Human Stress and the Environment

Environmental Topics

A series of books edited by J. Rose

Volume 1
Environmental Health
The Impact of Pollutants
Edited by J. Rose

Volume 2
Environmental Concepts, Policies and Strategies
Edited by J. Rose

Volume 3
Water and the Environment
Edited by J. Rose

Volume 4
Acid Rain
Current Situation and Remedies
Edited by J. Rose

Volume 5
Human Stress and the Environment
Health Aspects
Edited by J. Rose

Volume 6
Water Quality for Freshwater Fish
Further Advisory Criteria
Edited by Gwyneth Howells

This book is part of a series. The publisher will accept continuation orders which may be cancelled at any time and which provide for automatic billing and shipping of each title in the series upon publication. Please write for details.

Human Stress and the Environment

Health Aspects

Edited by

J. Rose

GORDON AND BREACH SCIENCE PUBLISHERS
Switzerland ● Australia ● Belgium ● France ● Germany ● Great Britain
India ● Japan ● Malaysia ● Netherlands ● Russia ● Singapore ● USA

Gordon and Breach Science Publishers S.A.
Y-Parc
Chemin de la Sallaz
1400 Yverdon, Switzerland

Library of Congress Cataloging-in-Publication Data

Human stress and the environment : health aspects / edited by J. Rose.
p. cm. -- (Environmental topics, ISSN 1046-5294 ; v. 5)
Includes bibliographical references and index.
ISBN 2-88124-851-9
1. Stress management. 2. Stress (Psychology) 3. Stress (Physiology) 4. Environmental health. I. Rose, John, 1917- II. Series.
RA785.H86 1994
155.9'042--dc20 94-2603
CIP

To my late brother Isaac.
A dedicated physician, devoted to his patients.

Contents

Preface

This volume is devoted to a study of the relationship between stress and health. The aim of this work is to present a fair and sober account of this complex and difficult problem without the sensationalism and slogans so beloved of today's extreme proponents of environmental remedies. In this age of distressed cities, environmental pollution and human follies of various kinds, it is necessary to preserve a sense of balance in order to deal with the serious problems of modern society before we can endeavour to deal with their deleterious effects on health.

The first chapter, concerned with the medical implications of stress, considers how stress may contribute to the development of a number of physical illnesses and major depressive conditions: the latter impair the immune response to various illnesses such as cancer, infections and cardiac disorders. This chapter deals with various aspects of the problem of stress and its effects, including etiology, cardiovascular system, risk factors, stress and the heart, the anxiety component, hypertension, psychiatric disorders, depression, sleep, menstrual disorders, sexual aspects, old age and the skin. Some of the psychological and psychiatric aspects of stress are elaborated on in Chapter 5, whilst the relationship between stress and cardiovascular diseases is treated in detail in the second chapter. Dr Wheatley concludes that there appears to be a correlation between stress and the deterioration of health for men: he suggests that in the case of women this association may even be stronger.

This theme is continued by Dr Rosenman in Chapter 2 on cardiovascular diseases. The author concentrates on emotions that relate to heart diseases, e.g., anxiety and anger. It appears that while such factors do not play a part in the pathogenesis of hypertension, they do in ischemic heart disease. An extensive literature review underpins the author's conclusions in this field of behavioural medicine.

A different aspect is reviewed in Chapter 3 by Professors Wong and Peacock, namely ageing. The authors discuss both theoretical and empirical findings and stress the importance of measuring both psychological and physical dimensions of the environment for the elderly. The chapter then considers atmosphere and societal stress that could predict psychological well-being. They conclude that elderly people with a high IQ tend to fare much better under environmental stress than those with a low IQ, in accordance with Lawton's docility hypothesis.

Professors Abe and Egashira consider the relationship between suicide and seasonal factors in various countries. They find that in spite of geographical differences, a common pattern underlies the occurrence of suicides during different seasons. Suicide and environmental stress factors are studied in detail using statistical data related to meteorological factors. In particular, sunshine appears to be one of the stress factors that may affect suicidal behaviour.

Chapter 5 addresses the complex problem of mental health and its relationship to stress, in particular the imbalance between the demands of the environment and the inability of the individual to adjust to them. Professor Freeman continues to discuss urbanization and the resulting psychosocial stress. He concludes that current urban programmes are not conducive to mental well-being and that extensive research is now required to demonstrate the dangers of urbanization. An alliance between the mental health professions and city planners is advocated to "humanize" the environment.

Stress in the work environment is discussed in Chapter 6. The authors believe that the modern environment is potentially very stressful and deleterious to society, economy and health. Sources of stress in the workplace are described and examples are cited to support the idea that stress affects the population in certain occupations. To cope with the problem of stress at work, a stress management system involving awareness, analysis and action is recommended.

Chapter 7 deals with stress within the office environment. The introduction of new technology and the design of more energy efficient office buildings have radically altered the office landscape and resulted in new stresses on employees. Dr Dorn reviews the extensive literature on the relationship between health and two potential sources of stress among office workers: the visual display unit and "sick building syndrome". The author stresses that care should be taken in the design of both technology and the office building itself in order to reduce ill-health among office workers.

The topic considered in Chapter 8 is stress and public concern in the management system over hazardous waste. The author presents case studies of specific UK hazardous waste facilities which have caused concern and cites different strategies for ameliorating stress. Health effects are analysed extensively, as well as public response and risk management. The chapter emphasizes the need to understand the risks and to involve communities in decisions about managing the risks to health.

The final chapter explores the theme of stress and transport, especially with regard to the impact of the motor vehicle on modern society. The author presents the situation caused by pollutants, which are responsible for accelerating other forms of environmental-related stress. Road transport is claimed to be a major factor in creating public disturbance, noise nuisance and ill-health, and various remedial measures in different countries are studied. Ms McHarry asserts that road pricing, traffic calming and improved public transport systems can be only partial solutions in the face of the rising number of vehicles on the roads.

This volume considers several aspects of the relationship between stress and the environment, ranging from ageing to stress in the workplace and from somatic diseases to mental health. The literature cited is extensive and various suggestions are made for remedial measures. Though it is impossible to describe completely all the facets of this complex problem, this work endeavours to present a realistic view of current conditions in order to guide both public opinion and those working in the field towards methods of reducing the effects of stress in modern life.

Finally, I wish to thank the authors for their patience and goodwill, as well as my wife Fay for her cooperation and valuable assistance.

List of Contributors

ABE, Dr. K.	Department of Psychiatry, University of Occupational and Environmental Health, Iseigaoka, Yahatanishi, Kitakyushu, 807 Japan.
COOPER, Professor C.L.	Organisational Psychology, Manchester School of Management, UMIST, PO Box 88, Manchester M60 1QD.
DORN, Dr. L.	Unilever Research, Colworth House, Sharnbrook, Bedford MK44 1LQ.
EGASHIRA, Dr. K.	Department of Psychiatry, University of Occupational and Environmental Health, Iseigaoka, Yahatanishi, Kitakyushu, 807 Japan.
FREEMAN, Professor H.	21 Montagu Square, London W1H 1RE.
McHARRY, Ms. J.	Environmental Research and Information, PO Box 893, London E5 9RU.
PEACOCK, Dr. E.J.	University of Toronto, 215 Huron Street, Toronto, Ontario M5S 1A1, Canada.
PETTS, Dr. J.	Centre for Extension Studies, University of Technology, Loughborough, Leicestershire, LE11 3TU
ROSENMAN, Dr. R.H.	Health Sciences Program, SRI International, Menlo Park, California 94115, United States of America.
SUTHERLAND, Dr. V.J.	Organisational Psychology, Manchester School of Management, UMIST, PO Box 88, Manchester M60 1QD.
WHEATLEY, Dr. D.	5 Warblers Green, Cobham, Surrey KT11 2NY.
WONG, Professor T.P.	Department of Psychology, Trent University, Peterborough, Ontario K9J 7B8, Canada.

1. MEDICAL IMPLICATIONS OF STRESS

DAVID WHEATLEY*

Stress at the present time is both subtle and prolonged, affecting the psyche rather than the soma. Adverse effects may ensue in a number of physical illnesses and major depressive illness may supervene. There are important associations with disordered sleep and sexual problems, whilst the elderly are particularly susceptible to increasing isolation from the community. Stress and depression consistently impair the immune response to illness as in cardiac disorders, infections and cancer.

1. ETIOLOGY

Living creatures, including man, are endowed with the autonomic "fight or flight" reaction that maintains the body in a state of constant readiness to respond to danger.[1] In this state, mental alertness is the basic requisite and the response to stress is manifested in animals many times a day by the release of catecholamines. This results in instantaneous diversion of cardiac output from the gastro-intestinal tract to the skeletal muscles. This state of mental and physical awareness is a normal physiological component of living creatures and is entirely appropriate to life in the wild, seemingly producing no adverse morbid effects. But what of civilised man?

Acute stress situations comparable to the constant fear of predators are, fortunately, relatively rare, but they have been replaced by other stresses of a more chronic nature, the effects of which are mental rather than physical. The changing nature of these stresses that challenge man's existence in our modern civilization,[2] have sorely taxed his coping mechanisms.[3] Stress is indeed a normal component of life,[4] but at some times and in some individuals, it can produce devastatingly adverse effects, which in turn may constitute a threat to health.

1.1. *Contemporary Themes*

At the present time, the individual lives against a constant background of sensual stresses,[5] including: urbanization, crime and boredom, changing sexual mores with associated problems concerning contraception and sexually transmitted diseases (particularly the human immune virus), the computer invasion, isolation of the aged, and drugs, alcohol and tobacco abuse. Increases in noise levels assault the individual in all walks of life and may have adverse effects on health and well-being.[6]

The technological revolution that has taken place in the last forty years has led to complacent forecasts of reduced work demands on the individual, to be

*5 Warblers Green, Cobham, Surrey, KT11 2NY, UK.

Table I Etiological and symptomatological components of stress

	Etiology	Symptomatology
S	Sorrow	Syncope
T	Terror	Tachycardia
R	Rage	Rigor
E	Elation	Extrasystole
S	Sleep Loss	Sweating
S	Sin	Shaking

accompanied by greater opportunities for leisure pursuits. This, in turn, should have led to the development of an expanded leisure industry, thereby achieving an optimal balance between work and pleasure. Unfortunately, the reverse has happened, and unemployment now represents one of the major stressors of the present decade.[7] Overpopulation too, particularly in underdeveloped countries, is a major problem.[8]

1.2. *Emotional Responses*

The etiology of stress-induced illness is multi-faceted but six important components and six relevant symptoms can be derived from the alphabetical constitution of the word itself.

Any combination of etiological and symptomatological factors can occur and those illustrated here are but some particularly relevant examples (Table I).

Stress may play an important adverse role in both initiating, maintaining and aggravating a number of physical and mental disorders, which will now be considered in more detail.

2. CARDIOVASCULAR SYSTEM

As in the "fight or flight" response, the main reaction to acute stress is manifested through the cardiovascular system. The response to an acute stress situation results in the following effects:[9] tachycardia, ECG changes (including ST segment depression and arrhythmias), hypertension, pulmonary oedema, and mobilization of free fatty acids, a putative factor in the development of atheroma.

In the acute stress situation these changes rapidly revert to normal with cessation of the challenge, but with long continued stress, they may become perpetuated and result in adverse effects on the cardiovascular system.

3. RISK FACTORS

Many important cardiac risk factors have psychogenic components that may be affected by stress. These are illustrated in Table II.

3.1. *Smoking*

The association between heavy smoking and coronary heart disease (CHD) has been established beyond doubt.[10] Nicotine may also act as a tranquillizer, and

Table II CHD risk factors with psychogenic components

Tobacco smoking	Alcohol consumption
Caffeine containing drinks	Sleep disturbances
Sexual problems	Obesity
Personality factors	Genetic factors
Racial characteristics	

excessive smoking may constitute a response to prolonged stress, as evidenced by Horowitz and his co-workers[11] who found significant differences in stress levels between smokers and non-smokers.

3.2. *Alcohol*

Although a number of reports have demonstrated that moderate alcohol consumption has a protective action[12,13] excess is undoubtedly dangerous.[14,15] Wine may have some specific effect in this context, since St. Leger *et al.*[16] have shown a strong negative association between CHD deaths and alcohol, wholly attributable to wine consumption.

In a study from Finland,[17] 2,671 doctors were assessed in relation to their consumption of alcohol and drinking behaviour. Selected variables related to work, stress and coping were correlated with the information recorded. The study concluded that alcohol consumption among doctors was higher than that of the general population and that heavy drinking seemed to be associated with stress and burn-out. Many people turn to alcohol as a means of coping with stress.[18,19]

3.3. *Caffeine*

A number of studies have shown that excessive consumption of stimulant drinks (tea, coffee, cola) may contribute to the physiological effects of stress and its adverse effects on the cardiovascular system.[20–22] It is a time-honoured custom, in times of stress, to brew a "cup of tea" in the mistaken belief that this will help to calm the individual. Illogical though the habit undoubtedly is, tea or coffee may also be consumed before going to bed, in order to induce sleep!

3.4. *Sleep*

Lack of sleep is a subtle form of stress that may insidiously undermine the sufferer's health,[23] influencing not only his or her whole life but also exerting an adverse effect in almost every illness known to man. Thus, changes in the pattern of sleep constitute potent stress factors and these can influence CHD. For example, in a large cross-sectional study of 5,419 Finnish adult men, a greater incidence of CHD was found among those who slept for more than nine hours or for less than six hours per night, as compared with those sleeping the more usual seven to eight hours.[24]

The restorative function of physiological sleep is not in doubt,[25] but sinus arrhythmia, sinus pauses and A-V blocks may occur during rapid eye-movement (REM) sleep, during which an increase in the variability of blood pressure and heart rate occurs.[26] The benzodiazepine (BDZ) hypnotics which are most commonly used to treat insomnia decrease both REM sleep (which may be beneficial) and deep sleep (which may have an adverse effect). However, new hypnotic drugs are being developed that selectively increase deep or slow-wave sleep (SWS).[27]

Sleep disturbance is, of course, a common accompaniment of anxiety states and also occurs in depressive illness. Hence the importance of treating these psychogenic factors that can increase the cardiovascular risk due to insomnia.

3.5. *Sex*

Sexual activity may exert an adverse effect in CHD as a result of the haemodynamic changes that take place during intercourse.[28,29] On the other hand, several studies have suggested that the risk of a "coital coronary" has been considerably exaggerated. Against any possible risks must be balanced the adverse effects of abstinence not only on the patient but also on the patient's partner, with resultant mental stress that may far outweigh the risk involved in the act itself. This factor will be discussed more fully in a later section.

3.6. *Obesity*

An adverse association between obesity and CHD is well documented.[30] In a recent study from Finland,[31] 3,786 men and 4,120 women were randomly selected and observed prospectively for seven years. Independent of age, men with a body mass index of 28.5 or more had a significantly higher incidence of acute myocardial infarction (MI), although this was not the case with women. The authors concluded that a body mass index of 29.0–31.0 or more is not only a marker for coronary risk factors, but is also a predictor of acute myocardial infarction in men. Another recent study has shown that fat loss achieved through diet or exercise produces comparable favourable changes in plasma lipoprotein concentrations.[32] There are many people who eat more under the minor stresses of everyday life, and motivation to maintain dietary restriction is also psychologically determined.

3.7. *Character*

The concept of the type A/B dichotomy in relation to CHD has become firmly established since the pioneering hypothesis of Friedman and Rosenman in 1959, supporting the concept that type A individuals (ambitious, aggressive, hyperactive) suffered from more CHD than type B (easy-going, placid, and unambitious).[33] On the other hand, Raglan and Brand have produced evidence showing that type A individuals may actually have *less* CHD than type B.[34] Nevertheless, personality is an important stress factor in the etiology of CHD and clearly the anxious type of individual will be more affected by the cardiovascular effects of anxiety.

3.8. *Heredity*

Of the non-preventable factors in CHD, heredity is undoubtedly the most important. It was shown in a study from Finland that the probability of CHD developing in the brother of a man experiencing a heart attack is 52–65% depending upon the region. Comparable figures for the brothers of healthy controls were

5–31% respectively.[35] The psychological effects of a family history of CHD may be of considerable importance. Fear of succumbing to the same ailment that killed one's forebears may constitute a potent causal factor.

3.9. *Race*

Considerable differences in the incidence of CHD occur between different races, extremes being the very low incidence in Eskimos as compared to the increased susceptibility of Jews to CHD. These are not necessarily due to genetic factors, and cultural differences may account for some of them.[36] For example, marked differences in mortality from CHD are seen in Japanese people living in the United States, Hawaii and Japan respectively, and between black people of similar race living in Africa and the Caribbean. The anomaly of the much higher incidence of CHD in Scotland, as compared to other parts of the U.K., and the lower incidence in France has not been properly explained,[37] as other factors, such as wine consumption, use of garlic and fresh food, may be involved.

4. STRESS AND THE HEART

The Stress Clinic at the Maudsley Hospital, London, has been established to evaluate and quantify stress occurring in nine important areas: social habits, social stress, life events, sexual, sleep, psychiatric, old age, menstrual in females, and stress and the heart.[38] In ensuing sections of this chapter, reference will be made in more detail to these stress areas as relevant to the text.

Stress in relation to the cardiovascular system is assessed in seven sections, although only four of these are scored for the purpose of assessing severity. These sections are shown in Table III: most items are scored on a five-point severity scale, ranging from 0 (absent) to 4 (very severe).

Whether or not there is any underlying cardiovascular etiology, cardiophobia can be a potent stress factor, the principal expression of which may culminate in panic disorder.[39]

4.1. *Family History*

Details are recorded of: cardiac disorders, hypertension or diabetes. Even although the patient may not worry about a family history, it remains a permanent stress factor and its severity has been related to the number of relatives

Table III Severity scores for the stress and the heart section of the stress profile

1. Personal history	Not scored
2. Family history	0–4
3. Physical measurements	
(a) diastolic B.P. (<95, 95–115, >115)	0–2
(b) pulse (<100, 100–121, >120 or irreg)	0–2
4. History of hypertension	Not scored
5. Cor. Hrt. Dis. (suspicious pain, dyspnoea, ECG proven, M.I.)	0–4
6. Other cardiac conditions	Not Scored
7. Cardiophobia (none–extreme)	0–4

affected, with corresponding severity grades up to a maximum of four or more affected family members.

4.2. *Blood Pressure and Pulse Rate*

Blood pressure is measured in the sitting position after three minutes at rest and the stress severity is rated on the diastolic reading. If below 95 mm.Hg. this is assessed as 0, 95–115 mm.Hg. scores 1, and 116 mm.Hg. and over scores 2. Pulse rate is recorded in similar manner, a rate of 100–120 scoring 1, and 121 and over, or an irregular pulse a score of 2.

4.3. *Coronary Heart Disease (CHD)*

Suspicious pain on exertion, (i.e. going up stairs, running for a bus, "tight band," referred to left shoulder/arm, jaw or back, relieved by rest, etc.) is scored as 1, whilst undue shortness of breath on exercise is scored as 2. If CHD has been proven by ECG, this is scored as 3 and if there has been a previous myocardial infarction, then this is scored as 4.

4.4. *Cardiophobia*

Worry over heart or blood pressure may be an important stress factor and so this is quantified as shown in Table IV.

5. THE ANXIETY COMPONENT

Our research group has undertaken a number of studies of anti-anxiety drugs compared to placebo as an adjunct to therapy in patients suffering from CHD.[40] In the first of these, the benzodiazepine (BDZ) anxiolytic clorazepate (Tranxene) was compared to placebo over an eight week period and both the anginal attack rate and glyceryl trinitrate requirement to relieve attacks, were reduced in the patients on the active drug as compared to the placebo, the differences being statistically significant in respect of trinitrate requirements (Figure 1).

However, in subsequent studies involving larger numbers of patients, we were not able to confirm these findings, although changes in trial design may have been responsible for these differences.

Despite these conflicting results, one of the most important aspects of the work was to determine whether prophylaxis with anti-anxiety drugs may reduce the

Table IV Cardiophobia definitions

Cardiophobia

How much does the patient worry about blood pressure or cardiac condition?

0 = Not at all
1 = Slight (occasionally)
2 = Moderate (when not otherwise mentally occupied)
3 = Marked (most of the time)
4 = Extreme (dominates patient's life)

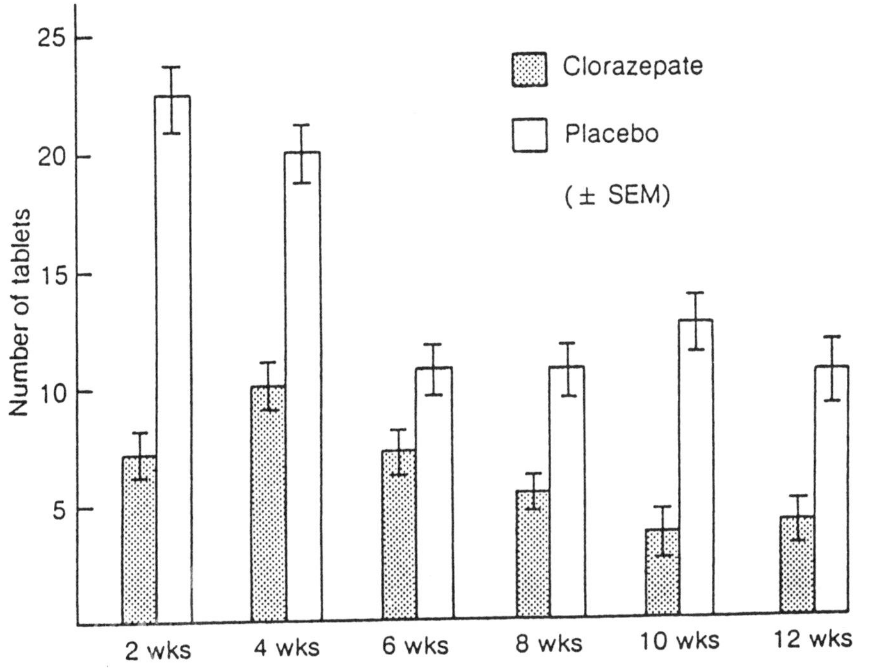

Figure 1 Mean nitroglycerin requirements per 2-week period (0.5 mg. per tablet) in a trial to compare clorazepate to placebo.

incidence of subsequent myocardial infarction (MI). In fact there were five cases of MI in a total of 77 patients treated with placebo (6%), as compared to no such cases in 81 patients treated with one of the anti-anxiety drugs (0%), a difference which is in fact statistically significant ($p < 0.05$).

6. HYPERTENSION

Since temporary blood pressure elevation is a normal physiological response to stressful stimuli, it is tempting to postulate a causal relationship between hypertension and external stresses of various kings. Although it is relatively easy to produce elevation of blood pressure in experimental animals exposed to various stresses, it is much more difficult to establish the relationship between repeated or prolonged stress and the pathogenesis of chronic hypertension.[41] Nevertheless, there are many examples that suggest, almost overwhelmingly, that this is the case.

Following the seminal work of Cobb & Rose on stress in air-traffic controllers in relation to the development of hypertension,[42] a number of other studies have demonstrated the particular vulnerability of such individuals to contract the illness. Such reports have reached their natural culmination in a study on the effects of occupational conflict superimposed on the "normal" stresses of the job.[43] Another occupation particularly suited to this type of research is provided by prison inmates, in which setting a number of external stressors may contribute to elevated blood pressure.[44]

In an era in which raised blood pressure can be effectively controlled by a variety of drugs, the quality of the patient's life is not always considered and well-being may be impaired as a result of the treatment itself.[45] An increasing awareness of the role that stress may play in the etiology of sustained hypertension might well lead to a better control of stress factors in the patient's life. Thus, he or she may be spared the "life sentence" of antihypertensive drug therapy, which itself may constitute a potent stress factor.

7. PSYCHIATRIC DISORDERS

A number of psychiatric disorders may be precipitated by stress with which the patient is unable to cope, and likewise, an existing psychiatric illness generates stress, both for the patient and those in contact with him or her.[46] Since the immediate response to an acute stress situation is analogous to an anxiety state, generalised anxiety disorder and its variants, including phobias and panic disorder, are intimately implicated diagnoses. However, just as anxiety is associated with acute stress, so may major depression develop insidiously in response to prolonged chronic stress and, moreover, this serious fundamental illness may be concealed by the continuing overlay of anxiety symptoms.

7.1. *Diagnosis*

The importance of psychiatric diagnosis in the management of stress-induced problems is illustrated by an analysis of the first 100 patients attending the Maudsley Stress Clinic (Table V).

In interpreting these figures, it should be appreciated that in some cases, more than one principal stress was present, i.e. insomnia and anxiety, depression due to organic illness, etc. However, this analysis does illustrate the enormous potential for treatment of associated psychiatric disorders, as a means of stimulating the patient's ability to cope with his or her stress problems. In this context, many

Table V Stress diagnoses in the first 100 patients seen at the Maudsley Stress Clinic (more than one in some cases)

Principal stress	N.	Males	Females
General anxiety	24	14	10
Social phobia	8	4	4
Panic disorder	9	6	3
Agarophobia	1	—	1
Simple phobia	3	2	1
Major depression	49	21	28
Insomnia	20	8	12
Organic illness	9	6	3
Emotional	1	—	1
Situational	1	1	1

stresses are ever-present but their impact on the patient can be reduced as a result of appropriate diagnosis and treatment.

8. ANXIETY

A number of comprehensive studies have been undertaken to document the association between stress and anxiety symptoms. Reference has already been made to the stress of incarceration in prison and, in another study of 208 male prisoners on the 10th day after entering prison, it was found that anxiety occurred in no less than 50% of the subjects.[47]

The working conditions of junior medical staff in Great Britain provide many examples of the association between anxiety and stress. For example, in a study of 20 such individuals, significantly more tension and anxiety were recorded on stressful days as compared to others.[48] Loud noise is a particularly potent form of stress, particularly when it cannot be controlled. Thus, in a study on 10 healthy human volunteers, higher self-ratings of tension, stress and anxiety were recorded when the subjects were unable to turn off the noise, than when they could do so.

8.1. *Pharmacological Interventions*

The benzodiazepine (BDZ) anxiolytics have proved to be highly effective in the control of anxiety symptoms, but recently the emergence of dependence and withdrawal effects[49] have resulted in a precipitous decline in their use. The majority of patients with stress-induced anxiety are treated by their general practitioners and alternative psychological techniques[50] are not always appropriate in situations of inadequate time and training. Although new effective anxiolytic drugs are being developed without the hazard potential of the benzodiazepines,[51] it may be some time before these are fully evaluated and made available to the clinician. An alternative approach is to use the beta-adrenergic blocking drugs for their anxiolytic properties.[52]

Although the main effect of such drugs may be peripheral and on the somatic symptoms of anxiety, nevertheless there may be a central component, in which case a drug with high lipo-solubility is indicated.[53] Propranolol has the highest lipid solubility of all beta-blocking drugs available in the UK and is highly concentrated in the central nervous system.[54] Of the cardio-selective beta-blockers, betaxolol is the most highly lipid soluble, although much less so than propranolol. We have used both these beta-locking drugs in the clinic and Figure 2 shows the results recorded on the mean Stress Profiles for an initial sample of six patients.

Concurrent with the highly significant relief of anxiety in response to this form of treatment, the adverse severity of most of the other stress factors was also reduced, significantly so in the case of depression, sexual stress and sleep disorders.

9. DEPRESSION

As indicated, there is a strong association between many forms of stress and the onset of depressive illness, particularly in relation to many social stresses.[55] Depression has also followed: stroke,[56] sexual and physical abuse of women,[57] confinement to prison,[58] HIV infection,[59] noise,[60] unemployment,[7] and the ardours of house officers on night duty,[48] amongst other stress factors.

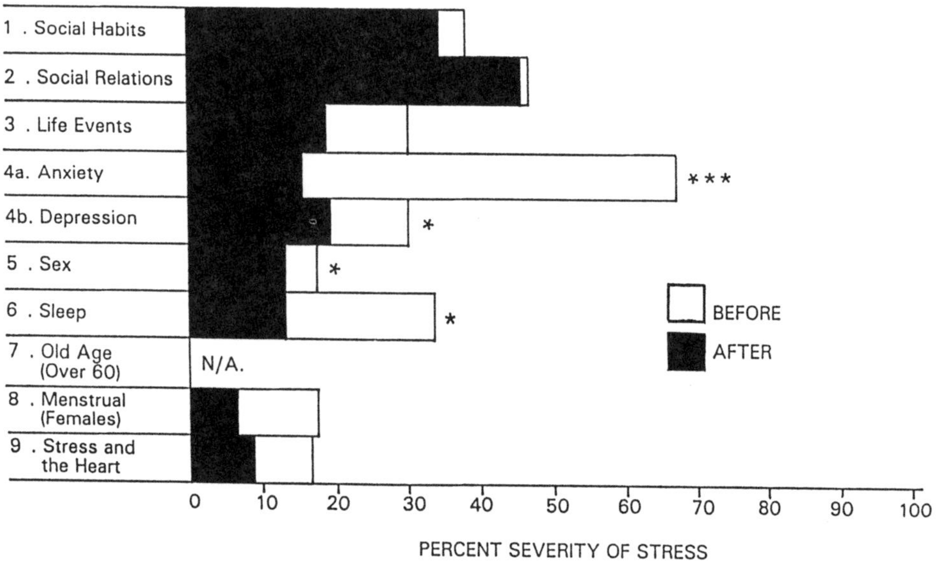

Figure 2 Mean stress profiles in a sample of six patients suffering from generalized anxiety before and after treatment with beta-blocking drugs. (*$p < 0.05$) (***$p < 0.001$)

9.1. *Anxiety Overlay*

In view of the considerable overlay of anxiety symptoms in a stress situation, at the clinic, both the Hamilton Anxiety and Hamilton Depression Rating Scales (HARS & HDRS respectively) are administered concurrently. The strong association between anxiety and depressive symptoms is emphasized by the repetition of a number of items in both scales, which may be difficult to interpret.[61] The mean scores on these two scales can be represented as percentage severities in relation to the maximum possible scores, namely 56 (HARS) and 52 (17-item HDRS). In a sample of 21 depressed patients treated with antidepressant drugs, the means of these percentages are shown before and after treatment in Figure 3.

It is apparent that the levels of depression and anxiety and total stress were, in fact, very similar initially, and did not differ statistically. Furthermore, the severities of all three were significantly reduced to similar extents as a result of treatment.

This finding applies also to the other components of the Stress Profile as shown in Figure 4.

Following the significant relief of depression ($p < 0.01$), as in the case of anxiety, there was concomitant relief in severity of stress due to other factors also. This applied to: social habits, life events, sleep, anxiety, menstrual, and stress and the heart ($p < 0.01$) and social stress ($p < 0.05$).

10. SLEEP

Lack of sleep is a subtle form of stress that may insidiously undermine the sufferer's health,[62] influencing not only his or her whole life but also exerting an adverse effect in almost every illness known to man. There are a number of

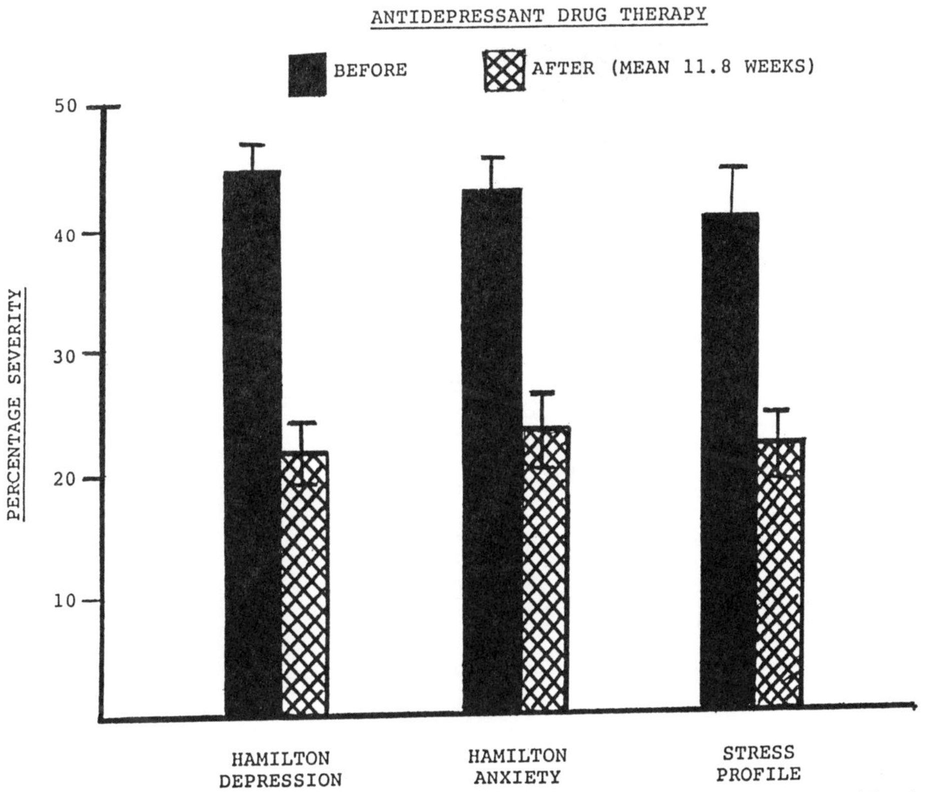

Figure 3 Severity of symptoms before and after antidepressant drug therapy, as measured by the Hamilton Depression and Anxiety Scales and by the Stress Profile

important aspects of sleep that need to be differentiated and the sleep questionnaire of the Stress Profile is concerned with these. Thus, the following aspects are quantified: time taken to fall asleep, number of awakenings during the night, early morning wakening, total nightly sleep duration and how refreshed the patient feels on final awakening.

It is also important to classify the etiology of insomnia[63] as this is also an important factor in determining the best form of treatment. Etiology is classified as follows:

10.1. *Physical*

There are many organic illnesses which may keep the patient awake because of the symptoms they produce. These include: pain, cough, pruritus, enuresis, asthma and cramp, to name a few common examples.

10.2. *Physiological*

Artificial disturbance of the normal sleep pattern may be experienced by night workers who sleep in the daytime, air crews who are continually adjusting to various time differences and individuals who sleep under conditions of unusual

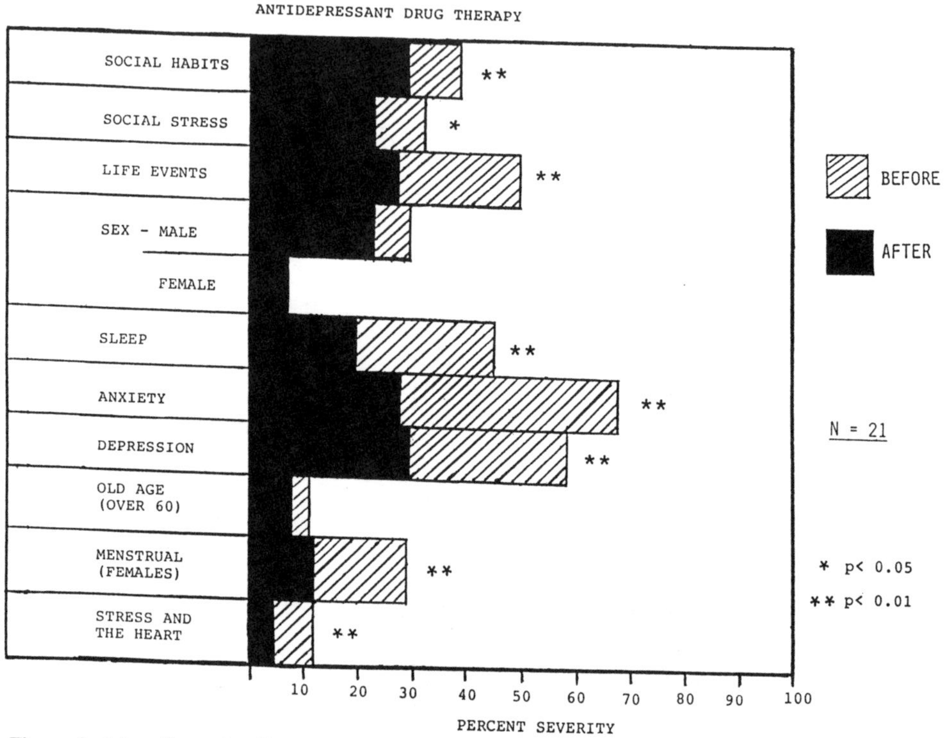

Figure 4 Mean Stress Profiles before and after antidepressant drug treatment in 21 patients suffering from major depression.

noise or light. The use of stimulants such as coffee and tea just before going to bed may also keep people awake, as may a restless or snoring bed companion.

10.3. *Psychological*

The commonest psychological causes of insomnia are emotional, often being concerned with the various life events already considered.

10.4. *Psychiatric*

Sleep disturbance is a regular component of anxiety states, depression and psychotic illnesses.

10.5. *Iatrogenic*

Certain drugs may keep the patient awake. These include: adrenergic compounds such as those used in the treatment of asthma, whether taken orally or by means of a nasal spray. Thiazide diuretics, if not taken first thing in the morning, may interfere with sleep because of the diuresis they produce, whilst beta-blocking drugs may cause vivid dreams that disturb sleep or even induce insomnia.

10.6 *Idiopathic*

There are many occasions when individuals are unable to obtain adequate sleep for any apparent reason.

Adequate assessment of these factors bears considerable implications for subsequent treatment. For example, changes in occupation, sleeping conditions, necessary drugs, etc., may all be considered when implicated in the etiology of the insomnia. Nevertheless, there will be many occasions when resort may have to be made to a hypnotic drug and the importance of the type of insomnia then becomes apparent from the point of view of choosing "the right drug for the right patient."

10.7. *Somnopharmacology*

The effectiveness of different hypnotic drugs on the various types of insomnia varies considerably in relation to rapidity of onset, duration of action and "hangover" effects the next morning. The specific indications for individual drugs are mainly concerned with their differing pharmacokinetic properties,[64] which in turn influence duration of action. These drugs can be broadly classified into four groups as follows: ultra short-acting (2–4 hours), short-acting (5–7 hours), medium (8–10 hours) and long (over 10 hours). However, the pharmacokinetic properties, and particularly the elimination 1/2-lives of individual drugs, can vary considerably from individual to individual and even in the same individual from time to time, particularly when illness supervenes.[64]

A clinical assessment of sleep can be achieved by means of somnography, a method which has been used in many trials undertaken by our research group.[65] The individual's sleep pattern can then be displayed graphically in the form of a "somnogram" and mean somnograms in relation to various hypnotic drugs and their elimination half-lives are shown in Figure 5.[51]

"Awake" periods are shown in white and "asleep" periods in black, the initial wake period representing the time taken to fall asleep after the patient has gone to bed ready to sleep. Individual periods awake during the night have been amalgamated to show a total period and, whatever the type of insomnia, the net result is reduction in total sleep time. This is illustrated by the figures shown for control (no treatment) periods and placebo, which also illustrate the lack of placebo effect when treating insomnia.

11. STRESS AND IMMUNITY

11.1. *Infectious Diseases*

The prime defense against infectious diseases is provided by the immune system and any impairment of this physiological mechanism must result in increased morbidity and even mortality of many infectious diseases. Upper respiratory tract infections provide an ubiquitous paradigm.

Thus, 29 college students suffering from sore throats were compared with a similar number of symptom-free students[66] from the point of view of antecedent stress factors. It was found that the infected students had suffered from signifi-

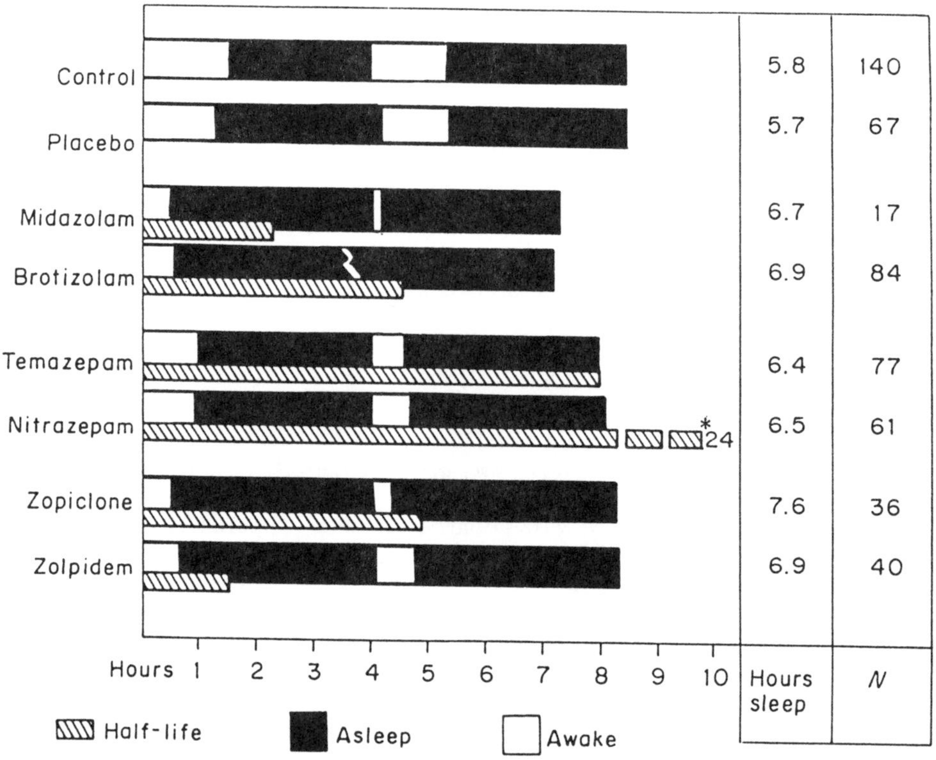

Figure 5 Mean somnograms for control, placebo and various hypnotic drugs. *Half-life 24 hours with nitrazepam.

cantly more disappointment, failure and role crisis in their lives as compared to the normal controls. In another study[67] subjects kept daily records of events and situations in their lives and colour photographs were taken of their noses and throats. The authors found that "life events that occasioned self-reported sadness were likely to be followed by an acute respiratory illness." The psychological factors involved in resistance to infectious disease have been comprehensively reviewed by Friedman and Glasgow.[68]

There are many reports demonstrating reduced cell-mediated immune function in patients suffering from major depression.[69] Impaired sleep, particularly when it results in reduction of deep sleep periods, will also adversely affect immunity.[70,71]

Not only is the patient under stress more susceptible to infectious diseases of various kinds as a result of stress-induced impairment of the immune system, but stress occurring during prolonged infections may not only delay recovery, but facilitate the onset of complications. Many important aspects of the effects of stress on the immune system have been reviewed by the author.[72]

11.2. *Neoplastic Disease*

Inexplicably, from time to time, reports appear of spontaneous regression of otherwise fatal forms of cancer. The factors controlling the initiation and progression of neoplastic disease are, in most cases, totally unknown, except where obvious carcinogens are involved as in the case of lung cancer. Survival times in patients suffering from the same type and stage of cancer may also vary considerably from individual to individual, and it is tempting to postulate that the mind may play some part in modulating the body's resistance to carcinogenesis.[69]

Breast cancer has been extensively studied in relation to stress[73] and, in a study by Ramirez *et al.*,[74] it was found that stressful adverse events occurring in the life of women with breast cancer, may affect the relative risk of recurrence following operation. Thus, the relative risk of relapse associated with severe life events was 5.67 times greater than in control patients without such events. Another study by Spiegel *et al.*[75] demonstrated a substantial difference in survival times between breast cancer patients who underwent intensive psychosocial treatment, and others who did not.

In a fascinating study by Goodwin *et al.*,[76] marital status was found to be an important factor associated with survival in cancer patients. In a population-based data study of 27,779 cancer cases, they found that, after adjustment for stage of cancer and treatment, unmarried persons had a significantly poorer survival than those who were married. They showed that being married is associated with an increase in 5 year survival that is comparable with being in an age category 10 years younger. Furthermore, poorer prognosis was associated with divorced persons, as compared with widowed or never married persons. Clearly, many factors may be involved, such as earlier diagnosis and treatment, and the likelihood that married individuals have a better socio-economic status than single individuals. However, it may well be that marriage provides social support, which has been postulated to buffer the effects of stressful events and this may be the factor involved. The authors quote a number of references in support of this hypothesis.

In the absence of specific therapy for neoplasia, and when surgical and other forms of treatment have failed, control of stress factors in the patients life may assume paramount importance in prolonging survival.

12. MENSTRUAL DISORDERS

Apart from childhood, there is no period of a woman's life that escapes the threat of ill-health in relation to her reproductive functions. These adverse effects, many of which are psychiatric, may occur at the menarche, proceeding to premenstrual manifestations in adult life, contraceptive problems, puerperal depression, and symptoms associated with the menopause and post-menopausal years. Finally, in old age, unless prevented, the morbid toll of osteoporosis frequently supervenes.

12.1. *Premenstrual Syndrome*

Premenstrual syndrome (PMS) provides a constant source of stress to many women and it has been estimated that only 5% of women appear to be totally symptom-free of PMS manifestations.[77] Vellacott and O'Brien have classified PMS into four groups.[77] In the first of these, effects are confined to physiological premenstrual changes only but these are not of sufficient degree to adversely affect life style. In the next stage, which they term primary premenstrual syndrome, psychological and/or somatic symptoms occur in the majority of cycles and these are of sufficient severity to cause disruption of the lives of both sufferers and their families. However, the symptoms are relieved by the end of menstruation, so that these women do have at least one totally symptom-free week during each cycle.

On the other hand, in secondary premenstrual syndrome, the third stage, similar changes occur not only in the premenstrual phase but, although they improve following menstruation, persist to a varying extent throughout the cycle. Finally, similar symptoms can occur that are not menstrually related and may be incorrectly labelled as PMS, although the underlying disorder is in fact a psychiatric one.

In assessing menstrual stresses for the Stress Profile,[78] the severity of various menstrual abnormalities, such as menorrhagia, dysmenorrhoea, hysterectomy, etc., is first recorded. This is followed by records of the severity of either premenstrual or menopausal symptoms as appropriate.

12.2. *Premenstrual Stress*

PMS is defined as the occurrence of symptoms at any time during 14 days before the menstrual period and/or on the first 1–2 days of the period itself. Symptoms are assessed in two categories, i.e. psychiatric and physical. The former includes: anxiety and depression, irritability, loss of energy and difficulty in thinking. Physical symptoms include: headaches, weight gain, swollen breasts, fluid retention, etc.

12.3. *Menopausal Symptoms*

These are defined as occurring from the age of 35 onwards, with or without menstrual changes such as heavy, missed, or irregular periods, amenorrhoea, or following hysterectomy. Such symptoms are divided into three categories, i.e. physical, sexual and psychiatric. Physical symptoms include the well-known hot flushes, sweats and headache; whilst sexual symptoms consist of: reduced libido, dry vagina, and difficult intercourse. Finally, psychiatric symptoms are similar to those with PMS: anxiety, tension, irritability and/or depression.[78]

12.4. *Iatrogenic*

Hormone replacement therapy is being increasingly used to treat the adverse manifestations of the menopause, with considerable improvement in the health of the majority of women who suffer from such sequelae to the climacteric. Menopause-induced psychiatric symptoms and, in particular depression, may also be relieved by this form of therapy.[79]

The majority of menopausal women have intact uteruses and so require additional treatment with progestogen to protect against putative endometrial cancer. Unfortunately, progestogen-induced depression is a well-recognized complication of such treatment.[80] The relief of menopausal misery has been revolutionized by the introduction of HRT, but it is not a universal panacea. When depression occurs at the menopause, whether idiopathic or iatrogenic, specific antidepressant drug therapy may be required in addition to HRT.[81]

It is apparent that menstrual stresses are the cause of a great deal of morbidity to women, both physical and psychiatric. Although the treatment of PMS is by no means very satisfactory, the elucidation of the problems involved may lead to improved coping on the part of the sufferers. Thus, for example, patients may be advised to try and arrange their work schedules to avoid stressful episodes in their life (if possible) at the time when these symptoms occur. This particularly applies to decision making and the avoidance of social contacts at these times. An explanation of the cause of the symptoms to the spouse, or even indeed to others who may be in intimate day-to-day contact with the patient, can be helpful in creating better understanding of this endocrine illness for which no specific treatment is at present available. In the menopause, on the other hand, pharmacological intervention in the form of HRT and psychopharmacology may be highly effective, and so an evaluation of menopausal stresses may be of positive help to the patient in coping with such problems.

13. SEX, STRESS AND THE PSYCHE

The sexual urge constitutes one of the most potent stress factors in human life and may produce numerous sequelae when the response to it is inadequate for whatever reason.[82] Since many patients are reluctant to mention sexual problems and many doctors omit to enquire about them, this is a particularly important aspect of stress assessment. The stress profile considers six aspects of sexual problems concerned with: libido, partner integration, physical, psychological, masturbation and problems peculiar to the female.

13.1. *Libido*

The amount of libido varies considerably from person to person, in the same person at different times, and in relation to age and circumstances. However, it is only when the individual is dissatisfied with his or her degree of libido, that this may constitute stress. Thus, the individual with excessive libido may worry over this, feeling that in some way it may be "abnormal." It may also create problems with the partner, and a common complaint encountered at the clinic is that the partner (usually the male in a heterosexual relationship) is "too demanding" and this results in friction between the couple. Lack of libido may similarly cause stress, as an individual may greatly desire to increase it and may suffer psychologically from failure to do so. Similarly, problems may arise with the partner.

Therefore, the section on libido in the questionnaire concerns satisfaction with sexual urges and, if unsatisfactory, the degree to which this causes distress to one, the other or both in the partnership. Similar remarks apply to homosexual relationships.

13.2. *Relationship with the Partner*

In addition to frictions that can be caused by one partner being over-demanding with regard to the frequency of sexual intercourse, other personality variables may be involved. Thus, the response of the partner is important, and one who is either too passive or too aggressive during union may cause stress to the other. Likewise, the amount of foreplay that may be acceptable to one partner may not please the other, or indeed, if too perfunctory, may result in orgasmic failure in the female. The partner's loving approach too is a factor involved, particularly in the avoidance of stress for the female.

Physical problems in relation to intercourse may provide a potent source of stress to both partners. For example, poor erection and premature ejaculation in the male may cause just as much, if not more, stress to the female; whereas anorgasmia in the female may or may not be communicated to her partner. Stress in relation to bodily inadequacy may be present, as for example the man who feels that his penis is too small; or the woman that her breasts may not be big enough. Guilt feelings may still be a cause of stress, and these may stem from childhood upbringing which may be affected by both family and social morals at the time. Guilt is a potent source of stress in illicit relationships and this can sometimes be so extreme as to affect the heart, when underlying cardiac disorder is present.

13.3. *Cardiovascular Effects*

In both sexes, intercourse culminating in orgasm may cause transient tachycardia, hypertension, and hyperventilation.[28] Heart rates of 180 beats/min and blood pressures of 230/130 mm.Hg have been recorded at orgasm in normotensive individuals.[84] Although ECG abnormalities have not been observed in normal men and women during intercourse,[84] in patients suffering from CHD, angina commonly occurs during coitus.[28,85] Indeed, it is well known that sudden deaths do occur during or shortly after sexual intercourse. It has been suggested, however, that these usually happen in clandestine circumstances, such as the bordello or the mistress's boudoir, or when the relationship has been between an older man and a younger woman.[29] It is perhaps only in the placid ambiance of established marriage that sex may be safely indulged in by the sufferer from CHD!

On the other hand, several studies have suggested that the risk of a "coital coronary" has been considerably exaggerated.[28,85,86] Against any possible risks must be balanced the adverse effects of abstinence,[87] not only on the patient but on the patient's partner also, with resultant psychological interaction, the adverse effects of which may far outweigh the risk involved in the act itself.

In an anecdotal example, when interviewing a 68 year-old woman at the Stress Clinic, my approach to the sexual section of the profile was somewhat diffident and apologetic. Her response, however, was indignant; to the effect that her sex life was not at all satisfactory! It transpired that her husband had had a myocardial infarction the year before and, since then, had been afraid to indulge in sexual intercourse for fear of adverse effects on his heart and, in particular, the fear of a possible further and fatal attack. He himself seemed not to be too much concerned about the lack of sex, but she was suffering considerable stress as a result of missing what had been a normal component of their life together for so long. Fortunately, reassurance that this bond to their relationship could be safely enjoyed, did much to relieve the woman's stress problems.

13.4. *Contraception*

The use of oral contraceptives by women may increase the risk of MI, particularly if associated with cigarette smoking.[88] The findings of this study suggested that the rate of MI is increased approximately fourfold among women using oral contraceptives. Furthermore, in combination with 25 or more cigarettes per day, the rate increases at least 20-fold and more probably about 40-fold. Further confirmation for the increased risk of CHD due to oral contraceptive use, is provided by a study undertaken by Wallace *et al.*[89] of women attending 10 North

American lipid research clinics. They found that, compared with controls, oral contraceptive users exhibited increased cholesterol, triglyceride, low density lipoprotein (LDL) and very low density lipoprotein (VLDL) levels, and that these levels were positively associated with the amounts of the oestrogen components.

13.5. *Other Factors*

Masturbation by both sexes is common, often as an adjunct to a normal sexual relationship. A common problem encountered at the clinic is that of a woman who is unable to achieve orgasm in an otherwise perfectly satisfactory sexual relationship, but who has no difficulty in climaxing as a result of masturbation. However, once again, guilt feelings can be present, particularly as a result of misinformation concerning the adverse effects of what should be regarded as a normal habit.

In the stress profile, there is an extra question applying to females, which enquires about dysparunia and fear of pregnancy. This question is confined to the female sex, since in the author's experience he has yet to encounter a male who experiences stress from either of these causes!

14. OLD AGE

The elderly are particularly vulnerable to the assaults of stress and, with increasing age, there is hardly any aspect of their life that is not so influenced.[90] These influences may often be of a subtle nature and are not always appreciated by those who strive to help them. One of the main problems for the elderly is increasing isolation from others in the community, and this is often aggravated by the failing senses. For example, an elderly person who seemingly shuns social contacts may merely be self-conscious of (or may not even appreciate) a hearing deficit. The provision of a hearing aid may be all that is required to treat such a major stress factor.

Indeed, in several studies Gilhome Herbst[91] found a significant relationship between not only deafness and depression, but with mental impairment also. She speculates on the danger of regarding, and therefore treating, elderly people as demented or confused when they are actually severely deaf, particularly as the manifestations of both these disorders are in many ways so similar. Furthermore, deafness increases fears of personal insecurity and emphasizes feelings of loneliness. Eyesight is of particular importance as its impairment threatens that last link with the world outside, the television screen. In consequence, depression often develops insiduously in the elderly and should be assiduously sought for. Physical defects such as arthritis limit mobility, self-sufficiency and social contacts.

Social stresses are particularly important to the elderly. A harmonious relationship between an elderly relative and younger members of the family, is rendered more difficult by the restrictions imposed by financial problems and limited living-space; aspects of modern life that affect most of the community in one way or another. No longer is the grandparent an honoured guest in the house, waited on hand and foot, in return for the many years that he or she gave in service to the children. Differences in outlook, the changed role of younger women in the community, the ubiquitous quest for material advantages, are some of the factors that tend to create a situation that may become strained in the extreme. And so, increasingly the old are isolated from their kith and kin, whilst their equally elderly friends die off around them, and they become cut off from contact with the outside world through their failing senses.

Alterations in the sleep pattern, although a normal physiological manifestation of the aging process, may cause considerable psychological stress to those who suffer from it. Thus, the psychological effects can be profound and should be taken into consideration when deciding whether or not to prescribe a hypnotic drug. There is no worse situation for the octogenarian widow or widower, retiring night after night, probably at an early hour as an escape from the boredom of sitting alone in a silent empty room, to awaken prematurely in the middle of the night when the safety of sleep was anticipated until the light of dawn. As with a child, so at night all things become magnified—the creak of the floorboards that might signal an intruder, the ticking of the clock that seems supernaturally loud, the squeaks and creaks of the building as it adapts to the colder night air that may signify the advent of marauding rodents, a displaced hair on the face felt as some obnoxious insect—and so on and so forth. No wonder that elderly people frequently turn on the lights during the night and, far from feeling refreshed from their short physiological sleep the next morning, spend a restless day worrying about what is going to happen on the next night, and the next night, and the next night after that. Decline in cognitive functioning may also cause stress, although it is not always appreciated as such.

14. STRESS AND THE SKIN

The skin provides an impressive milieu for the visible manifestations of stress. Individuals, very often young women, who flush embarrassingly under the stress of social encounters are wellknown. But stress may also play an important part in established skin disorders and in a review of the subject, Walton[92] has commented that the changes occurring in the skin as a result of stress are initially physiological and temporary. However, if repeated frequently enough, they eventually become permanent and so pathological. It has been estimated that, in 34% of cases of dermatological disease, stress was the precipitating factor. As he points out, the biological reaction to stress is very complicated, and the skin can react in a variety of ways.

Stress may affect: the neurological components, causing pruritus as in neurodermatitis; the vascular structure, as in rosacea and urticaria; the endocrine sweat glands, causing hyperhidrosis as seen in anxiety; the sebaceous gland structures, causing overactivity as in acne, seborrhoea and rosacea; and the hair follicles, causing piloerection as a result of fear. Increased capillary fragility and immunological changes may occur with emotional stress. Vasodilation and associated increased skin temperature are the concomitants of embarrassed and frustrated patients but, in contra-distinction with fear there is vasoconstriction and pallor. Although these are manifestations of the primitive "fight or flight" reaction to stress, if repeated frequently enough, they may eventually become permanent and therefore pathological.

Koblenzer[93] has classified "psychocutaneous disease" into three main groups: psychiatric conditions that manifest in the skin, conditions in which strong psychogenic factors are imputed, and conditions probably dependent on genetic or environmental factors, whose course is often significantly affected by emotional stress. These are shown in Table VI.

Table VI Classification of psychocutaneous disease (C. S. Koblenzer[93]).

I Psychiatric conditions that manifest in the skin

1. Dermatitis artefacta
2. Delusions and hallucinations as they relate to the skin
3. Psychogenic pain syndromes
4. Compulsive habits and obsessional concerns as they relate to the skin
5. The psychogenic purpura syndrome

II Conditions in which strong psychogenic factors are imputed

1. The urticarias
2. Pruritus; generalized and localized
3. Flushing reactions and rosacea
4. Psychogenic disorders of the sweat glands
5. Alopecia areata, telogen effluvium and hirsuitism

III Conditions probably dependent on genetic or environmental factors, whose course is often significantly affected by emotional stress

1. Psoriasis
2. Atopic dermatitis
3. Acne vulgaris
4. The spectrum of clinical dermatology

As is so often the case the manifestations of stress themselves may then serve as stress factors and produce profound psychological effects. Since physical appearance is so important to the majority of human beings (in contra-distinction to the animal world), it is hardly surprising that self-esteem may be diminished as a result of skin disorders. As Koblenzer comments: "...... patients often feel different, alienated, unacceptable or dirty. These feelings in turn may lead to social withdrawal, to poor job performance, and to impaired social and family relationships." As this author further observes, patients often feel that society does not understand their plight, and that they are discriminated against socially. Since, frequently, these patients already handle aggression poorly and have poor impulse control, their negative feelings may translate into inappropriate behaviours, that further increase their feelings of alienation.

The association between psychiatric disorder and dermatology has been further emphasized by Wessely & Lewis[94] who found that, in a random sample of 173 subjects attending a dermatology out-patient clinic, no less than 40% were classified as suffering from a psychiatric disorder. However, they found no correlation between psychiatric morbidity and the nature, severity or site of the skin disease. This led them to conclude that any interaction with psychological factors was probably of a general nature but that, nevertheless, diagnosis and treatment of associated psychiatric or emotional disorder is of considerable importance in prognosis of the skin condition.

16. CONCLUSION

In this chapter I have attempted to survey the many medical implications of stress and to emphasise the importance of diagnosing and quantifying stress factors that may be exerting adverse effects on a number of physical disorders.

The importance of the psychosomatic interface in relation to general health has been outlined by Vaillant[95] in a prospective study of 204 men, selected as adolescents for an interdisciplinary study of health. They were then followed up biennially and 185 remained in the study and in good health until 1964, when the mean age was 42 years. Over the next 11 years, 100 of these subjects remained in excellent physical health, 54 acquired minor problems, and 31 acquired serious chronic illness or died. Of 59 men with the best mental health, assessed from the age of 21 to 46 years, only two became chronically ill or died by the age of 53, whereas of the 48 men with the worst mental health from the age of 21 to 46, 18 became chronically ill or died. The relation between previous mental health and subsequent physical health remained statistically significant when the effects on health of alcohol, tobacco use, obesity, and longevity of ancestors, were excluded by multiple regression analysis. The author concluded that the data suggest that good mental health retards mid-life deterioration in physical health.

Therefore, at least as far as the male of the species is concerned, there would seem to be a correlation between the effects of stress and deterioration in health. Doubtless, if a similar study were to be undertaken on women, the associations might be even stronger, in view of the special stresses to which women are subject in relation to their physiological constitution. It is clear, therefore, that stresses of diverse nature may adversely influence the health of the individual in a variety of different ways.

References

1. D. J. Bruce Durie, "Sleep in Animals" **In**: *Psychopharmacology of Sleep* (ed. D. Wheatley) (Raven Press, New York, 1981) pp. 1–18.
2. A. Hordern and D. Wheatley, "Psychotropic Drugs in the Elderly" **In**: *Stress and the Heart* 2nd edition (ed. D. Wheatley) (Raven Press, New York, 1981) pp. 377–401.
3. P. G. F. Nixon, "Stress and the cardiovascular system" *The Practitioner* **226**, 1589–1598 (1982).
4. Hans Selye, *"Stress Without Distress"* (Lippincott, Phil. and New York, 1974).
5. A. Hordern, "The Spectrum of Stress" *Stress Medicine* **1**, 17–25 (1985).
6. S. Rehm, E. Gros and G. Jansen. "Effects of noise on health and well-being" *Stress Medicine* **1**, 183–191 (1985).
7. L. Fagin, "Stress and unemployment" *Stress Medicine* **1**, 27–36 (1985).
8. N. Morris, "Factors influencing population control. The 12th Jennifer Hallam Memorial Lecture" *J. Obstet. and Gynaecol.* **6**(Suppl. 2), 571–576 (1986).
9. P. Taggart and M. Carruthers, "Behaviour Patterns and Emotional Stress in the Etiology of Coronary Heart Disease: Cardiological and Biochemical Correlates" **In**: *Stress and the Heart*, 2nd edition (ed. D. Wheatley) (Raven Press, New York, 1982) pp. 25–37.
10. K. Ball and R. Turner, "Smoking and the heart. The basis for action" *Lancet* **ii**, 822 (1974).
11. M. J. Horowitz, R. Benfari, S. Hulley, *et al.*, "Life events, risk factors and coronary heart disease" *Psychosomatics* **20**(9), 586–592 (1979).
12. A. L. Klatsby, G. D. Friedman and A. B. Sigelaub, "Alcohol consumption before myocardial infarction" *Ann. Intern Med.* **81**, 294–301 (1974).
13. C. H. Hennekens, B. Robner and D. Cole, "Daily alcohol consumption and coronary heart disease" *Am. J. Epidemiol.* **107**, 196–200 (1978).
14. O. Paul, M. H. Lepper, W. H. Phelen, *et al.*, "A longitudinal study of coronary heart disease" *Circulation* **28**, 20–31 (1963).
15. L. Wilhelmsen, H. Wedel and G. Tibblin, "Multivariate analysis of risk factors for coronary heart disease" *Circulation* **48**, 950–958 (1973).
16. A. S. St. Leger, A. L. Cochrane and F. Morre, "Factors associated with cardiac mortality in developed countries with particular reference to the consumption of wine" *Lancet* **i**, 1017 (1979).
17. J. Juntunen, S. Asp, M. Olkinuora, *et al.*, "Doctors' drinking habits and consumption of alcohol" *Br. Med. J.* **297**, 951–954 (1988).

18. T. Stockwell, "Stress and alcohol" *Stress Medicine* **1**, 209–215 (1985).
19. M. King, "At risk drinking among general practice attenders: prevalence, characteristics and alcohol-related problems" *Br. J. Psychiat.* **148**, 533–540 (1986).
20. M. Kenny and A. Darragh, "Caffeine: an innocuous stimulant?" *Psychiatry in Practice* **4**(26), 15–18 (1985).
21. J.-P. Boulenger and T. W. Uhde, "Caffeine consumption and anxiety: preliminary results of a survey comparing patients with anxiety disorders and normal controls" *Psychopharmacol. Bull.* **18**(4), 53–57 (1982).
22. J. Onrot, M. R. Golfberg, I. Biaggioni, *et al.*, "Hemodynamic and humoral effects of caffeine in autonomic failure" *N. Engl. J. Med.* **313**(9), 549–554 (1985).
23. W. V. Mendelson, D. Garnett, J. C. Gilling, *et al.*, "The experience of insomnia and daytime and night-time functioning" *Psychiatry Res.* **12**, 235–250 (1984).
24. M. Partinen, "Sleep and coronary heart disease" *Stress Medicine* **1**, 135–141 (1985).
25. K. Adam and I. Oswald, "Protein synthesis, bodily renewal and the sleep-wake cycle" *Clin. Sci.* **65**, 561–567 (1983).
26. M. Partinen, "Stress and the heart: the sleep factor" *Stress Medicine* **4**, 253–263 (1988).
27. D. Wheatley, "Zolpidem and placebo. A study in general practice in patients suffering from insomnia" **In**: *Imidazopyridines in Sleep Disorders*, Volume 6 (eds. J. P. Sauvanet, S. Z. Langer and P. L. Morseli) (Raven Press, New York, 1988) pp. 305–316.
28. H. K. Hellestein and E. H. Friedman, "Sexual activity and the postcoronary patient" *Arch. Intern. Med.* **125**, 987–999 (1970).
29. Annotation, "Coitus and coronaries" *Br. Med. J.* **i**, 414, (1976).
30. R. Pelkonen, E. A. Nikkila, S. Koskinen, *et al.*, "Association of serum lipids and obesity with cardiovascular mortality" *Br. Med. J.* **ii**, 1185 (1977).
31. J. Tuomilehto, J. T. Salonen, B. Marti, *et al.*, "Body weight and risk of myocardial infarction and death in the adult population of eastern Finland" *Br. Med. J.* **295**, 623–627 (1987).
32. P. D. Wood, M. L. Stefanick, D. M. Dreon, *et al.*, "Changes in plasma lipids and lipoproteins in overweight men during weight loss through dieting as compared with exercise" *N. Engl. J. Med.* **319**, 1173–1179 (1988).
33. M. Friedman and R. H. Rosenman, "Association of specific overt behavior pattern with blood and cardiovascular findings: blood cholesterol level, blood clotting time, incidence of arcus senilis, and clinical coronary artery disease" *J.A.M.A.*, **169**, 1286 (1959).
34. D. R. Raglan and R. J. Brand, "Type A behavior and mortality from coronary heart disease" *N. Engl. J. Med.* **318**, 65–69 (1988).
35. A. M. Rissanen and E. A. Nikkila, "Coronary arterial disease and its risk factors in families of young men with angina pectoris and in controls" *Br. Heart J.* **39**, 875–883 (1977).
36. D. Wheatley. "Treatment of the anxiety factor" **In**: *Stress and the Heart*, 2nd edition (ed. D. Wheatley) (Raven Press, New York, 1981) pp. 39–62.
37. H. Tunstall-Pedoe, "Autres pays, autres moeurs" *Br. Med. J.*, **297**, 1559–1560 (1988).
38. D. Wheatley, "The Assessment of Stress" **In**: *Human Psychopharmacology. Measures and Methods.* Vol. 1. (eds. I. Hindmarch and P. D. Stoner) (John Wiley and Sons, 1987) pp. 47–67.
39. J. W. Jefferson, "Anxiety and the aching heart" *Stress Medicine* **3**, 33–39 (1987).
40. D. Wheatley, "Anxiolytic drug use in cardiovascular disease: An Overview" *Psychopharm. Bulletin* **20**, No. 4, 649–659 (1984).
41. R. Friedman, "Psychogenic hypertension" *Stress Medicine* **1**, 5–8 (1985).
42. S. Cobb and R. M. Rose, "Hypertension, peptic ulcer, and diabetes in air traffic controllers" *J.A.M.A.*, **224**, 489 (1973).
43. R. Mundal, J. Erikssen, R. Bjorklund and K. Rodahl, "Elevated blood pressure in air traffic controllers during a period of occupational conflict" *Stress Medicine* **6**, 141–144 (1990).
44. A. M. Ostfield, S. V. Kasl, D.A.D'Atri and E. F. Fitzgerald (eds.), *Stress, Crowding, and Blood Pressure in Prison*, (Lawrence Erlbaum Associates, New Jersey, 1987).
45. J. S. Gill and D. G. Beevers, "The relationship of hypertension and its treatment with stress and well-being" *Stress Medicine*, **1**, 279–286 (1985).
46. D. Wheatley, "The Wheatley Stress Profile" *Stress Medicine* **9**, 5–9 (1993).
47. J. Harding and E. Zimmermann, "Psychiatric symptoms, cognitive stress and vulnerability factors. A study in a remand prison" *Brit. J. Psychiat.* **155**, 36–43 (1989).
48. D. I. Orton and J. J. Gruzelier, "Adverse changes in mood and cognitive performance of house officers after night duty" *Brit. Med. J.* **298**, 21–23 (1989).
49. M. R. Trimble (ed), *Benzodiazepines Divided* (John Wiley, Chichester, 1983).
50. R. France, "The general practitioner and the anxious patient" **In**: *The Anxiolytic Jungle: where Next?* (ed. D. Wheatley) (John Wiley and Sons Ltd., Chichester, 1990) pp. 79–98.
51. D. Wheatley, "The new alternatives" **In**: *The Anxiolytic Jungle : Where Next?* (ed. D. Wheatley) (John Wiley and Sons Ltd., Chichester, 1990) pp. 163–183.

52. D. Wheatley, "Beta-blockers in anxiety: the stress connection" **In**: *The Anxiolytic Jungle: Where Next?* (ed. D. Wheatley) (John Wiley and Sons Ltd., Chichester, 1990) pp. 137–152.
53. D. Wheatley, "Beta-blocking drugs in anxiety" **In**: *Stress and the Heart*, 2nd edition (ed. D. Wheatley) (Raven Press, New York, 1981) pp. 73–89.
54. J. A. Street, B. A. Hemsworh, A. G. Roach and M. D. Day, "Tissue levels of several radio-labelled B-adrenoceptor antagonists after intravenous administration in rats" *Arch. Pharmacodyn. Ther.* **237**, 180 (1979).
55. G. W. Brown and T. O. Harris, "Social class, psychiatric disorder of mother, and accidents to children" *The Lancet* **i**, 378–381 (1978).
56. A. House, "Depression after stroke" *Brit. Med. J.* **294**, 76–78 (1987).
57. P. E. Mullen, S. E. Romans-Clarkson, V. A. Walton and G. P. Herbison, "Impact of sexual and physical abuse on women's mental health" *The Lancet* **i**, 841–845 (1988).
58. T. Harding and E. Zimmermann, "Psychiatric symptoms, cognitive stress and vulnerability factors. A study in a remand prison" *Brit. J. Psychiat.* **155**, 36–43 (1989).
59. H. N. Sno, J. G. Storosum and J. A. Swinkels, "HIV infection: Psychiatric findings in the Netherlands" *Brit. J. Psychiat.* **155**, 814–817 (1989).
60. A. Brier, M. Albus, D. Pickar, T. P. Zahn, O. M. Wolkowitz and S. M. Paul, "Controllable and uncontrollable stress in humans: alterations in mood and neuroendocrine and psychophysiological function" *Am. J. Psychiatry* **144**, No. 11, 1419–1425 (1987).
61. R. I. Wang and C. Kocher, "The choice of rating form for evaluating anxiolytics" *Psychopharm. Bull.* **25**(2), 291–5 (1989).
62. J. Horne, "Why we Sleep" Oxford University Press, Oxford (1988).
63. D. Wheatley, "Prescribing short-acting hypnosedatives" *Drug Safety* **7**(2), 106–115, (1992).
64. D. D. Breimer, R. Jochemsen, "Pharmacokinetics of hypnotics drugs" **In**: *Psychopharmacology of Sleep* (ed. D. Wheatley) (Raven Press, New York, 1981) pp. 135–152.
65. D. Wheatley, "Effects of drugs on sleep" **In**: *Psychopharmacology of Sleep* (ed. D. Wheatley) (Raven Press, New York, 1981) pp. 153–176.
66. M. A. Jacobs, A. Spilken and M. Norman, "Relationship of life change, maladaptive aggression, and upper respiratory infection in male college students" *Psychosomatic Medicine* **31**, No. 1, 31–44 (1969).
67. L. E. Hinkle, "The effect of exposure to cultural change, social change, and changes in interpersonal relationships on health" **In**: *Stressful life events: Their nature and effects* (eds. B. S. Dohrenwend and B. P. Dohrenwend) (Wiley, New York, 1974).
68. S. B. Friedman and L. A. Glasgow, "Psychologic factors and resistance to infectious disease" *Paediatr. Clin. N. Am.* **13**, 315–335 (1966).
69. Editorial, "Depression, stress, and immunity" *Lancet* **i**, 1467–1468 (1987).
70. H. Moldofsky, F. A. Lue, J. Eisen, E. Keystone and R. M. Gorczynski, "The relationship of interleukin-1 and immune functions to sleep in humans" *Psychosom. Med.* **48**, 309–318 (1986).
71. J. Palmblad, B. Petrini, J. Wasserman and T. Akerstedt, "Lymphocyte and granulocyte reactions during sleep deprivation" *Lancet* **i**, 834–836 (1979).
72. D. Wheatley, "Stress and the immune system" *Healthways Communications*, New York (1990).
73. A. Cheang and C. L. Cooper, "Psychosocial factors in breast cancer" *Stress Medicine* **1**, 61–66 (1985).
74. A. J. Ramirez, T. K. J. Craig, J. P. Watson, *et al.*, "Stress and relapse of breast cancer" *Brit. Med. J.* **298**, 291–293 (1989).
75. D. Spiegal, J. Bloom, H. C. Kraemer, *et al.*, "Effect of psychosocial treatment on survival of patients with metastatic breast cancer" *Lancet* **2**, 888–891 (1989).
76. J. S. Goodwin, W. C. Hunt, C. R. Key and J. M. Samet, "The effect of marital status on stage, treatment, and survival of cancer patients" *J.A.M.A.* **258**, No. 21, 3125–3130 (1987).
77. I. D. Vellacott and P. M. S. O'Brien, "Premenstrual stress" *Stress Medicine* **4**, 33–40 (1988).
78. D. Wheatley, "The stress profile" *Brit. J. Psychiat.* **156**, 685–688 (1990).
79. D. Gath and S. Iles, "Depression and the Menopause" *Brit. Med. J.* **300**, 1287–88 (1990).
80. J. Holst, T. Backstrom, S. Hammarback and B. von Schoultz, "Progestogen addition during oestrogen replacement therapy—effects on vasomotor symptoms and mood" *Maturitas* **11**, 13–20 (1989) (Elsevier Scientific Publications Ireland Ltd).
81. D. Wheatley, "Hormone replacement therapy in post-menopausal depression and the female climacteric" *Curr. Med. Res. and Opinion* **4**, Suppl. 3, 37–45 (1977).
82. D. Wheatley, (ed.), *Psychopharmacology and Sexual Disorders* (Oxford University Press, Oxford, 1983).
83. W. A. Littler, A. J. Honour and P. Sleight, "Direct arterial pressure, heart rate, and electrocardiogram during human coitus" *J. Reprod. Fertil.* **40**, 321–329 (1974).
84. W. B. Tuttle, W. L. Cook and E. Fitch, "Sexual activity in postmyocardial infarction patients" *Am. J. Cardiol.* **13**, 140 (1964).

85. L. R. Derogatis and K. M. King, "The coital coronary : a reassessment of the concept" *Arch. Sex. Behav.* **10**, 325–335 (1981).
86. G. Fletcher, B. Johnston and J. Cantwell, "Dynamic electrocardiographic monitoring during coitus in patients post myocardial infarction and revascularisation" *Circulation* **57**, 204 (1978).
87. J. Catalan, "Stress and sexual response" *Stress Medicine*, **2**, 45–53 (1980).
88. S. Shapiro, D. Slone, L. Rosenberg, *et al.*, "Oral contraceptive use in relation to myocardial infarction" *Lancet* **1**, 743–746 (1979).
89. R. B. Wallace, J. Hoover, E. Barrett-Connor, *et al.*, "Altered plasma lipid and lipoprotein levels associated with oral contraceptive and oestrogen use" *Lancet*, **2**, 111–114 (1979).
90. D. Wheatley, (ed.), *Psychopharmacology of Old Age* (Oxford University Press, Oxford, 1982).
91. K. R. Gilhome Herbst, "Deafness, dementia, and depression" **In**: *Psychopharmacology of Old Age* (D. Wheatley, ed.) (Oxford University Press, Oxford, 1982) pp. 145–155.
92. R. G. Walton, "Stress factors in dermatology" *Stress Medicine* **1**, 55–60 (1985).
93. C. S. Koblenzer, "Stress and the skin: significance of emotional factors in dermatology" *Stress Medicine* **4**, 21–26 (1988).
94. S. C. Wessely and G. H. Lewis, "The classification of psychiatric morbidity in attenders at a dermatology clinic" *Brit. J. Psychiat.* **155**, 686–691 (1989).
95. G. E. Vaillant, "Natural history of male psychologic health" *N. Engl. J. Med.* **301**, No. 23, 1249–1254. (1979).

2. SOME EFFECTS OF THE ENVIRONMENT ON EMOTIONS AND THEIR RELATIONSHIPS TO CARDIOVASCULAR DISEASES

RAY H. ROSENMAN*

This discussion deals with some effects of environmental stress on emotions that relate to cardiovascular function and diseases and which underlie the burgeoning field of behavioral medicine. Emotions such as anxiety and anger can affect cardiovascular function and are related to complications of cardiovascular diseases, some of which are discussed. The evidence fails to support a role of emotions on cardiovascular reactivity in the pathogenesis of hypertension. However, psychosocial and behavioral factors play a significant role in ischemic heart disease, and the causal relevance of Type A behavior pattern is reviewed. Particular importance is given to non-evolutionary Type A competitive behaviors, rather than to the hostility dimensions which appear to have a more superficial and unproven relationship.

1. INTRODUCTION

Consideration of environmental influences on human health usually calls to mind the effects of toxic materials and adverse ecological factors. However, environmental stressors also affect emotions that are related to cardiovascular function and diseases.

Ecology deals with relations of an organism to its environment and evolution with its ability to adapt to environmental changes. Basic to mammalian physiology is a need to maintain the internal milieu. Under threat, the organism is mobilized into adaptive efforts to regain its equilibrium by homeostatic adjustments. An early phylogenetic reaction involves a primitive sensory-motor system that is a simple neural reflex arc or coordinated in a primitive brain. Aside from primitive reactions to light and touch that are useful in nutrition and reproduction, touch stimulation usually evokes a flight response, the reaction of withdrawal that is commonly observed in invertebrates and lower vertebrates such as fishes.

Reptiles evolved another reaction, that of sustained snapping or the fight response. The response to turn and fight that is common to all living reptiles is an optional one in higher forms. In addition to a simple sensory-motor system, many vertebrates have a response choice that results from a limbic brain dispositional system in which sensory and motor reactions and responses are coordinated. This is primitive in reptiles, but evolved in higher forms, particularly in mammals. In addition to fight or flight, it permits a reaction of affection, as in

*Health Sciences Program, SRI International, Menlo Park, California, USA.

the licking and turning to grooming that is requisite for domestication of animals and pets and, perhaps, even of husbands, and underlies maternal-offspring relationships and the development of pack, family, clan, tribe, and communities. There is evidence that behavioral reactions have not only evolved, but that evolutionary somatic progression has been influenced by organized central nervous system (CNS) activity that coordinates the behavioral responses and changes that favor the organism's and species survival.

2. THE HYPOTHALAMIC DEFENSE REACTION AND FIGHT-FLIGHT RESPONSES

There are several closely coordinated neurohormonal changes that pattern instinctual responses to an environmental challenge. The hypothalamic defense reaction is described[1] as a highly differentiated response that overrides brainstem homeostatic mechanisms and adjusts the cardiovascular system in an anticipatory fashion that prepares it for vigorous physical activity to meet environmental challenges, and is thus highly suited for fight-flight. It is readily provoked in the conscious organism in response to any sudden or novel environmental stimulus and is characterized by increased cardiac output, vasodilatation in skeletal muscles, and vasoconstriction in the kidney, skin and particularly in the splanchnic area, with an acute rise of heart rate and blood pressure. The rise of cardiac output is directed to skeletal muscles and heart, and passively to the brain through a rise of perfusion pressure.[2] The defense reaction thus elicits changes closely similar to those during strenuous muscular exercise, in which active muscle vasodilatation is produced by vasoactive metabolites in response to increased oxygen need.[2]

The fight-flight response is an early part of the defense reaction. Cannon[3] found the sympathoadrenal system (SAS) involved in rapid animal fight or flight responses to life-threatening challenges and Selye[4] later found an importance for cortisol, and that humans respond differently, influenced by age, sex, genetics, and various environmental factors. His emphasis was on exogenous stressors, in contrast with an earlier one on endogenous personality. Lazarus[5] observed that neurohormonal responses depend upon individual stress perception, and that effective coping methods minimize distress. The conceptual evolution went from a general model of stress response to complex, dynamic interactions between environmental stressors and individual perception and coping. It is of interest that this was recognized about two millennia ago, when Marcus Aurelius noted that "If you are pained by any external thing, it is not the thing that disturbs you, but your judgment about it. And it is in your power to wipe out this judgment now."

The CNS integrates information from the organism and the environment and provides neurohormonal adjustments essential to blood pressure maintenance. A complex interaction between the heart and vasculature is involved in pressor responses to environmental challenges, mediated via autonomic nervous system (ANS) and local factors responsible for arteriolar autoregulation. The CNS is therefore highly involved in SAS responses to perceived stress and in blood pressure regulation. Mental stress elicits a circulatory response akin to that evoked by the defense reaction, and the SAS is activated by noise, crowding, other stress associated with industrialization and urbanization, and by emotions. Stress releases catecholamines[6] and the fight-flight response[3] is driven by them and the renin-angiotensin system.[1,7] However, physiological responses result from selective neuronal activation rather than from uniform or generalized sympathetic nervous system (SNS) response.[8] Moreover, the defense reaction results from an arousal of the central amygdala, linked with the locus ceruleus, and noradrenergic responses override epinephrine effects.[1] Although the SAS was emphasized by Cannon,[3] the parasympathetic nervous system also is involved, and an increased heart rate may result more from vagal withdrawal than SNS activity.

Some[7] believe that the Cannon's SAS responses can be separated from Selye's alarm response. In addition to the defense reaction activated by the fight-flight response, they find a defeat reaction, and that different response patterns involve the defense and defeat components of the neuroendocrine stress responses with which Cannon,[3] Selye,[4] and Mason *et al.*[9] were concerned. They distinguish two pathways for stress responses and believe that fight-flight SAS responses differ in emotional origins and pathophysiological consequences from the Selyean adrenocortical mode associated with passive stance, withdrawal, and defeat behavior, and differentiate the confident anger of a dominant reaction and the anxious sadness or depression of a subordinate one. The alarm response[4] involves arousal of the pituitary-adrenocortical axis system (PAS).[7] For example, humans with specific phobias react with PAS arousal during defense reactions elicited by viewing pictures of specific phobic but not other objects.[10]

Distinct responses to environmental stressors can be related to the extent of an organism's control over the environment.[7] Thus, when animals retain control over their environment during challenges, they are observed to respond with SAS and aggressive activity, but when this control is prevented, they become defeated, withdrawn, immobilized, and release cortisol. The perception of loss of control associated with a defeat response is marked by increased release of ACTH and adrenal hypertrophy.[7] The defeat response mode occurs when environmental situations elicit feelings of frustration and despair in mammals, and a struggle without perception of winning in animals can cause death in this struggle-defeat syndrome.[1,7] The SAS is primary in fight-flight, and the PAS in defeat-withdrawal reactions in which cortisol is released, and hormonal patterns may be accompanied by less defined changes in activity of the amygdala nuclei, tractus solitarius, and locus ceruleus.[7]

The defense reaction causes an acute rise of blood pressure that may hemodynamically be due to a rise of cardiac output and/or increase of total peripheral vascular resistance (TPR). The usually increased cardiac output suggests that it is integral, rather than a reflex consequence of peripheral vasodilatation.[2] A predominant TPR response was once associated with a suppressed hostility prone to occur in hypertensives,[11] and some subjects may have a combination of the two mechanisms, with predominant rise of cardiac output more common in normotensives.[12] Brod and associates[2] confirmed the dual pressor mechanism in defense reactions, but found this to occur variably from increased cardiac output or increased TPR. The defense reaction in both produces renal and splanchnic vasoconstriction and vasodilatation in muscles, without qualitative differences in response patterns found by others.[11,12] The net change in TPR thus reflects the balance between vasoconstriction in viscera and vasodilatation in muscles, with an increase of cardiac output and shift of blood flow from viscera to muscle that closely resembles circulatory changes that accompany strenuous muscular exercise, the only difference being in the balance between regional vasoconstriction and vasodilatation, and which predominates from relative regional intensity. Brod *et al.*[2] also found no difference in qualitative responses between normo-

and hypertensives, but that the latter tend to quantitatively respond with greater renal vasoconstriction, less skeletal muscle vasodilatation, and hemodynamic responses that may persist longer after cessation of an emotional stimulus.

It is unlikely that there are intrinsic fight-flight response differences in the defense reaction. In response to environmental threats, both appear similarly to activate the SAS, release catecholamines and testosterone, and increase heart rate, myocardial contractility, cardiac output, TPR, blood pressure, and platelet aggregation. However, a chronic state of vigilance activates the PAS, with increased release of ACTH and cortisol, and decreased release of testosterone. Consistent with findings in animal models,[1,7] both SAS and PAS systems are activated in humans when situations are either monotonous or involve aggressive, competitive responses that are accompanied by feelings of distress, while situational responses without perceived distress are found to activate the SAS and suppress PAS responses.[13]

The question arises whether the PAS defeat response mode is part of the defense reaction which is readily and rapidly provoked in conscious organisms even by minor environmental stimuli.[1] Fight-flight is a primitive, instinctual response that prepares for vigorous physical exertion, while the PAS response is associated with chronic mental vigilance, withdrawal, immobilization, and defeat,[7] and probably differs from the defense reaction and fight-flight responses both in emotional origins and temporal relationships to threats. It is also possible that the mammalian defense reaction may partly be subcortical and not involve cognitive perceptions that appear integral to the defeat response. An immediacy of the hypothalamic defense reaction[1–3] contrasts with delayed cognitive PAS responses,[1,7] and it may therefore be questioned whether the defeat response is a valid part of the hypothalamic, fight-flight defense reaction.

3. THE EMOTION OF ANXIETY

The defense reaction may particularly occur in situations that cause anxiety,[2] which is defined as "a specific unpleasurable state of tension which indicates the presence of some danger to the organism,"[14] as "the apprehensive tension or uneasiness which stems from the anticipation of imminent danger, in which the source is largely unknown or unrecognized,"[15] and "as a psychobiological emotional state or reaction that can be distinguished from other emotions such as anger or sadness by unique experiential qualities. An anxiety state consists of unpleasant feelings of tension, apprehension, nervousness, and worry, and activation of the ANS."[16] The unpleasantness of anxiety is invariably stressed, making it an important force in human existence. Moreover, there appear to be many reasons to equate perceived stress with anxiety.[17–19]

Anxiety is an affective reaction at the end of a long, evolutionary chain going back to protoplasmic irritability in protozoans. It is a primitive, basic, physiological, and affective response to the perception of danger that should not be misconstrued as abnormal. Anxiety serves an important biologic function that involves several brain neurotransmitters, but without an identified major brain dysfunction, consistent with it being a biologically useful reaction to danger that increases alertness and cardiac output, and otherwise prepares an organism for fight-flight. It is a universal emotion that has served as an evolutionary force, contributed to biological adaptation and reinforced processes of social bonding and communal living.[18,20]

Anxiety is thus a constructive force that spurs the organism to overcome threats to homeostasis, and refers to a signal derived from present or future threats that the organism uses to restore homeostasis, mobilize for defense and self-preservation, and make adaptive efforts to cope with future threats. When the ego's efforts fail to master threats, however, anxiety can handicap further adaptive efforts, and be worst when threats are nameless dreads which the ego cannot cope with, and which may be repressed, suppressed, or lead to psychosomatic conversion mechanisms. Extreme anxiety can lead to mental disorganization and ego-paralysis, and when habitually inappropriate, to generalized anxiety and panic disorders, psychosomatic and mental illness, and other dissociation from reality.[21]

A profundity of new and unique chronic threats to homeostasis are both natural and man-made problems that include increasing existential anxiety due to constant threat of environmental disasters and even nuclear extinction. Primitive dangers have been replaced by environmental stressors that are more insidious and associated with heightened fear and anxiety. It is said[22] that "Contemporary man, in much of the world, is faced every day with people and with situations about which there is uncertainty of outcome, where appropriate behavior is not prescribed and validated by tradition, where the possibility of bodily or psychological harm exists, where running or fighting is inappropriate, and where mental vigilance is called for." Humans are more preoccupied with power and destruction, and we have acquired the faculty of total destruction before learning how to prevent it. Man has ascended the ladder of power, with a descending ability to control a rapidly-expanding technology and without the knowledge needed for survival.

Lazarus[5] believed that Cannon's fight-flight[3] is a differential response to life-threatening situations in animals, while modern stress is very different in the urbanized and industrialized world. However, Folkow[1] pointed out that the same defense reaction is similarly activated, whether in response to major threats or when the organism is even slightly alerted in normal waking activities. Although many stressors that now evoke anxiety responses are different from those in which the emotion evolved,[23] we still use the same primitive defense reaction to face our new and more abstract stresses, ego threats and mental challenges, a carryover from lower animals and ancestral humans that still prepares the organism for maximum muscular activity for self-preservation. However, somatic responses may now be partly suppressed by social inhibitions.

Anxiety sets off other affective responses, such as fear and worry, which probably infrequently occur without some anxiety, and with no sharp line of demarcation. However, anxiety is a subjective response to less tangible, anticipated threats, while fear is a reaction to known and more objective ones. Thus, fear is more of the present, and anxiety more of the future. While anxiety is a primitive affective response, worry is an effort to cope at the intellectual level with justifiable anxiety and the apprehensive thoughts that are mobilized by anxiety in the organism's efforts to cope with anticipated dangers. Worry thus differs from anxiety, albeit often being set off by it.

Although the vertebrate limbic control system reacts similarly, whether a stimulus elicits anxiety or fear-worry, it determines whether there will be fight or flight, and the intensity of associated SNS and catecholamine responses. For example, in the acute reaction, the limbic system determines whether a person responds to sight of a rattlesnake by fleeing or turning to fight it. A poor grade may lead a student to react with anxiety, or with an intellectual and adaptive worry response that can either lead to coping with the threat by studying harder and thereby eliminating it, or to a new homeostatic equilibrium achieved by accepting the fact and setting sights more realistically. The system also participates in whether the student approaches another examination with anticipatory anxiety, fear, resentment, or optimism.

States of consciousness are accompanied by somatic expressions. An acute change can result in the vasoconstrictive pallor of anger or fear, or in the vasodilatory flush of pleasure or embarrassment. However, physiological changes that are associated with anxiety are due to increased beta-adrenergic and other ANS activity that reflect its secondary physiological expression rather than pathology.[24]

These are relatively acute responses. But what happens when consciousness is threatened by stimuli that are constant threats to homeostasis, whether real or imagined? Repetitive sensory input can evoke conditioned or behavioral responses and, when effected by reverberatory neural circuits, can participate in causalgic-like states. An excess of repetitive stimuli also can lead to emotional and mental problems. Attesting to the fragility of our consciousness is the increase of divorce, faddism, addictions, suicide, neurosis, chronic anxiety disorders, psychosomatic ailments, and mental illness that are common to industrialized, urbanized modern societies, but little known in more primitive civilizations. The same factors may play a role in an apparent widespread increase of hostility/anger dimensions and sense of alienation. The risk of psychosomatic illness and neurosis stemming from emotional causes results from our inability to cope with unknown threats to consciousness and partly from our ignorance of mechanisms.

Investigators attempted to differentiate physiological changes in the defense reaction that might be associated with different emotional stress responses. Some believed that Cannon's fight-flight response can be separated into fight (anger) or flight (fear) responses, and that their hemodynamics may differ.[25–28] Anger responses were likened to those during isometric exercise, with a combined norepinephrine-like and epinephrine-like peripheral response accompanied by increases of both cardiac output and TPR. Fear responses were likened to isotonic exercise, with increased cardiac output and predominant epinephrine responses that decreased TPR. However, although emotional arousal may raise blood pressure more during anger and anxiety than happiness, newer ambulatory monitoring methodology in natural environments finds only quantitative differences, and that even these are quite small.[29]

4. HOSTILITY AND ANGER EMOTIONS

Aside from relationships of anxiety, most attention is given to hostility/anger dimensions that are often related to environmental stress. It has been said[30] that man unceasingly hurries to increase productivity so as to consume larger amounts of useless things. Industrialized, urban areas and, to a lesser extent, even more remote areas are characterized by noise, deadlines, information overload, mobility, and the rapid pace of a civilization of transistors, mechanization, and "hot" nuclei, in which man has discovered and produced before developing the faculties of control. He must ceaselessly consume the manufactured goods which keep at work the machines which in part enslave him more and more.

Toynbee[31] emphasized that our stresses are uniquely new, and of a type never before witnessed in world history. Alexis de Tocqueville[32] was long ago amazed at the rapid pace of life in urban America. These uniquely new stresses doubtless contribute to anxiety neuroses, anger, and increasing divorce, addictions, suicide, and psychosomatic illness. There may be a smaller bulk of so-called adapted or normal humans, with an increasing fringe on either side, one being an increase of those who manifest a primitive reaction of withdrawal. Evolutionary changes require prolonged time, and humans have had insufficient time to adapt to ceaseless socioeconomic and psychosocial agitation. Until recently, there was greater tolerance of environments that supersede nature and constantly impose threats on homeostasis. However, we may be fortunate in having inadequate time for evolutionary development, since the more perfectly a species adapts to its environment, the closer it may come to extinction, as it becomes less able to adapt and survive major environmental changes.

Our knowledge and technology have not given adequate attention to man's affective and moral form, or to his inner life, character and aesthetic, or psychological needs and relationships to mental and spiritual environment. Science and industrial productivity have failed to increase intelligence, moral discipline, or to provide improved nervous equilibrium, security, or serenity. In our modern industrialized society, there is less belief in immortality, faith in a personal, protective God, and ability to rely on one's neighbors or even relatives. Our industrialized, geographic mobility has resulted in partial loss of family protection and social support networks, and the security that comes with a community of friends, and that does not induce the competitive feeling that often occurs when one seeks identity after moving to a new community. Carrel[30] believed that, as a whole, man has lost the clarifying power of human consciousness, which is the only real protection of human life against a world preoccupied with speed, mechanization, power, useless productivity, transistorized intelligence, destruction, and other factors that demoralize and lead to loss of our noblest traits.

This is a pessimistic outlook, but modern society has fostered conformity, socioeconomic insecurity, and loss of identity, family relationships and many true satisfactions. The so-called normal human is content with his lot, getting along with spouse or a companion, and almost as satisfied to have a job, despite increasing preoccupation with job stress, and is perhaps increasingly lacking in creativity, imagination, and spontaneity. His personality is often less stimulating, and aspirations for self and children even more limited, with a more elementary form of consciousness that is increasingly capable of doing only the relatively easy work which can assure survival in modern society. More humans produce, consume, ask few questions, and are satisfied with food, alcohol, cinema, television, spectator sports, and being rapidly transported while watching the fleeting scenery or TV enroute. The problems of modern education are increasingly emphasized.

The essential qualities that distinguish humans from all other forms have long been pondered. Some have used theology, saying that God created man in his own image, and others have used biology in calling him an unfettered biped or talking ape. Man is the only organism that simultaneously seeks the consolation of certainty in traditions of the past with that of hope in the future. The radical ignores the past, and the reactionary seeks to preserve it at all costs, despite its possible uselessness. Many others are only concerned with the present, and specifically their own present and the industrial bottom line. However, whereas primitive man was satisfied with adequate sustenance, modern man in an industrialized society faces a new competition in socioeconomic and psychological realms. This may be inappropriate from the evolutionary standpoint, since it contrasts with a truer biological competition that in lower forms is concerned with food and survival, and in primitive man with food, shelter, and survival.

Perhaps a major goal of existence for Zen Buddhists and others is serenity as a state of mystical expression, and as a high attainment of the spirit. But this may be the very opposite of modern man's so-called normalcy, in which many humans conform, and accept a loss of identity and security. However, we may be witnessing a swing of the pendulum, with increasing alienation and rejection. True serenity may be the opposite of passivity, apathy, and resignation, while being the refutation of disorder and hysteria, and depending upon a harmony with existence, lucidity, calmness, and active willingness to participate.

Cassirer[33] distinguished man from invertebrates and all other vertebrates as the only animal that is time-conscious and time-consuming. Our modern society has above all been marked by its rapid pace, fostered by every form of mechanization, and by a pervading accent on speed. Although some attempt to flee from it, the bulk of society tends to accept this. Some variously adapt to the new stresses of industrialized society, develop an excessive sense of competitiveness, and teach it to their children. They may develop an aggressive and ambitious approach to life, and are habitually immersed in a multiplicity of vocational and avocational pursuits that are subject to deadline pressures.

Among many ways to divide humans is to contrast the doers who like to get things done, with those who have to be driven. The doers aggressively grapple with all challenges perceived as relevant, compete for advancement, and seek recognition in a society that attempts to plunge them into an abyss of conformity and loss of identity. They are pleased to have at their disposal every form of timesaving, mechanized gadgetry, and are possessed with a sense of lack of time to do everything they wish to accomplish each day. They are impatient, detest waiting in lines and otherwise being delayed, and epitomize Cassirer's definition of hominid man.

Their motivations are probably many, sometimes stemming from a failure to derive adequate satisfactions from family, work, and other life-pursuits. It is probably not very often that a person who sells a boring product can derive a real sense of satisfaction from increased sales. Others derive inadequate gratification from their own past and more important achievements. As fast as one rung of the ladder is reached, the next is already being viewed, and they climb as fast as possible toward vague and ill-defined goals that can rarely be attained. Their truly important accomplishments are infrequent and creative ones rare. Such persons may develop enhanced hostility/anger dimensions.

It remains to be seen if these psychosocial, behavioral and emotional factors are causally related to cardiovascular diseases. A relationship of psychosocial and behavioral factors to hypertension will first be considered.

5. RELATIONSHIPS OF ANXIETY TO HYPERTENSION

The SNS is a major determinant of blood pressure homeostasis, as in circadian rhythmicity and in stress responses, and has an important role in hypertension. Borderline hypertensives often exhibit increased SNS activity and pressor responses to stressors, and tend to have an elevated cardiac output, and their normal TPR is inappropriate relative to their increased cardiac output.[34] They have a neurogenic hypertension that is strongly related to central control of the ANS, increased SNS activity and decreased parasympathetic cardioinhibition.[34]

Since the cardiac output remains essentially unchanged across the spectrum of hypertension, an age-related increase of TPR may explain the hemodynamic transition from borderline to established hypertension. Older established hypertensives tend to have lower cardiac output, heart rate, and stroke volume, and higher renal and TPR.[35] However, sustained hypertension does not result from a primary increase in TPR, without a concurrent abnormality that affects renal function.[36] In the person who develops hypertension with high cardiac output, the later increase of TPR is a secondary result of hypertension, rather than its primary cause, and structural vascular wall changes may be responsible for the hemodynamic transition.[37,38] The question is whether anxiety, stress or cardiovascular reactivity plays a causal role in these patterns.

Anxiety disorders are viewed in psychological terms that convey images of neurosis, and include generalized anxiety and panic disorders.[21,39] The latter may be due to genetic alterations of brain receptors or unidentified ligands that bind benzodiazepine receptors and make persons overly anxious even when not under stress.[40] Panic attacks are characterized by central alpha-adrenergic discharge involving the locus ceruleus,[39] and this major norepinephrine-containing area seems to be highly involved in anxiety,[24] and in the elicitation of defense reaction-like responses.[2] Chronic anxiety stimulates SNS activity and can induce tachycardia, increased myocardial contractility, and the vasoconstriction that may play a role in hypertension.

Acute anxiety elicits acute pressor responses, but most studies do not find heightened cardiovascular reactivity in anxious subjects or in patients with chronic anxiety disorders, correlations between psychometric scale ratings of anxiety and blood pressure,[41] or significant differences in anxiety levels between normo- and hypertensives. The weak correlations in some hypertensives may be due to selection bias and the labelling phenomenon, since negative correlations found in undiagnosed hypertensives indicate that they even have lower anxiety levels, compared to normotensives.[42]

Behavioral competitiveness may have seminal importance for such other Type A behaviors as aggressive drive, accelerated pace of activities, and increased potential for hostility and anger dimensions, and the underlying basis for this enhanced sense of competitiveness in severe Type A persons is often a deep-seated, covert anxiety that is independent of neuroticism in the latter's usual construct.[43,44] However, as a group, such Type As have no higher resting blood pressure or prevalence of hypertension than in converse Type Bs.[45]

These consistent findings do not support a causal role of anxiety in hypertension. Moreover, compared to controls, patients with generalized anxiety disorders exhibit low blood pressures and prevalence of hypertension.[46] A transient pressor response during acute anxiety can lead to diagnostic problems, as in many with elevation during screening, but who are later found to be normotensive without therapy,[47] a manifestation of "white coat hypertension." Repeated measurements lower blood pressure, but anxiolytics do not cause a sustained fall in those with hypertension-associated anxiety. Moreover, subjects with pressure elevated in the clinic but normal at home, do not differ in anxiety levels, compared to those elevated in both settings.[48]

The CNS integrates information from both organism and environment, provides neurohormonal adjustments essential to blood pressure homeostasis, and is highly involved in SNS responses to perceived stress. Blood pressure is raised by activating neural sites with neuroanatomic connections to the ANS and that connect with higher centers that appraise the environment, and mental stress elicits a circulatory response pattern akin to the hypothalamic defense reaction.[2] Although hypertensives may have heightened SNS responses to mental stressors, physiological responses result from selective neuronal activation rather than from a generalized SNS response.[8] External stressors can elevate pressure via neurohormonal mechanisms but, when the stressor is removed, long term changes are the exception. Thus, a considerable stress load is needed to induce a sustained elevation of blood pressure.

Changing cultural norms from more primitive to more complex also can raise blood pressure in association with anxiety and anger dimensions. However, this involves interactions with genetic and other relevant variables. Genetic factors place an individual at risk for hypertension, but environmental factors act to modulate the prevalence and severity of hypertension.[49,50] There is a strong relationship of hypertension to weight, body fat distribution, and intake of ethanol and sodium.[50] Migration from more primitive to urbanized areas, and the cultural transition to an industrial state may be associated with a rise of blood pressure, but the changes are of small magnitude,[51] and an increased prevalence of hypertension,[49] may be unrelated to social class,[52] low social support,[53] or geographic mobility.[54] Some of the increase is due to false positive hypertension,[55] and much of it to changes of weight and body mass, and to increased ethanol and salt intake.[56]

There is a vast literature on stress and hypertension. However, the Task Forces at a major Conference[57] found no compelling evidence that repeated behavioral pressor elevations can produce sustained hypertension in animal models or in humans in the absence of predispositions engendered, for example, by genetic factors, salt, or loss of renal mass. Recent studies in undiagnosed hypertensives even find inverse relationships between self-reported stressors and hypertension, similar to those with anxiety.[42]

A higher prevalence of hypertension in air traffic controllers has often been ascribed to occupational stress. However, the large prospective study found that future hypertensives had gained more weight and ingested more ethanol, but had significantly less distressing life change events, and were less hard-driving personalities, compared to the normotensives.[58] Moreover, although higher blood pressure may occur in air traffic controllers during occupational conflicts, there is a marked fall after their resolution.[59]

6. RELATIONSHIPS OF CARDIOVASCULAR REACTIVITY TO HYPERTENSION

Some believe that persons with heightened pressor responses to stress are predisposed to hypertension, and that their summated occurrence in natural environments is involved in its pathogenesis. This is based on acute responses in the laboratory during performance of mental and physical psychomotor tasks in some offspring of hypertensives.[60] A corollary would be that individual reactivity patterns generalize to the individual's natural environment.

The many problems associated with laboratory stress testing include lack of standardization for procedures and for definition of heightened responses, and reactivity studies tend to disregard potential influences of a wide variety of confounding variables.[61] Test-retest reliability for laboratory reactivity is low for commonly used stressors and also for ambulatory measures.[62] Responses in children generalize across laboratory procedures that involve qualitatively different behavioral responses,[63] suggesting that individual patterns are genetically determined rather than being related to differences of cognitive perceptions of the laboratory tasks.

Some relate SNS overactivity to hypertension because the hyperkinetic state of borderline hypertensives may be neurogenically-mediated,[34] and offspring of hypertensives are at increased risk.[60] However, their heightened pressor reactivity largely occurs in responses to mental stressors, but is usually of small magnitude, varies widely, and doesn't generalize to dynamic or static exercise, cold pressor testing, orthostatic stress, tilt, or blood volume expansion.[64] It is sometimes disregarded that they do not exhibit generalized autonomic dysregulation, and that blood pressure regulation in hypertensives is normal.[64–66]

Complex interactions between the heart and vasculature are involved in pressor responses mediated through the SNS and baroreceptor function is a homeostatic mechanism that plays a role, but varies considerably.[8] Stressor tasks may affect systolic and diastolic pressures differently, due to different effects on alpha- and beta- adrenoreceptors and neurogenic activity and, during more intense stress, on circulating catecholamines. Moreover, heightened responses may variably be due to increased release or decreased uptake of norepinephrine, to increased peripheral receptor density or responsiveness, or to increased central command over SNS activity. The notion of "hot" versus "cold" cardiovascular reactors is clearly simplistic.

Ambulatory monitoring finds that wide blood pressure fluctuations reflect reactive responses to various physical and mental stressors. If reactivity is causal for hypertension, borderlines should exhibit heightened pressor responses and variability in both the laboratory and natural environment. However, variability in borderlines is no greater than in normotensives, and spontaneous milieu fluctuations are similar in normo- and hypertensives. Moreover, compared to low laboratory reactors, high reactors at best show only small increases of systolic pressure, and no enhanced ambulatory variability in natural milieus. The increased variability accounted for by laboratory stress test reactivity is small and fails to predict either milieu variability or the changes that occur across daily activities.[62]

The results of stress testing thus provide no support for the hypothesis that hypertension is causally related to summated pressor responses over time. Heightened pressor responses in a subset of borderlines to mental stressors is said to be improperly interpreted to mean that behaviorally-induced blood pressure reactivity is the mechanism by which hypertension can develop.[64] The increased risk for hypertension that is associated with heightened reactivity in borderlines is small, and there are no population-based studies that find hyperreactors more likely to develop hypertension over time.[34,64]

Thus, there is little evidence that behaviorally-induced pressor responses or variability can explain either the development of hypertension or a shifting hemodynamic transition to established hypertension,[64,67,68] and critical reviews[67,68] find inadequate evidence to document the hypothesis. It is still possible that repetitive heightened pressor elevations or delay in return to baseline might contribute to structural wall damage in resistance vessels and the hemodynamic transition from borderline to sustained hypertension, and also play a role in cardiovascular complications of hypertension.

Many Type A persons have heightened SNS responses in competitive laboratory challenges and in their occupational milieus,[43] but do not differ from Type Bs in resting pressure or prevalence of hypertension,[45] suggesting that they have no intrinsic reactivity differences, but only that more competitive Type As have a heightened perception of relevant stressors that induces a more active coping style known to enhance SNS activity.[43]

The effects of antihypertensive treatment with a wide variety of drugs are relevant, since they do not decrease pressor responses to mental or physical stressors in the laboratory or diminish reactive pressor responses or ambulatory variability in natural environments.[61] Recent anatomic evidence of a partly independent regulation of resting and reactive blood pressures[69] may explain their ability to lower resting pressures, while having insignificant effects on reactive pressor responses. The dual regulation may also explain why normotensive and borderline and established hypertensives generally have the same circadian patterns, spontaneous fluctuations and ambulatory variability of blood pressure in natural milieus, and why stressful sensory stimuli affect its variability, but not the mean blood pressure itself.

Cardiovascular responses constantly occur in the waking state, whether or not associated with perceived stress. Any type of emotional arousal similarly raises the blood pressure, and differences during all types of emotions are only quantitative and small,[29] with no specificity of SNS response to different stressors or emotional responses.[70] Reactive responses can thus be viewed as physiological and homeostatic, rather than as psychologically reactive,[61] and the findings clearly fail to support old hypotheses about causal roles of anxiety, stress and cardiovascular reactivity in the pathogenesis of sustained hypertension.

7. RELATIONSHIPS OF PERSONALITY TO HYPERTENSION

There has long been emphasis on hostility and difficulties with anger management and expression in hypertensives.[3,71] Various hostility dimensions were related to heightened reactivity,[29] higher blood pressure,[34] increased ambulatory variability,[72] and to prediction of hypertension.[73,74] There may be an enhanced risk among genetically-predisposed persons who have the greatest difficulty in anger expression.[60] However, ongoing controversy is shown by finding blood pressure related to expressive hostility,[75,76] unrelated to anger expression[29,77] or inversely related to experienced hostility.[76,78] Similarly, hostility dimensions are related to heightened cardiovascular reactivity,[78] unrelated,[79] or even inversely related.[80]

Some found hypertensives to be submissive persons with low assertiveness and unexpressed hostility and anger.[34] Anger acutely raises blood pressure,[29] with greater fall in those who vent it.[81] However, any type of emotional arousal similarly raises blood pressure,[70] including anxiety and happiness,[29] and there is no specificity for SNS emotional responses.[70] Indeed, pressor response to anxiety may exceed that to anger,[82] and anger-induced elevations may only involve a systolic response in males.[70] Studies that relate personality to hypertension have used a confusing multitude of methods, with inconsistent findings and selection bias in choice of subjects for investigation,[71] and the same behavioral patterns are found both in normotensives and in patients with many psychosomatic problems, with little evidence for an antegrade specificity for hypertension.[71] For example, no differences in hostility were found in comparing normo- and undiagnosed hypertensives,[83] and observed findings may be mediated by differences of body mass.[75,76]

8. RELATIONSHIPS OF ANXIETY TO MITRAL VALVE PROLAPSE

Although the ability of the psyche to cause hypertension remains unproven, anxiety is related to functional prolapse of the mitral heart valve (MVP).[84] The long clinical interest in relationships of anxiety with the cardiovascular system is exemplified in MVP.[85] Although independent, there is an increased prevalence of anatomic, functional MVP in anxiety disorders.[86] Hemodynamic states can induce functional prolapse in normal mitral valves of healthy individuals, the common denominator being the pathophysiology of hypovolemia, orthostatic intolerance, chronic hyperadrenergic activity and veno- and vasoconstriction.[86] Chronic anxiety enhances SNS activity,[87] and thereby induces a self-sustaining vasoconstriction, hypovolemia and increased left ventricular contractility,[88] which combine to decrease left ventricular preload and intracavity size and thereby induce a functional MVP.[84]

Sodium is critical in regulation of the body's fluid volume (ECF). A common denominator of functional MVP is the hypovolemia that underlies orthostatic intolerance and stimulates SNS activity. Sodium depletion may be involved in pathogenesis of functional MVP in subjects predisposed by anthropometric habitus. It may also play a major role in the fatigability and other symptoms and clinical manifestations that are related to the chronic hyperadrenergic activity engendered by hypovolemia and orthostatic intolerance, rather than ascribing them to cardiovascular manifestations of chronic anxiety.[88]

The primary integrator of volume-regulation and relationships of cardiac output and TPR is baroreceptor activation of the SNS. When lowered blood pressure is sensed by arterial baroreceptors, SNS stimulation increases peripheral vasoconstriction and activates the renin-angiotensin (RAA) system and arginine vasopressin, thus stimulating the three major vasoconstrictor systems.[89] Autoregulatory mechanisms normally maintain a rather precise balance between sodium intake and excretion, regardless of load size. Changes in ECF are detected by volume sensors in the cardiopulmonary area, vascular system, kidney and other organs. A perceived change in normal relationships between intravascular volume and circulatory capacity activates mechanisms that effect physiologically appropriate changes of renal sodium excretion. This adjusts ECF volume to a quantity appropriate for the vascular capacitance and to maintain normal circulatory function. Proper functioning of the system uses a neural mechanism for regulation that parallels a classic reflex neural arc.

SNS activity is inversely related to sodium intake[90] and is particularly stimulated during short-term sodium restriction and depletion,[91] Dietary restriction markedly stimulates central SNS activity and also influences peripheral adrenoreceptor density and function.[92] Chronic restriction stimulates SNS activity, but may not increase myocardial contractility because of induced dysfunction at the receptor level.[93] It also increases plasma renin activity, which is important in maintaining homeostasis during sodium restriction and in blood pressure responses. Renal SNS activity is an important physiological regulator of sodium excretion since it decreases sodium excretion by enhancing renal antinatriuresis.

The quantity of sodium determines ECF volume, and the steady state urinary sodium excretion is closely matched to dietary intake to ensure its near-constancy for optimal circulatory performance.[94] However, individual sensitivity varies widely and is closely related to SNS activity. Salt-sensitive persons may have higher resting plasma norepinephrine, although their pressor responses are poorly influenced by altered sodium intake, and SNS activity is decreased in those who are less sodium-sensitive.[95] Sodium-sensitive subjects have increased forearm vascular resistance during high intake, accompanied by decreased responsiveness of the RAA system during sodium depletion.[96] Sodium-sensitivity in younger persons may partly be due to higher SNS venous tone which serves to increase cardiac output and release a plasma $Na+K+$ pump inhibitor.[97]

Although dietary restriction is useful in antihypertensive therapy in salt-sensitive subjects, a wide misunderstanding ignores the fundamental role of sodium in normal body economy and cardiovascular function and in homeostasis. A pejorative belief has led to its near elimination from prepared foodstuffs and often to inappropriate rigid dietary salt restriction. Younger females may chronically restrict salt intake and add diuretics to achieve slimness. Sodium-sensitivity is genetically related to anthropometric variables and SNS activity. Younger females at highest risk for MVP exhibit leanness for height, poor muscular development, hypovolemia, reduced cardiac output, and relative hypotension, thus being at the opposite end of the spectrum that characterizes subjects prone to hypertension. Sodium-sensitivity predisposes to anatomic MVP, that is enhanced when such persons restrict dietary sodium and stimulate the SNS activity that augments hypovolemia and poor orthostatic tolerance and thereby induces complex central and peripheral cardiovascular dysregulation. Sodium-repletion expands ECF volume and cardiac preload, enlarges left ventricular intracavity size, reduces orthostatic SNS stimulation and thereby decreases central SNS activity and beta-adrenergic responses. Repletion can normalize low blood pressure and markedly improve the poor exercise tolerance and fatigability, cardiac ectopy, and cold extremities often found in such persons. Their mid-systolic mitral clicks tend to disappear and lowered SNS activity may have salutary benefits for anxiety and associated hyperadrenergic manifestations. However, response to salt-loading is only successful when ECF volume is expanded, and this may not be possible until chronic vasoconstriction can be effectively interrupted.[86]

Variable findings in anatomic MVP indicate an association with diverse underlying disorders that share a reduction of left ventricular intracavity size relative to the fixed mitral valve orifice, the dominant factor that induces functional leaflet prolapse, whether idiopathic or secondary to other cardiovascular disorders. The evidence suggests that MVP and associated hyperadrenergic manifestations are not directly inherited, constitutional neuroendocrine disorders, but that genetic relationships prevail only for the anthropometric habitus that predisposes to MVP, and which is characterized by leanness for height, narrow antero-posterior chest diameter, straight back, and variable pectus excavatum.[84]

9. RELATIONSHIPS OF HOSTILITY/ANGER DIMENSIONS TO CARDIOVASCULAR DISEASES

Hostility and anger dimensions may be causally related to cardiovascular diseases.[98] Many emotions induce cardiovascular responses and it is primarily perceptions that determine emotional responses and psychophysiological consequences.[99] Anger is a cognitive response associated with personal appraisal and interpretation,[100] and it induces an active coping response in the susceptible individual.[27]

Some believe humans to be genetically and instinctively aggressive from an evolutionary heritage. The limbic system has a three-fold evolutionary development in which the amygdala is identified with aggression. However, its existence does not indicate that the brain is instinctually programmed for aggression. Its stimulation in animals elicits aggressive behaviors only when these were previously learned, and variously induces anxiety, depression, fright, horror, or anger in humans, but not an attack behavior. Behaviors arise out of an interaction of genetic potentials and environmental factors, but attitudes, behaviors, and personality traits probably far more result from environmental than inherited variables. A child partly learns aggression by enhanced status, approval, or rewards for such behavior. Montagu[101] points out that the drive for preservation of species and self leads to a type of aggression that is biologically adaptive, life-serving, and phylogenetically programmed, and that is common to both animals and humans. Most other forms of aggression may be biologically maladaptive and experiential, since they are almost exclusively found in humans. Indeed, the principal factor operating in animal evolution is identified as cooperation rather than conflict.[101,102] Aggressive behavior in animals is largely for acquisition and retention of food and establishment of a dominance relationship, rather than for inflicting injury. Life or death fights other than for food and survival are infrequent and conflicts consist mainly of teeth-baring and horn-butting that are relatively harmless trials of strength and mutual appraisal for hierarchical dominance and social order, and which serve to minimize divisiveness, aggression, and fighting.[101] Animals kill for food, defense, or dominance, but only rarely for other aggressive purposes. Most fighting in lower animals and even in primates is a ritualized process designed for self and species preservation and different from innate aggression that has killing as the goal. Thus, man does not fight wars merely because animals are territorial or have aggressive goals for domination which do not involve destruction. There is no innate aggressiveness in chimpanzees, since they do not fight among themselves, nor do males strike females, and they prevent fighting among their offspring merely by a stylized stare.[103]

The belief that man became man when he became aggressive and learned to hunt is discarded by anthropologists who observed that the key to man's evolution is cooperative culture.[101,102] The successful evolution of *Homo sapiens* appears to stem from an ability to live in effective groups, compared with the Australopithecines and other protohominids who failed in this regard. In *The Brothers Karamazov* Dostoyevsky has Ivan paint a Nietzschean picture of human aggression in his discussion of cruelty to children, but resents the reference to man as sometimes being bestial, pointing out that humans, but not animals exhibit such behaviors. He observes in *Notes from the Underground* that it is civilization rather than evolution that has increased instrumental aggression in humans. Montagu[101] also points out that drive is different from instinct, which is a phylogenetically programmed and fixed action pattern designed to react to a specific stimulus in a biologically adaptive manner that is characteristic for any given species, and that aggressive behavior is not necessarily instinctive in humans since it is not stereotyped or inevitable and does not appear in vacuo. By analogy, he notes that sexual drive may be instinctively fueled by hormonal actions on the brain. However, although born with a capacity for sexual activity, humans must learn how to perform. He questions whether we correctly describe Neanderthals as bestial, since they were the first to bury their dead on a bed of flowers, surrounded by objects placed there for their final journey. Cooperation remains a key behavior in many primitive societies.[101,102]

Cultures do not differ by reason of genetic factors but from variable experience. A genetic contribution probably underlies most human behaviors, but socialization, conditioning, and learning from infancy on are the major determinants. There are doubtless genetic influences on the limbic system's neuronal and receptor distributions, variably influenced in interactions with cognitive appraisals and somatic feedbacks. Morphology, biochemistry, neuroanatomy, and behavior are intertwined in evolutionary development, but there is increasing evidence for a major effect of behavioral changes on the other parameters.[104] Darwinian tradition is challenged by evidence that evolution is not fully responsible for shaping human traits, which may not only be transmitted by genes, but also influenced by important cultural factors. For example, Tanner[105] hypothesizes that progressive reduction of canine teeth size resulted from preferential female selection of males with smaller canines associated with less aggressive behavior, an influence that accordingly could be held responsible for shaping the development of a physical trait. She also hypothesizes that more social and less aggressive, disruptive males may have more often been selected as sexual partners during human evolution, and that change in mating patterns may therefore have increased sociability between males and females, and even between the sexes. Although born with a potential for aggression, there is no brain center that compels us to become aggressive. Therefore aggression is not inevitable and the behavior must be learned. Darwin taught that there is a competition for existence, but never stated that this involved aggression rather than cooperation, and he did not confuse aggressive drive with instinct.[101]

Reviews[106,107] point out how parental abuse and neglect are associated with alcoholism in youth, and parental crime and abuse with more aggression in children. Observing hostile aggression on the sporting fields can increase aggressive behavior in spectators who identify with the players,[103] and it is the culture that encourages bone-jarring football tackles, rituals of ice hockey duels, unregulated industrial competition, and the concept of the strongest being the most fit. Comparison of cultures find that rape is neither universal nor stems from a biological drive, but is a response to the way certain kinds of societies are organized.[108] Sanday found certain behaviors common to rape-prone societies, including tolerance for violence and the encouragement of males of all ages to be aggressive, tough, and competitive. The typical rapist is not sexually deprived but is a hostile and aggressive male who enjoys violence toward females.[109] Rape is thus not inherent in the sex relationship and remains unknown in many simpler African societies whose people live in harmonious groups that place high value on the nurtures and fertility of the female. Rape-free societies do not have clubs or special political and religious groups that exclude females, and they do not have machos, which means "male of animal species." Their absence of male aggression and violence may be related to the fact that such societies also have no lack of land and resources, and no need for the competitive survival that, for example, characterized trigger-happy American frontiersman, Southern planters outnumbered by restive slaves, Israelites who approached Canaan, or Azande conquerors of neighboring African tribes.[101] A major point in examining relationships of hostility to cardiovascular disease is to realize that these dimensions are largely acquired and related to environmental stressors.

10. RELATIONSHIPS OF THE SNS TO VARIOUS CARDIOVASCULAR DISEASES

The SNS is highly involved in cardiovascular disorders and complications, and may be a final common pathway for relating nutritional, environmental, and other risk factors to ischemic heart disease (IHD).[110–112] Anxiety and anger can increase food, ethanol, and nicotine consumption, with adverse consequences for hypertension and IHD. Ethanol stimulates SNS activity,[113] is strongly related to blood pressure in hypertensives, and can induce coronary vasoconstriction and cardiac arrhythmias.[114] Smoking is strongly related to sudden cardiac death and to incidence of myocardial infarction, and by SNS stimulation, can induce coronary vasoconstriction[115,116] even in asymptomatic subjects without coronary artery disease,[117] and it increases potential for myocardial ischemia and cardiac arrhythmias. It also increases platelet aggregability and responses to other aggregating agents, thus playing a potential role in thrombosis.[118] Smoking may increase locus ceruleus function and thus be anxiogenic and emotions can thus induce coronary vasoconstriction and silent myocardial ischemia in patients with coronary artery disease,[119] it may have additive SNS effects during stress responses. Anxiety and anger affect sleep directly, and indirectly by influences on consumption of food, alcohol, caffeine and nicotine. In subjects with IHD, an acute coronary event may occur at times of sleep deprivation, fatigue and exhaustion.[120]

SNS cardiac nerves are driven by preganglionic neurons in the spinal cord, under control of descending brain pathways. SNS traffic modulates development of atrial and ventricular arrhythmias by releasing norepinephrine into the myocardium, and a precipitating role of emotions is well known.[121–124] The final common pathway is via vagal and SNS neural fibers that act on electro-physiologic mechanisms, and SNS neural interactions regulate myocardial electrical stability in the conscious state.[123] Sudden cardiac death is usually due to ventricular fibrillation and myocardial ischemia, often silent,[119] lowers the threshold.[122] The pathophysiology of sudden cardiac death is best comprehended as an electrophysiologic abnormality manifested as myocardial electrical instability (MEI).[122,123] The vulnerable period of the cardiac cycle occurs in the brief interval at the apex of the upright T wave in the electrocardiogram, when there is a normal patchwork of inhomogeneous electrical activity due to variable fiber recovery time during repolarization. MEI provides the substrate for SNS stimulation to cause fibrillation,[122–124] and MEI and catecholamine surges are related to central SNS activity,[122–124] hence to emotional responses to perceived stress, and coronary vasospasm and platelet aggregation may contribute.[125] The role of higher neural traffic in genesis of lethal arrhythmias is now firmly based on evidence that links emotionally stressful events and sudden death in both animals and humans.[117,121–126]

Considerable attention is now given to the SNS in congestive heart failure (CHF) and its fatal complications,[127] which can be adversely affected by anxiety and anger. The high prevalence of potentially lethal arrhythmias in CHF is the most important predictor of mortality.[128] An inadequate stroke volume stimulates SNS activity that contributes to the CHF,[129] exerts a toxic effect on the myocardium,[130] and can induce arrhythmias.[128]

Mental arousal, anxiety, and anger are well known to cause myocardial ischemia[131] and symptomatic angina in patients with IHD.[132,133] However, we recognize a spectrum of angina pectoris, with angina at one end due to obstructive coronary artery disease, and in which ischemic symptoms occur at relatively fixed exercise levels. At the other end is angina that unpredictably occurs while sedentary or during usually well tolerated activities, due to transiently impaired coronary blood flow, and many patients have a mixed form.[134] Silent myocardial ischemia occurs frequently in patients with symptomatic or asymptomatic disease,[135] and the substantial majority of their ischemia is silent.[135,136] This frequently occurs during daily activities that are not perceived as stressful.[119] However, an important pathogenetic factor may be SNS stimulation and patients may exhibit silent ischemia with a change in their emotional state.[136]

Normal coronary arteries vasodilate with exercise, while diseased segments vasoconstrict due to endothelial dysfunction associated with unopposed increase in alpha-adrenergic tone.[137] The vasomotor activity of resistance vessels determines the autoregulatory mechanism that matches oxygen demand to adequate coronary blood flow. Opposing actions mediated by alpha- and beta- adrenergic inputs normally lead to automatic dilatation of resistance arterioles in response to oxygen demand. Excessive vasomotor tone interferes with autoregulatory mechanisms, with a deficient ability of resistance arterioles to vasodilate distal to points of stenosis. In coronary artery disease, there is increased sensitivity to alpha-adrenergic stimulation induced by exercise, cold, hyperventilation, agents which are mediated by endothelial injury and platelet deposition, and by soluble blood-borne mediators that include serotonin, histamine, thromboxane A2 and prostacyclin.[138] Thus, although alpha-adrenergically mediated tone is a determinant of coronary flow and ischemia, other mechanisms prevail in normal subjects,[139] to prevent excess vasoconstriction and allow compensatory responses not observed in patients with coronary disease, in whom there is no vasodilatory reserve, since resistance vessels are already maximally vasodilated in the basal state.[140]

The arterial lesions associated with spontaneous silent ischemia suggest that lipid-engorged foam cells rupture and expose surface endothelial collagen, a nidus for deposition of platelets and fibrin to form microthrombi that elaborate vasoactive substances and initiate vasoconstriction. Focal vasoconstriction causes endothelial fragmentation and desquamation, and platelet microthrombi adhere to subendothelial tissues and release vasoconstrictor mediators.[141] The local stimulus is a breach of endothelium and, when a crack, rupture or ulceration occurs in an atherosclerotic plaque, increased platelet accumulation leads to more severe vasoconstriction and thrombosis[142] and, in patients with coronary vasospasm, to acute myocardial infarction.[141] In patients with coronary disease, anger may stimulate SNS activity that can trigger a fissure or rupture of an atherosclerotic plaque, the primary cause of acute transmural myocardial infarction, and which also plays an important role in crescendo angina and sudden cardiac death.[143]

Proliferation of smooth muscle and deposition of cholesterol and calcium produce a focal atherosclerotic injury that propagates itself and enhances local sensitivity to vasoconstrictor mediators in a pattern that helps to explain coronary artery disease progression.[144] Endothelial injury and smooth muscle proliferation are thus important in atherogenesis.[145,146] Platelets that adhere to subendothelial tissues in areas of endothelial cell desquamation release a platelet mitogenic factor that stimulates proliferation of intimal smooth muscle involved in early atherogenesis, and also stimulate collagen organization of platelet-fibrin thrombi. The rate of progression and severity of coronary atherogenesis are strongly related to silent ischemia,[145,147–149] which partly explains the episodic nature of disease progression.[150]

Although silent ischemia mainly occurs with coronary disease,[139,151] it also occurs in Syndrome X, with exertional angina and silent ischemia occurring in the absence of large vessel coronary artery disease,[152] or impaired vasodilatory reserve.[153] Patients tend to have high anxiety.[154] Anger can readily trigger silent ischemia and increase cigarette consumption, and there is a high proportion of smokers in patients with vasospastic angina.[155]

SNS activity is involved in serum cholesterol regulation,[110,156] and in the variability that occurs in response to emotional arousal.[157–161] Type A persons may have higher serum cholesterol than Type Bs,[162] particularly more hostile Type As,[163] and there may be correlations of anger dimensions with serum cholesterol.[164]

11. RELATIONSHIPS OF TYPE A BEHAVIOR PATTERN TO ISCHEMIC HEART DISEASE

The classical risk factors are important for IHD but do not explain historical changes, gender or wide differences of geographic IHD prevalence in different populations. They account for under half of the numerical IHD incidence in prospective studies,[165,166] and provide relative levels of risk that do not explain the individual specificity of risk for IHD that occurs at similar risk factor levels. They do not account for the much higher IHD rate in males in the U.S.A., compared to Europe[167] or, for example, the much higher incidence in males in the Framingham Heart Study than in similar studies in Yugoslavia, Hawaii, Puerto Rico, and Paris.[168–170] They also do not correlate with a highly variable rate of progression of coronary artery disease.[171–175] It appears that relationships of biological variables to IHD are modulated by psychosocial factors that characterize different individuals, and particularly by certain behaviors.[176]

It was personal observation of patients that led to the concept of the Type A behavior pattern (TABP),[177] and of interest that these behaviors were long ago described by Von Dusch[178] and Osler,[179] and later by psychiatrists who observed patients to be ambitious, hard-driving, aggressive, and compulsively striving to achieve goals incorporating power and prestige.[180–185] Their consistent descriptions did not consider the interaction with western-style cultural milieus that was recognized by Stewart[186] and the TABP concept.[177,187]

TABP was originally defined as an action-emotion complex which individuals use to confront their environmental milieus and challenges. It involves behavioral dispositions such as aggressiveness, competitiveness and impatience; specific behaviors such as muscle tenseness, alertness, rapid and emphatic vocal stylistics, and accelerated pace of activities; and emotional responses such as irritation and increased potential for anger/hostility dimensions, and many studies have given strong construct validation for these behaviors.[177]

The belief that TABP doesn't equate with stress or neuroticism[177] is now questioned.[188–190] A meta-analysis of the literature found TABP correlated with psychological disorders.[189] Enhanced competitiveness has seminal importance for most overt Type A behaviors, including aggressive drive, accelerated pace of activities, and increased potential for hostility/anger dimensions.[43] The underlying basis for this Type A competitiveness may be a deep-seated, covert anxiety that is associated with insecurity and fear of failure, and different from neuroticism in the latter's usual construct.[43,44,188–190]

TABP was found to be related to IHD prevalence[177] and then prospectively associated with higher IHD incidence in the Western Collarborative Group Study (WCGS),[166] and other studies.[191–194] Although this was not confirmed in later studies,[195–197] it was pointed out[198,199] that later studies were done during a period of decline of IHD mortality, wide use of protective beta-adrenergic blockers, and low rates in association with risk factors, as exemplified in the controls in the MRFIT study.[200] A meta-analysis of the literature on psychosocial variables found stronger TABP/IHD associations in cross-sectional than in intervention studies.[201] However, the latter may be flawed by problems. For example, volunteers for risk screening are biased by health consciousness,[202,203] and often are the "worried well" who have rish factors and no disease.[204] Intervention study dropouts are more often smokers and blue collar workers,[205] with higher incidence in dropouts, compared to remaining participants.[206] Moreover, the results of different studies are applicable only to their participants and not representative of other populations.[207]

Various studies found TABP predictively related to recurring coronary events.[208–210] This was not confirmed by others,[211–217] possibly due to use of poor TABP assessment measures.[218] More relevant is that risk factors for long-term postinfarction survival reflect the amount of myocardial damage, rather than those for first infarction.[211,219]

Various studies also found TABP related to severity of coronary artery disease,[220–227] while others failed to confirm this.[288–233] Angiography studies may not properly evaluate behavioral risk factors because of small sample sizes, lack of normal controls, selection bias in referral of patients, different methodologies, poor correlation of disease severity with all risk factors, and lack of appropriate endpoints.[171–175,201,221,234,235] There are more risk factors in IHD suspects who are not confirmed at angiography, compared to those without suspected disease.[234] Regardless of this, these problems were obviated in the large series of 2,289 patients referred for diagnostic angiography, and in which TABP was found to be significantly related to disease severity.[226]

Higher IHD incidence is strongly associated with urbanization, population density, and industrialization, which enhance the competitiveness that underlies TABP in susceptible individuals.[43] However, as already pointed out, this type of competitiveness is not an evolutionary development.[101]

12. RELATIONSHIPS OF HOSTILITY/ANGER DIMENSIONS TO ISCHEMIC HEART DISEASE

TABP has diverse behavioral components that include an enhanced potential for hostility.[98,177] Psychometric hostility measures discriminate Type A/B behaviors,[177] and it is now popular to focus on this component.[236,237] An initial impetus was finding that Type A behavioral hostility was significantly related to intake prevalence of silent myocardial infarction in the WCGS.[238] Bortner compared WCGS intake interviews of subjects who developed IHD during follow-up with paired controls, and later analysis of his blinded ratings found competitive drive and potential for hostility to be the most important predictive, coronary-prone Type A behaviors.[239] A different method of scoring these intake interviews confirmed a relationship of competitiveness and hostility to CHD incidence during the WCGS follow-up.[240]

Megargee[241] noted that those who attempt to relate dimensions of anger, hostility or aggression to cardiovascular disease, may operationally define different constructs by using a confusing array of dissimilar technics in their studies, often interchangeable and without appropriate differentiation. He and others[242] point out the considerable ambiguity and inconsistency in how these constructs are defined, separated, or overlap, and the lack of agreement on how they are measured. For example, Type A hostility refers to a potential for anger/hostility dimensions[243] that reflects an antagonistic interpersonal style, rather than intrapsychic affect.[44,189,241] This potential correlates with expression of hostile behaviors, but not with feelings of resentment, guilt, or mistrust,[240] and these are independent personality dimensions.[241] Type As with high potential for hostility do not often exhibit another form of neurotic hostility that is related to anxiety, and characterized by feelings of annoyance, resentment, or rage.[189]

The Cook and Medley "Ho" Scale of the MMPI[244] was found predictively related to IHD incidence in a 25-year follow-up of 255 physicians who took the MMPI as students,[245] and in a 10-year follow-up of 1,877 men.[246] However, IHD incidence was not linearly related to Ho scores in either study. In one[245] it related only to scores above a cutoff point, and below which scores were unrelated and, in the other,[246] although lowest IHD rates were associated with lowest scores and highest with middle range socres, the rates were intermediate in those in higher score quintiles.

Moreover, no relationship of Ho Scale scores were found with IHD incidence or mortality in a 33-year follow-up of 1,399 university students,[247] in a 3-year follow-up of healthy, middle-aged Finnish men,[248] in a 25-year follow-up of 478 medical students,[249] in a large Japanese study in which the major proponent of the Ho Scale participated,[250] or in The Framingham Heart Study.[251] In one study,[248] Ho scores were only related to IHD morbidity and mortality in subjects with intake hypertension and IHD, thus only to triggering acute IHD events in subjects with coronary artery disease.

Angiographic severity of coronary disease was found to be related to Type A behavioral hostility.[225–228,252] It also was related to Ho scale scores in two studies,[225,226] but this was not confirmed by others[250,253] and regardless of whether the subjects had IHD.[250]

The recognized problems with anger/hostility constructs[241,242] are particularly relevant for the Ho Scale.[241] Its original correlations with hostility were made in teachers, adults convicted of violent crimes, and sucidal outpatients, and do not generalize to the normal population.[254] The measure has been correlated with anger, cynicism, mistrust, psychosocial and physical distress, social maladjustment, ineffective coping style, and social supports,[245,255–260] thus appearing to be a measure of neuroticism and general psychopathology.[241,254,261] However,it is not a reliable measure of hostility or overtly aggressive behavior[241,245,258,262] and is negatively correlated with MMPI and other psychometric measures of hostility and paranoia.[241,254,261] Type A behavioral hostility is not associated with psychopathology,[254] or related to the Ho Scale,[225,228,254] with which it shares only a small part of its variance.[228,254] Thus, TABP is correlated with psychometric measures of tolerance, vigor, self-confidence, achievement via independence and dominance,[254,261] while the strongest association for Ho is with the MMPI scale that measures social desirability, and high Ho scores characterize neurotic persons with attributes of psychopathology.[241] Megargee[241] authoritatively states that "Most distressing is the failure of Ho to measure hostility. All in all, the evidence for the construct validity of the Ho Scale in minimal. Thirty years after its derivation it is difficult to say with any confidence what Ho measures."

Unlike Ho, TABP is not a predictor of general illness,[263,265] or of all cause mortality.[265] However, Ho scale scores were similarly related to IHD, cancer, and all cause mortality in the "positive" studies.[245,246] Aside from the fact that Ho does not predict IHD incidence or coronary disease severity in most studies, the latter also find it unrelated to all cause mortality. Thus, Ho scores lack specificity for IHD prediction even in the two studies,[245,246] and these cannot be used to document a causality of hostility for IHD.

These findings suggest that the Ho Scale is a measure of psychopathology that is associated with increased all cause mortality, and possibly with precipitation of fatal coronary events in subjects with IHD,[248] consistent with other measures of "suspiciousness" that relate to cardiac mortality.[266] An increased all cause mortality associated with Ho scores in the two studies[245,246] may be due to low levels of social support, physological distress, and anger variables[266–272] assessed by the Ho Scale.[254] However, although Ho strongly correlates with psychopathology, this is not related to IHD or severity of coronary artery disease.[273,274] Some use the Ho Scale inappropriately to support a belief that hostility is the "toxic" behavior that relates TABP to IHD, and continue to assume that the Ho Scale measures hostility,[275] even though their own studies do not support this.[245]

Although sometimes more related to fatal events,[272] TABP appears to be an independent risk factor for coronary artery disease,[226] and for incidence of initial myocardial infarction. This is confirmed by findings of the 27-year WCGS follow-up, in which global TABP, but not its hostility component, was found to be significantly related to the prevalence of symptomatic myocardial infarction in the 1,796 surviving participants.[276] As expected, TABP was much less related to recurring coronary events and long-term postinfarction mortality, since these are primarily related to the severity of myocardial damage rather than to the classical risk factors for primary events.[211,219]

Considerable data relate IHD to psychosocial factors,[176,277,278] the strongest being for behavioral factors such as the TABP.[176] However, in contrast to an ill-defined hostility construct, competitiveness looms as the so-called "toxic" factor, since it has seminal importance for Type A aggressive drive, accelerated pace, impatience, and greater potential for hostility.[43] Although useful for psychometric questionnaires, it is highly improbable that highly intercorrelated emotional and behavioral aspects of a human being can ever satisfactorily be subdivided into component parts such as "hostility."

References

1. B. Folkow, "Physiology of behavior and blood pressure regulation in animals" **In**: (S. Julius and D. R. Bassett, eds.), *Behavioral Factors in Hypertension, Handbook of Hypertension, Vol. 9* (Elsevier, Amsterdam, 1987), pp. 1–18.
2. J. Brod *et al.*, "Circulatory changes underlying blood pressure elevation during acute emotional stress (mental arithmetic) in normotensive and hypertensive subjects" *Clin. Sci.* **18**, 269–279 (1956).
3. W. B. Cannon, *Bodily Changes in Pain and Rage* (2nd ed., Appleton-Century Crofts, New York, 1936).
4. H. Selye, "A syndrome produced by diverse nocuous agents" *Nature* (London) **138**, 32–38 (1936).
5. R. S. Lazarus, *Psychological Stress and the Coping Process* (McGraw Hill, New York, 1966).
6. J. E. Dimsdale, "Wet Holter monitoring: techniques for studying plasma responses to stress in ambulatory subjects" **In**: (T. M. Dembroski, T. H. Schmidt and G. Blumchen, eds.), *Biobehavioral Bases of Coronary Heart Disease* (Karger, Basel, 1983) pp. 175–184.
7. J. P. Henry and C. E. Grim, "Psychosocial mechanisms of primary hypertension" *J. Hyperten.* **8**, 783–793 (1990).
8. P. Hjemdahl and K. Eliasson, "Sympatho-adrenal and cardiovascular response to mental stress and orthostatic provocation in latent hypertension" *Clin. Sci.* **57**, 189s–191s (1979).
9. J. W. Mason *et al.*, "Selectivity of corticosteroid and catecholamine responses to various natural stimuli" **In**: (G.Serban, ed.), *Psychopathology of Human Adaptation* (Plenum, New York, 1976) pp. 147–171.
10. M. Fredrikson *et al.*, "Cortisol excretion during the defense reaction in humans" *Psychosom. Med.* **47**, 313–319 (1985).

11. S. Wolf and H. G. Wolff, "A summary of experimental evidence relating life stress to the pathogenesis of essential hypertension in man" **In**: (E. T. Bell, Ed.), *Hypertension, A Symposium* (University of Minnesota Press: Minneapolis, 1951), p. 288.
12. Z. Hejl, "Changes in cardiac output and peripheral resistance during simple stimuli influencing blood pressure" *Cardiol.* **31**, 375–380 (1957).
13. M. Frankenhaeuser, "The sympathetic-adrenal and pituitary-adrenal response to challenge: comparison between the sexes" **In**: (T. M. Dembroski, T. H. Schmidt and G. Blumchen, Eds.), *Biobehavioral Bases of Coronary Heart Disease* (Karger, Basel, 1983), pp. 91–105.
14. E. Weiss and O. English, *Psychosomatic Medicine* (Saunders, Philadelphia, 1957).
15. H. P. Laughlin, *Neuroses in Clinical Practice* (Saunders, Philadelphia, 1956).
16. C. D. Spielberger, "Anxiety as an emotional state" **In**: (C. D. Spielberger, Ed.), *Anxiety: Current Trends in Theory and Research, Vol. 1* (Academic Press: New York, 1972) pp. 24–49.
17. R. S. Lazarus and S. Folkman, *Stress, Appraisal, and Coping* (Springer, New York, 1984).
18. P. McReynolds, "The concept of anxiety" **In**: (D. G. Byrne and R. H. Rosenman, Eds.) *Anxiety and the Heart* (Hemisphere, Washington, D.C., 1990) pp. 3–28.
19. R. W. Winters *et al.*, "The neurobiology of anxiety" **In**: (D. G. Byrne and R. H. Rosenman, Eds.) *Anxiety and the Heart* (Hemisphere, Washington, D.C.) pp. 187–212.
20. R. M. Jeresaty, "The syndrome associated with midsystolic click and/or late systolic murmur: analysis of 32 cases" *Chest* **59**, 643–647 (1971).
21. D. V. Sheehan and S. Soto, "Diagnosis and treatment of pathological anxiety" *Stress Med.* **3**, 21–32 (1987).
22. A. M. Ostfeld and R. B. Shekelle, "Psychological variables and blood pressure" **In**: (J. Stamler, R. Stamler and T. N. Pullman, Eds.) *Epidemiology of Hypertension* (Grune and Stratton, New York, 1967) pp. 321–331.
23. G. L. Klerman, "Understanding anxiety: Normal emotion, symptom, and clinical disorder" (Special Issue) *Int. Med.* **6**, 3–12 (1985).
24. D. F. Klein *et al.*, "Antidepressants, anxiety, panic and phobias" **In**: (M. A. Lipton, A. DiMascio and K. F. Killam, Eds.), *Psychopharmacology: A Generation of Progress* (Raven Press: New York, 1978) pp. 1401–1410.
25. A. R. Ax, "The physiological differentiation between fear and anger in humans" *Psychosom. Med.* **15**, 433–442 (1953).
26. J. Schacter, "Pain, fear and anger in hypertensives and normotensives" *Psychosom. Med.* **19**, 17–29 (1957).
27. G. E. Schwartz *et al.*, "Cardiovascular differentiation of happiness, sadness, anger and fear following imagery and exercise" *Psychosom. Med.* **43**, 343–364 (1981).
28. T. C. Weerts and R. Roberts, "The physiological effects of imagining anger-provoking scenes" *Psychophysiol.* **13**, 174 (1976).
29. G. R. James *et al.*, "The influence of happiness, anger, and anxiety on the blood pressure of borderline hypertensives" *Psychosom. Med.* **48**, 502–508 (1986).
30. A. Carrel, *Man the Unknown* (Harper, New York, 1935).
31. A. J. Toynbee, *A Study of History* (Oxford University Press, New York, 1946).
32. A. de Tocqueville, *Democracy in America* (Appleton, New York, 1904).
33. E. Cassirer, *Substance and Function, and Einstein's Theory of Relativity* (Dover, London, 1953).
34. S. Julius and E. H. Johnson, "Stress, autonomic hyperactivity and essential hypertension: an enigma" *J. Hyperten.* **3**(Suppl 4), S11–S17 (1985).
35. F. H. Messerli *et al.*, "Essential hypertension in the elderly: hemodynamics, intravascular volume, plasma renin activity, and circulating catecholamine levels" *Lancet* **2**, 983–986 (1983).
36. A. C. Guyton, "Essential cardiovascular regulation—the control linkages between bodily needs and circulatory function" **In**: (C. J. Dickinson and J. Marks, Eds.), *Developments in Cardiovascular Medicine* (University Park Press, Baltimore, 1982) pp. 265–302.
37. B. Folkow, "Structural factor in primary and secondary hypertension" *Hyperten.* **16**, 89–101 (1990).
38. S. Julius, "Transition from high cardiac output to elevated vascular resistance in hypertension" *Am. Heart J.* **116**(Part 2), 600–606 (1988).
39. E. Hollander and M. R. Leibowitz, "Generalized anxiety disorders and panicogenic syndromes" **In**: (D. G. Byrne and R. H. Rosenman, Eds.), *Anxiety and the Heart* (Hemisphere, Washington, D.C., 1990) pp. 53–72.
40. C. Braestrup and M. Nielsen, "Anxiety" *Lancet* **74**, 1030–1034 (1982).
41. M. W. Ward, "Anxiety and cardiovascular reactivity" **In**: (D. G. Byrne and R. H. Rosenman, Eds.), *Anxiety and the Heart* (Hemisphere, Washington, D.C.), 1990, pp. 347–365.
42. M. A. Winkleby *et al.*, "Self-reported stressors and hypertension: evidence of an inverse association" *Am. J. Epidemiol.* **127**, 124–134 (1988).

43. R. H. Rosenman, "Type A behavior pattern: a personal overview" **In**: (M. Strube, Ed.), Type A Behavior (Special Issue). *J. Soc. Behav. Pers.* 1–24 (1990).
44. J. Suls, "Type A behavior pattern: the role of anxiety, self-awareness, and denial" **In**: (D. G. Byrne and R. H. Rosenman, Eds.), *Anxiety and the Heart* (Hemisphere, Washington, D.C., 1990), pp. 259–276.
45. R. H. Rosenman, "Type A behavior and hypertension" **In**: (S. Julius and D. R. Bassett, Eds.), *Handbook of Hypertension, Vol. 9: Behavioral Factors in Hypertension* (Elsevier, Amsterdam, 1987) pp. 141–149.
46. R. B. Devereux *et al.*, "Association of mitral valve prolapse with low body-weight and low blood pressure" *Lancet* **2**, 792–795 (1982).
47. Australian Therapeutic Trial in Mild Hypertension, *Lancet* **1**, 1261–1267 (1980).
48. R. H. Schneider *et al.*, "Anger and anxiety in borderline hypertension" *Psychosom. Med.* **48**, 242–248 (1986).
49. R. Cruz-Coke, "Correlation between prevalence of hypertension and degree of acculturation" *J. Hyperten.* **5**, 47–50 (1987).
50. B. Falkner *et al.*, "Effect of salt loading on the cardiovascular response to stress in adolescents" *Hyperten* **3**(Suppl II), II-195–II-199 (1981).
51. N. R. Poulter *et al.*, "Higher blood pressures of urban migrants from an African low-blood pressure population are not due to selective migration" *Am. J. Hyperten.* **1**, 143S–145S (1988).
52. A. G. Shaper *et al.*, D. Ashby and S. J. Pocock, "Blood pressure and hypertension in middle-aged Bristish men" *J. Hyperten.* **6**, 367–375 (1988).
53. D. S. Strogatz and S. A. James, "Social support and hypertension among blacks and whites in a rural, southern community" *Am. J. Epidemiol.* **124**, 949–956 (1986).
54. C. L. Broman, "Social mobility and hypertension among blacks" *J. Behav. Med.* **12**, 123–134 (1989).
55. T. Lang *et al.*, "Social, professional conditions and arterial hypertension: an epidemiological study in Dakar, Senegal" *J. Hyperten.* **6**, 271–276 (1988).
56. R. C. Patrick *et al.*, "Relationship between blood pressure and modernity among Ponapeans" *Intl. J. Epidemiol.* **12**, 36–44 (1983).
57. Conference on Behavioral Medicine and Cardiovascular Disease, *Circ.* **76**(Part 2, Suppl), I-1–I-227 (1987).
58. R. M. Rose *et al.*, "Air traffic controller health change study" *Report of the FAA* (1978).
59. R. Mundal *et al.*, "Elevated blood pressure in air traffic controllers during a period of occupational conflict" *Stress Med.* **6**, 141–144 (1990).
60. B. Falkner, "Reactivity to mental stress in hypertension and pre-hypertension" **In**: (S. Julius and D. R. Bassett, Eds.), *Handbook of Hypertension. Vol. 9: Behavioral Factors in Hypertension* (Elsevier: Amsterdam, 1987) pp. 95–103.
61. R. H. Rosenman and M. W. Ward "The changing concept of cardiovascular reactivity" *Stress Med.* **4**, 241–251 (1988).
62. L. F. van Egeren and A. W. Sparrow, "Laboratory stress testing to assess real-life cardiovascular reactivity" *Psychosom. Med.* **51**, 1–9 (1989).
63. K. A. Matthews *et al.*, "Are cardiovascular responses to behavioral stressors a stable individual difference variable in childhood?" *Psychophysiol.* **24**, 464–473 (1987).
64. S. Julius, "The blood pressure seeking properties of the central nervous system" (editorial review). *J. Hyperten.* **6**, 177–185 (1986).
65. H. Liebau and J. Manitius, "Diurnal and daily variations of plasma renin activity, plasma catecholamines and blood pressure in normotensive and hypertensive man" **In**: (J. Bahlmann and H. Liebau, Eds.), *Stress and Hypertension. Contributions to Nephrology, Vol. 30* (Karger, Basel, 1982) pp. 57–63.
66. G. Mancia, "Blood pressure variability at normal and high blood pressure" *Chest,* **83**, (Suppl), 317–320, 1983.
67. G. Mancia and G. Parati, "Reactivity to physical and behavioral stress and blood pressure variability in hypertension" **In**: (S. Julius and D. R. Bassett, Eds.), *Handbook of Hypertension. Vol. 9: Behavioral Factors in Hypertension.* (Elsevier, Amsterdam, 1987) pp. 104–122.
68. S. Julius *et al.*, "Does behaviorally induced blood pressure variability lead to hypertension?" **In**: (K. Matthews *et al.*, Eds.), *Handbook of Stress, Reactivity, and Cardiovascular Disease,* (Wiley, New York, 1986) pp. 71–84.
69. D. J. Reis *et al.*, "The C1 area of the rostral ventrolateral medulla oblongata; a critical brainstem region for control of resting and reflex integration of arterial pressure" *Am. J. Hyperten.* **2**, (Part 2), 363S–374S, 1989.
70. J. E. Dimsdale *et al.*, "Anger and norepinephrine: a classic hypothesis reexamined" *Psychosom. Med.,* **50**, 192–109, 1988.

71. C. Cottier *et al.*, "Personality traits and hypertension: an overview" **In**: (S. Julius and D. R. Bassett, Eds.), *Handbook of Hypertension. Vol. 9: Behavioral Factors in Hypertension.* (Elsevier: Amsterdam, 1987) pp. 123–140.
72. G. A. Harshfield *et al.*, "Do laboratory tests of blood pressure reactivity predict blood pressure changes during everyday life?" *Am. J. Hyperten.*, **1**, 168–174, 1988.
73. J. H. Medalie *et al.*, "The incidence of hypertension and associated factors: the Israeli ischemic heart study" *Am. Heart J.*, **84**, 171–180, 1972.
74. D. C. McClelland, "Inhibited power motivation and high blood pressure" *J. Abnorm. Psychol.*, **88**, 182–190, 1979.
75. J. M. Siegel and C. J. Leitch, "Behavioral factors and blood pressure in adolescence: The Tacoma Study" *Am. J. Epidemiol.*, **31**, 287–300, 1981.
76. F. A. Treiber *et al.*, "The relationship between hostility and blood pressure in children" *Ann. Behav. Med.*, 173–178, Winter 1989.
77. E. H. Johnson *et al.*, "Emotional and familial determinants of elevated blood pressure in black and white adolescent males" *J. Psychosom. Res.*, **31**, 287–300, 1987.
78. T. W. Smith and K. D. Allred, "Blood-pressure responses during social interaction in high- and low-cynically hostile males" *J. Behav. Med.*, **12**, 135–143, 1989.
79. J. F. Sallis *et al.*, "The relationship between cynical hostility and blood pressure reactivity" *J. Psychosom. Res.*, **31**, 111–116, 1987.
80. B. K. Houston, "Psychological variables and cardiovascular and neuroendocrine reactivity" **In**: (K. A. Matthews *et al.*, Eds.) *Handbook of Stress, Reactivity, and Cardiovascular Disease* (Wiley, New York, 1986) pp. 207–229.
81. J. E. Hokanson *et al.*, "Effects of displaced aggression on systolic blood pressure" *J. Abnorm. Soc. Psychol.*, **67**, 214–218, 1963.
82. W. E. Whitehead *et al.*, "Anxiety and anger in hypertension" *J. Psychom. Res.*, **21**, 383–389, 1977.
83. R. Cochrane, "Hostility and neuroticism among unselected essential hypertensives" *J. Psychosom. Res.*, **17**, 215–218, 1973.
84. R. H. Rosenman, "Pathogenesis of the relationship of anxiety to mitral valve prolapse" **In**: (D. G. Byrne and R. H. Rosenman, Eds.), *Anxiety and the Heart.* (Hemisphere, Washington, D.C., 1990) pp. 295–346.
85. C. F. Wooley and H. Boudoulas, "Mitral valve prolapse: a classification" **In**: (H. Boudoulas and C. F. Wooley, Eds.), *Mitral Valve Prolapse and the Mitral Valve Prolapse Syndrome.* (Futura, Mount Kisco, NY, 1988) pp. 3–10.
86. F. A. Gaffney and C. G. Blomqvist, "Mitral valve prolapse syndrome: autonomic dysfunction" *Primary Cardiol.*, 132–145, 1981.
87. N. Weissman *et al.*, "Contrasting patterns of autonomic dysfunction in patients with mitral valve prolapse and panic attacks" *Am. J. Med.*, **82**, 880–888, 1987.
88. J. S. Turkkan *et al.*, "Animal studies of stressful interactions; a behavioral-physiological overview" **In**: (L. Goldberger and S. Breznitz, Eds.), *Handbook of Stress: Theoretical and Clinical Aspects.* (Macmillan, New York, 1986) pp. 153–180.
89. R. W. Schrier, "Pathogenesis of water and sodium retention in high-output and low-output cardiac failure, nephrotic syndrome, cirrhosis and pregnancy" *New Engl. J. Med.*, **319**, 1065–1072, 1988.
90. D. P. Henry *et al.*, "Norepinephrine in urine and plasma following provocative maneuvers in normal and hypertensive subjects" *Hyperten.*, **2**, 20–28, 1980.
91. B. Folkow and D. L. Ely, "Dietary sodium effects on cardiovascular and sympathetic neuroeffector functions as studied in various rat models" (editorial review). *Hyperten.*, **5**, 383–395, 1987.
92. F. Skrabal *et al.*, "Up-regulation of alpha-2 adrenoreceptors and down-regulation of beta-2 adrenoceptors by high-salt diet in normotensive men: enhanced up-regulation of operative (alpha-2: beta-2) adrenoreceptor ratio predicts salt sensitivity" *J. Hypertension*, **6**, 196–199, 1986.
93. T. Nii *et al.*, "Changes in hemodynamics and myocardial contractility during chronic sodium depletion in conscious dogs" *Hyperten.*, **9** (Part 2), III-176–III-180, 1987.
94. K. L. Skorecki and B. M. Brenner, "Body fluid homeostasis in man" *Am. J. Med.*, **70**, 77–88, 1981.
95. M. I. Koolen and P. van Brummelen, "Adrenergic activity and peripheral hemodynamics in relation to sodium sensitivity in patients with essential hypertension" *Hyperten.*, **6**, 820–825, 1984.
96. J. M. Sullivan and T. E. Ratts, "Sodium sensitivity in human subjects: hemodynamic and hormonal correlates" *Hyperten.*, **11**, 717–723, 1988.
97. J. M. Krzesinski and G. L. Rorive, "Is the decrease of central venous compliance the factor responsible for salt sensitivity?" *J. Hyperten.*, **6**, 68–70, 1986.
98. R. H. Rosenman, "Health consequences of anger and implications for treatment" **In**: (M. A.

Chesney and R. H. Rosenman, Eds.), *Anger and Hostility in Behavioral Medicine.* (Hemisphere, New York, 1985) pp. 103–126.

99. M. Frankenhaeuser, "Psychobiological aspects of life stress" *Coping and Health.* (Plenum, New York, 1980).
100. C. D. Spielberger *et al.*, "Assessment of anger: the State-Trait Anger Scale" **In**: (J. N. Butcher and C. D. Spielberger, Eds.) *Advances in Personality Assessment, Vol. 2* (Erlbaum, Hillsdale, NJ, 1983).
101. A. Montagu, *The Nature of Human Aggression.* (Oxford University Press, Oxford, 1976).
102. C. M. Turnbull, "The politics of non-aggression" **In**: (A. Montagu, Ed.), *Learning Non-Aggression: the Experience of Non-Literate Societies* (Oxford University Press, Oxford, 1978).
103. J. E. Pfieffer, "The apish origins of human tension: the case of the amiable chimps and the nervous baboons" *Harvard Mag.*, 55–60, 1981.
104. H. F. Judson, "An empirical presence" *The Sciences,* 20–23, (May, 1983).
105. N. M. Tanner, *On Becoming Human.* (Harvard University Press, Cambridge, 1981).
106. M. Konner, *The Tangled Wing.* (Holt, New York, 1982)
107. W. Mandell, *Med. World News,* (October 11, 1982) p. 40.
108. B. L. Benderly, "Rape-free or rape-prone societies" *The Sciences,* **221**, 40–43, 1982.
109. O. Olmert, "Medieval mayhem" *Smithsonian,* 34–36 (1982).
110. R. H. Rosenman, "Results of the multicenter antihypertensive treatment trials. Therapeutic implications and the role of the sympathetic nervous system" **In**: (J. Izzo and R. H. Rosenman, Eds.), *Hypertension: the noradrenergic factor. Am. J. Hyperten.*, **2** (Suppl. Part 2) 313S–338S, 1989.
111. J. B. Young and L. Landsberg, "Catecholamines and intermediary metabolism" *Clin. Endocrinol. Metab.*, **6**, 599–631, 1977.
112. J. B. Young and L. Landsberg, "Catecholamines and the regulation of hormone secretion" *Clin. Endocrinol. Metab.*, **6**, 657–695, 1977.
113. D. B. Beevers *et al.*, "Alcohol and hypertension" **In**: (J. Bahlmann and H. Liebau, Eds.), *Stress and Hypertension: contributions to Nephrology.* Vol. 30, (Karger, Basel, 1982) pp. 92–97.
114. A. Takizawa *et al.*, "Variant angina induced by alcohol consumption" *Am. Heart J.*, **107**, 25–27, 1984.
115. J. E. Deanfield *et al.*, "Direct effects of the heart on smoking: silent ischemic disturbances of coronary flow" *Am. J. Cardiol.*, **57**, 1005–1009, 1986.
116. M. D. Winniford *et al.*, "Smoking-induced coronary vasoconstriction in patients with atherosclerotic coronary artery disease; evidence for adrenergically-mediated alterations in coronary artery tone" *Circ.*, **73**, 662–667, 1986.
117. M. A. Brodsky *et al.*, "Ventricular tachyarrhythmia associated with psychological stress: the role of the sympathetic nervous system" *JAMA,* **257**, 2064–2067, 1987.
118. T. W. Meade, "Effects of changes in smoking and other characteristics on clotting factors and the risk of ischemic heart disease" *Lancet,* **2**, 986–988, 1987.
119. A. Rozanski *et al.*, "Mental stress and the induction of silent myocardial ischemia in patients with coronary artery disease" *New Engl. J. Med.*, **318**, 1005–1012, 1988.
120. P. G. F. Nixon, "Are there clinically significant prodromal signs and symptoms of impending sudden death" *Pract. Cardiol.*, **8**, 173–183, 1982.
121. P. Reich *et al.*, "Acute psychological disturbances preceding life-threatening ventricular arrhythmias" *JAMA,* **246**, 233–235, 1981.
122. J. E. Skinner, "Psychosocial stress and sudden cardiac death: brain mechanism" **In**: (R. Beamish, P. Singal and N. Dhalla, Eds.), *Stress and Heart Disease,* (Martinus Nijhoff, Boston, 1985) pp. 44–59.
123. B. Lown, "Sudden cardiac death: biobehavioral perspective" *Circ.*, (Suppl. 1) **76**, I-186–I-196, 1987.
124. P. J. Schwartz, "Stress and sudden cardiac death: the role of the autonomic nervous system" *J. Clin. Psych.*, (Suppl. 2), **45**, 7–13, 1984.
125. P. R. Kowey *et al.*, "Influence of coronary platelet aggregation on ventricular electrical properties during partial coronary artery stenosis" *Am. J. Cardiol.*, **51**, 595–602, 1983.
126. C. D. Brackett and L. H. Powell, "Psychosocial and physiological predictors of sudden cardiac death after healing of acute myocardial infarction" *Am. J. Cardiol.*, **61**, 979–983, 1988.
127. J. N. Cohn, "Sympathetic nervous system activity and the heart" **In**: (J. Izzo and R. H. Rosenman, Eds.), *Hypertension: The noradrenergic factor. Am. J. Hyperten.*, (Suppl. Part 2), **12**, 353S–356S, 1989.
128. J. G. F. Cleland and M. J. Dargie, "Arrhythmias, catecholamines and electrolytes" *Am. J. Cardiol.*, **62**, 55A–59A, 1988.
129. G. S. Francis *et al.*, "The neuro-humoral axis in congestive heart failure" *Ann. Int. Med.*, **101**, 370–377, 1984.

130. R. Garcia and J. M. Jennings, "Pheochromocytoma masquerading as cardiomyopathy" *Am. J. Cardiol.*, **29**, 368–371, 1972.
131. L. J. Freeman *et al.*, "Psychological stress and silent myocardial ischemia" *Am. Heart J.*, **114**, 417–424, 1987.
132. J. Hunter, *Treatise on the Blood, Inflammation and Gun-shot Wounds* (Richardson, London, 1794).
133. W. Heberden, *Commentaries on the History and Cure of Disease* (Wells and Lilly, Boston, 1818).
134. A. Maseri *et al.*, "Mixed angina pectoris" *Am. J. Cardiol.*, **56**, 30–33, 1985.
135. P. F. Cohn *et al.*, "Characteristics of silent myocardial ischemia occurring during out-of-hospital activities in asymptomatic coronary patients" *Circ.*, **74**, Part 2, 226 1986.
136. A. Maseri *et al.*, "Mechanisms of ischemic cardiac pain and silent myocardial ischemia" *Am. J. Med.*, **79**, (Suppl 3A), 7–11, 1985.
137. J. B. Gordon *et al.*, "Coronary constriction with exercise: possible role for endothelial dysfunction and alpha tone" *Circ.*, **74**(Part 2), II-481, 1986.
138. F. Crea *et al.*, "Different susceptibility to myocardial ischemia provoked by hyperventilation and cold pressor test in exertional and variant angina pectoris" *Am. J. Cardiol.*, **56**, 18–22, 1985.
139. A. P. Selwyn and P. Ganz, "Myocardial ischemia in coronary disease" *New Engl. J. Med.*, **318**, 1058–1060, 1988.
140. G. H. Mudge *et al.*, "Reflex increase in coronary vascular resistance in patients with ischemic heart disease" *New Engl. J. Med.*, **295**, 1333–1337, 1976.
141. S. D. Gertz *et al.*, "Endothelial cell damage and thrombus formation after partial arterial constriction: relevance to the role of coronary artery spasm in the pathogenesis of myocardial infarction" *Circ.*, **63**, 476–486, 1981.
142. J. Y. Lam *et al.*, "Is vasospasm related to platelet deposition? Relationship in a porcine preparation of arterial injury *in vivo*" *Circ.*, **75**(Suppl 1), 243–249, 1987.
143. M. J. Davies and A. C. Thomas, "Plaque fissuring—the cause of acute myocardial infarction, sudden ischemic death, and crescendo angina" *Br. Heart J.*, **53**, 353–373, 1985.
144. A. A. Bove, "Coronary spasm—etiology and diagnosis" *Learning Center Highlights. Am. Coll. Cardiol.*, **2**, 1–5, Fall 1986.
145. B. Lown and R. A. DeSilva, "Is coronary arterial spasm a risk factor for coronary atherosclerosis?" *Am. J. Cardiol.*, **45**, 901–903, 1980.
146. H. Yasue *et al.*, "Induction of coronary artery spasm by acetylcholine in patients with variant angina: possible role of the parasympathetic nervous system in the pathogenesis of coronary artery spasm" *Circ.*, **74**, 955–963, 1986.
147. S. Goldberg, "Unstable angina and progressive coronary obstruction" *The Complicated Cardiovascular Patient.* 22–25, Fall 1986.
148. A. Moise *et al.*, "Unstable angina and progression of coronary atherosclerosis" *New Engl. J. Med.*, **309**, 685–689, 1983.
149. M. Marzilli *et al.*, "Some clinical considerations regarding the relation of coronary vasospasm to coronary atherosclerosis: a hypothetical pathogenesis" *Am. J. Cardiol.*, **45**, 882–886, 1980.
150. A. P. Selwyn *et al.*, "Clinical problems in coronary disease are caused by a wide variety of ischemic episodes that affect patients out of hospital" *Am. J. Med.*,. **79**(Suppl 3A), 12–17 (1985).
151. J. E. Deanfield *et al.*, "Clinical evaluation of transient myocardial ischemia during daily life" *Am. J. Med.*, **79**(Suppl 3A), 18–24 (1985).
152. A. N. De Maria *et al.*, "The anginal syndrome with normal coronary arteries: etiologic and prognostic considerations" *JAMA*, **244**, 826–828, 1980.
153. J. C. Kaski *et al.*, "Transient myocardial ischemia during daily life in patients with syndrome X" *Am. J. Cardiol.*, **58**, 124–147, 1986.
154. R. P. Sprafkin *et al.*, "Cardiovascular and psychological characteristics of patients with chest pain" *Behav. Med. Update*, **6**, 11, 1984.
155. A. Maseri *et al.*, " 'Variant' angina: one aspect of a continuous spectrum of vasospastic myocardial ischemia. Pathogenetic mechanisms, estimated incidence and clinical and coronary arteriographic findings in 138 patients" *Am. J. Cardiol.*, **42**, 1019–1035, 1978.
156. L. G. Howes *et al.*, "Plasma cholesterol levels are dependent on sympathetic activity" *J. Hyperten.*, **5**(Suppl 5); S361–S363, 1987.
157. J. E. Dimsdale and A. Herd, "Variability of plasma lipids in response to emotional arousal" *Psychosom. Med.*, **44**, 413–430, 1982.
158. M. Friedman *et al.*, "Changes in the serum cholesterol and blood-clotting time in men subjected to cyclic variation of occupational stress" *Circ.*, **17**, 852–861, 1958.
159. R. Grundy and A. Griffin, "Relationship of periodic mental stress to serum lipoprotein and cholesterol levels" *JAMA*, **171**, 1793–1796, 1959.
160. C. D. Jenkins *et al.*, "Psychological traits and serum lipids. Part I. Findings from the California Psychological Inventory" *Psychosom. Med.*, **31**, 115–128, 1969.

161. L. O'Donnell *et al.*, "Plasma catecholamines and lipoproteins in chronic psychological stress" *J. Roy. Soc. Med.*, **80**, 339–344, 1987.
162. R. H. Rosenman and M. Friedman, "Neurogenic factors in pathogenesis of coronary heart disease" *Med. Clin. No. Am.*, **58**, 269–279, 1974.
163. G. Weidner *et al.*, "The role of Type A behavior and hostility in an elevation of plasma lipids in adult women and men" *Psychosom. Med.*, **49**, 136–145, 1987.
164. U. Lundberg *et al.*, "Type A behavior in healthy males and females as related to physiological reactivity and blood lipids" *Psychosom. Med.*, **51**, 113–122, 1989.
165. T. Gordon and J. Verter, "Serum cholesterol, systolic blood pressure, and Framingham relative weight as discriminators of cardiovascular disease" **In**: (W. B. Kannel and T. Gordon, Eds.), *The Framingham Study: an Epidemiological Investigation of Cardiovascular Disease.* (U.S. Government Printing Office, Washington, D.C., 1969).
166. R. H. Rosenman *et al.*, "Multivariate prediction of coronary heart disease during 8.5 year follow-up in the Western Collaborative Group Study" *Am. J. Cardiol.*, **37**, 903–910, 1976.
167. A. Keys *et al.*, "Probability of middle-aged men developing coronary heart disease in 5 years" *Circ.*, **45**, 815–872, 1972.
168. D. Kozarevic *et al.*, "The Yugoslavia cardiovascular study. II. Factors in the incidence of coronary heart disease" *J. Chron. Dis.*, **29**, 405–414, 1976.
169. T. Gordon *et al.*, "Differences in coronary heart disease in Framingham, Honolulu, and Puerto Rico" *J. Chron. Dis.* **27**, 329–37, 1972.
170. P. Ducimetière *et al.*, "Coronary heart disease in middle-aged Frenchmen" *Lancet*, **1**, 1346–1349, 1980.
171. A. V. G. Bruschke *et al.*, "The anatomic evolution of coronary artery disease demonstrated by coronary arteriography in 256 non-operated patients" *Circ.*, **63**, 527–36, 1981.
172. J. R. Kramer *et al.*, "Progression and regression of coronary atherosclerosis: relation to risk factors" *Am. Heart J.*, **105**, 134–139, 1983.
173. B. Marchandise *et al.*, "Angiographic evaluation of the natural history of normal coronary arteries and mild coronary atherosclerosis" *Am. J. Cardiol.*, **41**, 216–220, 1978.
174. D. R. Holmes *et al.*, "Association of risk factor variables and coronary artery disease documented with angiography" *Circ.*, **63**, 293–299, 1981.
175. R. E. Vlietstra *et al.*, "Factors affecting the extent and severity of coronary artery disease in patients enrolled in coronary artery surgery" *Atheroscler.*, **2**, 208–215, 1983.
176. C. D. Jenkins, "Epidemiology of cardiovascular diseases" *J.Consul. Clin. Psychol.*, **56**, 324–332, 1988.
177. R. H. Rosenman, "Current and past history of Type A behavior pattern" **In**: (T. H. Schmidt, T. M. Dembroski and G. Blumchen, Eds.), *Biological and Psychological Factors in Cardiovascular Disease* (Springer-Verlag, Heidelberg, 1986) pp. 15–40.
178. T. von Dusch, *Lehrbuch der Herzkrankheiten* (Verlag von Wilhelm Engelman, Leipzig, 1868) p. 334.
179. W. Osler, "The Lumleian lectures on angina pectoris" *Lancet*, **1**, 829–844, 1892.
180. K. A. Menninger and W. C. Menninger, "Psychoanalytic observations in cardiac disorders" *Am. Heart J.*, **11**, 10–26, 1936.
181. H. F. Dunbar, *Psychosomatic Diagnosis* (Hoeber, New York, 1943).
182. J. A. Arlow, "Identification of mechanisms in coronary occlusion" *Psychosom. Med.*, **7**, 195–207, 1945.
183. C. Kemple, "Rorshach method and psychosomatic diagnosis: personality traits of patients with rheumatic disease, hypertension, cardiovascular disease, coronary occlusion, and fracture" *Psychosom. Med.*, **7**, 85–89, 1945.
184. E. Gildea, "Special features of personality which are common to psychosomatic disorders" *Psychosom. Med.*, **11**, 273–277, 1949.
185. M. Gertler and P. D. White, *Coronary Disease in Young Adults: A Multidisciplinary Study* (Harvard University Press, Cambridge, 1954).
186. I. M. G. Stewart, "Coronary disease and modern stress" *Lancet*, **2**, 867–878, 1950.
187. C. G. Helman, "Heart disease and cultural construction of time: Type A behavior pattern as a western culture-bound syndrome" *Soc. Sci. Med.*, **25**, 969–979, 1987.
188. D. G. Byrne and R. H. Rosenman, "The Type A behaviour pattern as a precursor to stressful life events: a confluence of coronary risks" *Br. J. Med. Psychol.*, **59**, 75–82, 1986.
189. J. Suls and C. K. Wan, "The relationship between Type A behavior pattern and chronic emotional distress: a meta-analysis" *J. Pers. Soc. Psychology*, **57**, 503–512, 1989.
190. M. J. Strube, "A self-appraisal model of the Type A behavior pattern" **In**: (R. Hogan and W. Jones, Eds.), *Perspectives in Personality Theory (Vol. 2)* (JAI Press, Greenwich, Ct, 1987) pp. 201–50.

191. G. De Backer *et al.*, "Behavior, stress, and psychological traits as risk factors" *Prev. Med.,* **12**, 32–36, 1983.
192. French-Belgian Collaborative Group. Ischemic heart disease and psychological patterns. Prevalence and incidence studies in Belgium and France. **In**: (H. Denolin, Ed.), *Psychological Problems Before and After Myocardial Infarction. Advances in Cardiology (Vol. 29)* (Karger, Basel, 1982) pp. 25–31.
193. E. A. Eaker *et al.*, "Frequency of uncomplicated angina pectoris in Type A compared with Type B persons (The Framingham Study)" *Am. J. Cardiol.,* **63**, 1042–1045, 1989.
194. S. G. Haynes and M. Feinleib, "Type A behavior and the incidence of coronary heart disease in the Framingham Heart Study" **In**: (H. Denolin Ed.), *Psychological Problems Before and After Myocardial Infarction. Advances in Cardiology (Vol 29).* (Karger, Basel, 1982) pp. 85–95.
195. J. B. Cohen and D. Reed, "Type A behavior and coronary heart disease among Japanese men living in Hawaii" *J. Behav. Med.,* **8**, 343–352, 1985.
196. T. M. Dembroski *et al.*, "Components of hostility as predictors of sudden death and myocardial infarction in the Multiple Risk Factor Intervention Trial" *Psychosom. Med.,* **51**, 514–522, 1989.
197. R. B. Shekelle *et al.*, "The MRFIT Behavior pattern Study. II. Type A behavior and incidence of coronary heart disease" *Am. J. Epidemiol.,* **122**, 559–570, 1985.
198. K. A. Matthews and S. G. Haynes, "Type A behavior and coronary disease risk" *Am. J. Epidemiol.,* **123**, 923–960, 1986.
199. S. G. Haynes and K. A. Matthews, "Review and methodologic critique of recent studies on Type A behavior and cardiovascular disease" *Ann. Behav. Med.,* **10**, 47–59, 1988.
200. Multiple Risk Factor Trial Research Group, "Multiple risk factor intervention trial: risk factor changes and mortality results" *JAMA,* **248**, 1465–1477, 1982.
201. S. Booth-Kewley and H. S. Friedman, *Psychol. Bull.*, **101**, 343–362, 1987.
202. B. E. Edye *et al.*, "Evaluation of a worksite program for the modification of cardiovascular risk factors" *Med. J. Aust.,* **150**, 574–581, 1989.
203. T. P. Gill *et al.*, "Risk factors for coronary heart disease in a self-referred population compared with a general population" *Med. J. Aust.,* **151**, 518–525, 1989.
204. M. H. Criqui *et al.*, "The effect of non-response on risk ratios in a cardiovascular disease study" *J. Chron. Dis.,* **32**, 633–638, 1979.
205. N. B. Oldridge *et al.*, "Predictors of dropout from cardiac exercise rehabilitation—Ontario Exercise-Heart Collaborative Study" *Am. J. Cardiol.,* **51**, 70–74, 1982.
206. E. M. Bruce *et al.*, "Comparison of active participants and dropouts in CAPRI cardiopulmonary rehabilitation programs" *Am. J. Cardiol.,* **37**, 53–62, 1976.
207. L. Werko, "Risk factors and coronary heart disease—facts or fancy?" *Am. Heart J.,* **91**, 87–98, 1976.
208. R. H. Rosenman *et al.*, "Recurring and fatal myocardial infarction in the Western Collaborative Group Study" *Am. J. Cardiol.,* **19**, 771–775, 1967.
209. C. D. Jenkins *et al.*, "Type A behavior pattern and mortality after recurrent myocardial infarction: preliminary results from a follow-up study of five years" *Psychother. Psychosom.,* **46**, 653–665, 1986.
210. C. D. Jenkins *et al.*, "Risk of new myocardial infarction in middle-aged men with manifest coronary heart disease" *Circ.,* **53**, 342–47, 1976.
211. R. B. Case *et al.*, "Multicenter Post-Infarction Research Group. Type A behavior and survival after acute myocardial infarction" *New Engl. J. Med.,* **312**, 734–741, 1985.
212. D. De Leo *et al.*, "Type A behavior pattern and mortality after recurrent myocardial infarction: preliminary results from a follow-up study of five years" *Psychother. Psychosom.,* **46**, 653–665, 1986.
213. J. E. Dimsdale *et al.*, "Predicting cardiac morbidity based on risk factors and coronary angiographic findings" *Am. J. Cardiol.,* **47**, 73–76, 1981.
214. E. D. Eaker and W. P. Castelli, "Type A behavior and mortality from coronary disease in the Framingham Study" *New Engl. J. Med.,* **319**, 1480–1481, 1988.
215. M. Friedman *et al.*, "Alteration of Type A behavior and its effect on cardiac recurrences in post-myocardial infarction patients: summary results of the Recurrent Coronary Prevention Project" *Am. Heart J.,* **11**, 653–665, 1986.
216. D. R. Ragland and R. J. Brand, "Type A behavior and mortality from coronary heart disease" *New Engl. J. Med.,* **318**, 65–69, 1988.
217. R. B. Shekelle *et al.*, "Type A score (Jenkins Activity Survey) and risk of recurrent coronary heart disease in the aspirin myocardial infarction study" *Am. J. Cardiol.,* **56**, 221–225, 1985.
218. D. G. Byrne *et al.*, "Consistency and variation among instruments purporting to measure the Type A behavior pattern" *Psychosom. Med.,* **7**, 242–261, 1985.

219. Multicenter Postinfarction Research Group, "Risk stratification and survival after myocardial infarction" *New Engl. J. Med.*, **309**, 331–336, 1983.
220. J. A. Blumenthal *et al.*, "Type A behavior and angiographically documented coronary disease" *Circ.*, **58**, 634–639, 1978.
221. C. A. Derby and T. A. Pearson, "Type A behavior: associated with coronary disease or with arteriography" *Circ.*, **80**(Suppl.), II-613, 1989.
222. K. A. Frank *et al.*, "Type A behavior pattern and coronary atherosclerosis" *JAMA*, **240**, 761–763, 1978.
223. M. Friedman *et al.*, "The relationship of behavior pattern A to the state of coronary vasculature" *Am. J. Med.*, **44**, 525–537, 1968.
224. D. S. Krantz *et al.*, "Psychological correlates of progression of atherosclerosis in men" *Psychosom. Med.*, **41**, 467–475, 1979.
225. R. B. Williams *et al.*, "Type A behavior, hostility, and coronary atherosclerosis" *Psychosom. Med.*, **42**, 539–549, 1980.
226. R. B. Williams *et al.*, "Type A behavior and angiographically documented coronary atherosclerosis in a sample of 2,289 patients" *Psychosom. Med.*, **50**, 139–52, 1988.
227. S. J. Zyzanski *et al.*, "Psychological correlates of coronary angiographic findings" *Archiv. Int. Med.*, **136**, 1234–1237, 1976.
228. T. M. Dembroski *et al.*, "Components of Type A hostility, hostility, and anger-in: relationships to angiographic findings" *Psychosom. Med.*, **47**, 219–233, 1985.
229. J. E. Dimsdale *et al.*, "Type A personality and extent of coronary atherosclerosis" *Am. J. Cardiol.*, **42**, 583–586, 1978
230. J. E. Dimsdale *et al.*, "Type A behavior and angiographic findings" *J. Psychosom. Res.*, **23**, 273–276, 1979.
231. J. M. MacDougall *et al.*, "Components of Type A hostility, and anger-In: further relationship to angiographic findings" *Health Psychol.*, **4**, 137–152, 1985.
232. M. Kornitzer *et al.*, "Angiographic findings and the Type A behavior pattern assessed by means of the Bortner Scale" *J. Behav. Med.*, **5**, 313–319, 1982.
233. L. Scherwitz *et al.*, "Type A behavior, self-involvement, and coronary atherosclerosis" *Psychosom. Med.*, **45**, 47–57, 1983.
234. T. Pearson *et al.*, "Selection bias in persons undergoing angiography" *Am. J. Epidemiol.*, **116**, 568, 1982.
235. T. G. Pickering, "Should studies of patients undergoing coronary angiography be used to evaluate the role of behavioral risk factors for coronary heart disease?" *J. Behav. Med.*, **8**, 203–213, 1985.
236. T. M. Dembroski *et al.*, "Components of the Type A behavior pattern and cardiovascular responses to psychomotor performance challenge" *J. Behav. Med.*, **1**, 159–176, 1978.
237. R. B. Williams, Jr. *et al.*, "The health consequences of hostility" **In**: (M. A. Chesney and R. H. Rosenman, Eds.), *Anger and Hostility in Cardiovascular and Behavioral Disorders* (Hemisphere, Washington, D.C., 1985) pp. 173–86.
238. C. D. Jenkins *et al.*, "Components of the coronary-prone behavior pattern: their reaction to silent myocardial infarction and blood lipids" *J. Chron. Dis* **19**, 599–606, 1966.
239. K. A. Matthews *et al.*, "Competitive drive, pattern A, and coronary heart disease: a further analysis of some data from the Western Collaborative Group Study" *J. Chron. Dis.*, **30**, 489–498, 1977.
240. M. H. L. Hecker *et al.*, "Coronary-prone behaviors in the Western Collaborative Group Study" *Psychosom. Med.*, **50**, 53–64, 1988.
241. E. I. Megargee, "The dynamics of aggression and their application to cardiovascular disorders" **In**: (M. A. Chesney and R. H. Rosenman, Eds.), *Anger and Hostility in Cardiovascular and Behavorial Disorders* (Hemisphere, Washington, D.C., 1985) pp. 31–38.
242. C. D. Spielberger *et al.*, "The experience and expression of anger: construction and validation of an anger expression scale" **In**: (M. A. Chesney and R. H. Rosenman, Eds.), *Anger and Hostility in Cardiovascular and Behavioral Disorders* (Hemisphere, Washington, D.C., 1985) pp. 5–30.
243. P. T. Costa *et al.*, "Agreeableness versus antagonism: explication of a potential risk factor for CHD" **In**: (A. Siegman and T. M. Dembroski, Eds.), *In Search of Coronary-prone Behavior: Beyond Type A* (Erlbaum, Hillsdale, NJ, 1989) pp. 41–63.
244. W. Cook and D. Medley, "Proposed hostility and pharisaic-virtue scales for the MMPI" *J. Appl. Psychol.*, **38**, 414–418, 1954.
245. J. C. Barefoot *et al.*, "Hostility, CHD incidence, and total mortality: a 25-year follow-up study of 255 physicians" *Psychosom. Med.*, **45**, 59–63, 1983.
246. R. B. Shekelle *et al.*, "Hostility, risk of coronary heart disease, and mortality" *Psychosom. Med.*, **45**, 109–114, 1983.

247. M. D. Hearn *et al.*, "Hostility, coronary heart disease, and total mortality: a 33-year follow-up study of university students" *J. Behav. Med.,* **12**, 105–121, 1989.
248. M. Koskenvuo *et al.*, "Hostility as a risk factor for mortality and ischemic heart disease in men" *Psychosom. Med.,* **50**, 330–340, 1988.
249. E. W. McCranie *et al.*, "Hostility, coronary heart disease incidence, and total mortality: lack of association in a 25-year follow-up study of 478 physicians" *J. Behav. Med.,* **9**, 119–125, 1986.
250. T. Matsushima *et al.*, "Studies on Type A behavior pattern and hostility in Japanese male subjects with special reference to CHD" *Shinshin-Iqaku,* **213**, 321–328, 1983.
251. G. R. Leon *et al.*, "Type A behavior and mortality from coronary disease in the Framingham Study" *New Engl. J. Med.,* **319**, 1480–1481, 1988.
252. A. W. Siegman *et al.*, "Components of hostility and severity of coronary artery disease" *Psychosom. Med.,* **49**, 127–135, 1987.
253. C. C. Tennant *et al.*, "Anger and other psychological factors in coronary atherosclerosis" *Psychol. Med.,* **17**, 425–431, 1987.
254. G. Swan *et al.*, "Cook and Medley hostility and the Type A behavior pattern; psychological correlates of two coronary prone behaviors" **In**: (M. Strube, Ed.), *Type A Behavior* (Special issue) *J. Soc. Behav. Pers.,* **5**, 89–106, 1990.
255. J. A. Blumenthal *et al.*, "Psychological correlates of hostility among patients undergoing coronary angiography" *Br. J. Med. Psychol.,* **60**, 349–355, 1987.
256. P. T. Costa *et al.*, "Cynicism and paranoid alienation in the Cook and Medley HO Scale" *Psychosom. Med.,* **48**, 283–285, 1986.
257. J. D. Hardy and T. W. Smith, "Cynical hostility and vulnerability to disease: social support, life stress, and physiological response to conflict" *Health Psychol.,* **7**, 447–459, 1988.
258. T. W. Smith *et al.*, "Components of the Cook-Medley HO Scale" *Proc. of 9th Annual Soc. of Behavioral Medicine,* Boston, p. 155, 1988.
259. T. W. Smith and K. D. Frohm, "What's so unhealthy about hostility? Construct validity and psychosocial correlates of the Cook and Medley HO Scale" *Health Psychol.,* **4**, 503–520, 1985.
260. F. L. Watkins *et al.*, "Comparison of hostility and Type A behavior as cardiovascular disease risk factors within a worksite screening program" *Proc. of 9th Annual Society of Behavioral Medicine,* Boston, p. 148, 1988.
261. R. H. Rosenman *et al.*, "Some recent findings relative to the relationship of Type A behavior pattern to coronary heart disease" **In**: (S. Maes, C. D. Spielberger, P. Defares and I. Sarason, Eds.), *Topics in Health Psychology* (Wiley, Chichester, 1988) pp. 21–29.
262. W. G. Shipman, "The validity of MMPI Hostility scales" *J. Clin. Psychol.,* **21**, 186–190, 1965.
263. J. Suls and G. S. Sanders, "Type A behavior as a general risk factor for physical disorder" *J. Behav. Med.,* **11**, 201–209, 1988.
264. P. J. Woods and J. Burns, "Type A behavior and illness in general" *J. Behav. Med.,* **7**, 411–414, 1984.
265. I. S. Yakubovich *et al.*, "Type A behavior pattern and health status after 22 years of follow-up in the Western Collaborative Group Study" *Am. J. Epidemiol.,* **128**, 579–588, 1988.
266. J. C. Barefoot *et al.*, "Suspiciousness, health and mortality: a follow-up study of 500 older adults" *Psychosom. Med.,* **49**, 450–457, 1987.
267. L. F. Berkman and S. L. Syme, "Social networks, host resistance, and mortality: a nine year follow-up study of Alameda County residents" *Am. J. Epidemiol.,* **109**, 186–204, 1978.
268. A. Menotti *et al.*, "Prediction of all causes of death as a function of some factors commonly measured in cardiovascular population surveys" *Prev. Med.,* **12**, 318–325, 1983.
269. P. D. Somervell *et al.*, "Psychologic distress as a predictor of mortality" *Am. J. Epidemiol.,* **130**, 1013–1023, 1989.
270. W. Ruberman *et al.*, "Psychosocial influences on mortality after myocardial infarction" *New Engl. J. Med.,* **311**, 522–529, 1984.
271. S. Julius *et al.*, "Anger-coping types, blood pressure, and all-cause mortality: a follow-up in Tecumseh, Michigan (1971–1983)" *Am. J. Epidemiol.,* **124**, 220–233, 1986.
272. A. Appels *et al.*, "A prospective study about vital exhaustion as a precursor of myocardial infarction and other serious diseases" *Nederlands Tijd. Psychol.,* **44**, 122–128, 1989.
273. C. Bass *et al.*, "Unexplained breathlessness and psychiatric morbidity in patients with normal and abnormal coronary arteries" *Lancet,* **1**, 605–609, 1983.
274. P. T. Costa *et al.*, "Is neuroticism a risk factor for CAD? Is Type A a measure of neuroticism?" **In**: (T. H. Schmidt, T. M. Dembroski and G. Blumchen, Eds.), *Biological and Psychological Factors in Cardiovascular Disease* (Springer-Verlag, Heidelberg, 1986), pp. 85–95.
275. R. B. Williams Jr., "Refining the Type A hypothesis: emergence of the hostility complex" *Am. J. Cardiol.,* **60**, 27J-32J, 1987.

276. R. H. Rosenman *et al.*, "A 27-year follow-up of coronary heart disease in the Western Collaborative Group Study" (submitted for publication).
277. L. H. Sigler, "The mortality from arteriosclerotic and hypertensive heart diseases in the United States. Part I. Possible relation to distribution of population and economic status" *Am. J. Cardiol.*, **1**, 606–614, 1959.
278. M. Marmot, "Socioeconomic and cultural factors in ischemic heart disease" **In**: (H. Denolin, Ed.), *Psychological Problems Before and After Myocardial Infarction* (Karger, Basel, 1982) pp. 68–76.

3. ENVIRONMENT, STRESS AND AGING

PAUL T. P. WONG* and EDWARD J. PEACOCK†

This paper first discusses the need for measuring both psychological and physical dimensions of environment and reviews various theoretical perspectives. Data from a study that investigated environment, stress and the well-being of community-based elderly are presented. The results show that perceived atmosphere of the environment is the only predictor of everyday stress, while perceived controllability is the only predictor of societal stress. Both atmosphere and societal stress predict psychological well-being. The results generally support Lawton's environmental docility hypothesis in that high IQ elderly tend to fare better than low IQ elderly under high environmental press (i.e. bad atmosphere, low predictability).

1. INTRODUCTION

That people are affected by the environment they live in is a truism. Surely, geographical features and climate affect the lifestyle of human inhabitants. The human-made environment, such as buildings and communities, can also affect our activities. Similarly, our perceptions of living conditions can affect how we feel and behave. However, in spite of the pervasiveness of environmental influences in human life, much remains unknown regarding the complex relationships between environment and behaviour.

Research on the relationship between environment and aging has gained increasing attention in recent years. Progress has been slow because of conceptual and methodological difficulties. Environment is difficult to define and measure, because it is a global, all encompassing concept; in fact, it includes everything external to our bodies. Environment can be studied at the macro or micro level. Macro-environment refers to the region, the community, and the neighbourhood, while micro-environment refers to the characteristics of the individual's dwelling unit. One can also differentiate between the physical, cultural and psychological aspects of environment.

Environment and stress are treated as distinct constructs here for two reasons. First, environmental factors are not always stressful. Some aspects of environment, such as beautiful landscapes and comfortable homes can be benign and uplifting. Furthermore, environment can also function as a moderator against the negative impact of stress. Words like retreat, haven, sanctuary readily come to mind when we think about places where we can find temporary refuge from the storms of life. Second, stress does not always reside in the external world, even though stress in everyday living often originates from environmental factors. We need to keep in mind that there are both internal and external sources of stress.[1]

---

*Department of Psychology, Trent University, Peterborough, Ontario K9J 7B8, Canada.
†University of Toronto and Queen's University at Kingston, Ontario, Canada.

In this chapter, we begin with a discussion of the components of environment and theoretical perspectives on person-environment interactions. Second, we review selected areas of research findings on environmental influences in the aging process from the perspective of person-environmental congruence. Third, we review the literature on environment, stress and aging. Finally, we present new data to illustrate how certain dimensions of the environment and stress are related to well-being in a sample of community dwelling elderly.

2. COMPONENTS OF ENVIRONMENT

2.1. *Dimensions of Environment*

In view of the enormous complexity of variables subsumed under the rubric of environment, there is the need to develop an environmental taxonomy.[2] When environment is treated as an ecosystem involving person-environmental transactions, the problem of classification becomes even more complex. Lawton[3] writes that "a major lack in our ability to deal with the environmental aspects of the behavioral system comes from the absence of an environmental taxonomy" (p.40). A comprehensive taxonomy of person-environment interactions will facilitate systematic research in environmental psychology.

Little[4] provides a good overview of the conceptual and methodological issues in measuring person-environment relations, particularly the contributions by Craik[5,6] and McKechnie.[7,8] Drawing upon historical, literary and other resources for items reflecting various environmental dispositions, and using factor analytical techniques, McKechnie developed a 184-item Environmental Response Inventory; it consists of various scales, such as Pasturalism, Urbanism, Environmental Adaptation, and Need for Privacy.

Focusing on studies of how the elderly respond to the environment, McPherson[9] has identified 17 different components. These are: 1) the type, variety, and quality of housing, 2) the place of residence, 3) the degree and type of institutionalization, 4) the availability of and access to social networks, transportation and services, 5) the physical characteristics, spatial configuration, and size of the living area, 6) the age, ethnic, and racial composition of the neighbourhood, 7) the amount of actual or imagined crime in the neighbourhood, 8) the size and quality of private space especially within institutionalized settings, 9) the objective indicators and subjective meaning of the quality of the environment, 10) the availability and use of sensory, perceptual, and cognitive resources, 11) the degree and state of urban renewal, 12) the mobility and migration patterns of the elderly, 13) the degree of familiarity with the physical environment, 14) the degree of physical danger in the environment (poor lighting, deteriorating buildings), 15) the degree of personal control within the home and community, 16) the degree of choice in relocation, 17) the degree of actual or perceived change in the local physical and social environment over time. This list includes both physical and psychological components of the environment.

2.2. *Subjective Perceptions of Environment*

Golant[10] emphasizes the subjectively defined environment. Perceptions, feelings and evaluations of the residents rather than the objectively defined physical characteristics of the environment are the focus of interest. Environment is conceptualized from the experience of residents and not from the observations of a detached scientist.

Koffka's[11] "behavioural environment," Lewin's[12] "psychological environment," and Murray's[13] "beta environmental press" are all forerunners of this phenomenological approach. For example, Koffka[11] wrote: "Let us therefore distinguish between a geographical and a behavioural environment . . . our difference between the geographical and the behavioural environment coincides with the difference between things as they 'really' are and things as they look to us, between reality and appearance. And we see also that appearances may deceive, that behaviour well adapted to the behavioural environment may be unsuited to the geographical" (pp. 28, 33). Murray[13] conceived of the individual as striving to meet certain needs and classified environment in terms of how well it meets individuals needs. The objects and situations in the environment can exert either a positive or negative impact with respect to an individual's needs; such impact is conceptualized as "environmental press."

As a result of a factor analysis of questionnaire items completed by an elderly sample in Evanston, USA, Golant[10] obtained 15 interpretable factors of environmental experiences. Nine of these factors are related to environmental stress; these are: concern and worry about crime; annoyance at neighbour's care of property; accessibility difficulties and restrictions; bothered by noise and dirt; bothered/restricted by bad weather; feeling bored and lonely; feelings of restriction by dwelling; discomforts in dwelling; feelings of lack of privacy. Some of the factors obtained by Golant[10] overlap with those identified by McPherson.[9]

Golant has provided a composite picture of the many facets of the place where people grow old. It cuts across both macro and micro-environments. Furthermore, it is a mixed bag of feelings, needs, concerns and perceptions. Golant's results, though informative, are too complex and unwieldy to be useful for a more detailed analysis of environmental influences.

A few years ago, when the senior author and G. Reker began to study the psychosocial factors in successful aging, we could not find any instrument that measured important psychological dimensions of the environment, such as predictability, control, and the general atmosphere (e.g. friendliness). As a result, we developed our own instrument which we call the Living Environment Survey. The scales that measure the psychological aspects of living conditions were theoretically driven rather than empirically based. Some physical aspects of the environment known to impact the well-being of the elderly were also measured in the questionnaire. In short, we have chosen the strategy of measuring selective components of environment that have clear implications for adaptation in the elderly.

3. THEORETICAL PERSPECTIVES

Theoretical considerations play a major role in how we conceptualize and investigate environmental influences. Basically, there are three approaches to the study of environmental impact on well-being. The first approach focuses on the objectively definable characteristics of the environment, such as noise, temperature, and the physical aspects of the residence. The focus is on the properties and dispositions of the physical environment; individuals tend to be treated as passive recipients of environmental demands.

The second approach emphasizes the resident's phenomenological experiences. The focus is on subjective perceptions rather than objective realities. Individuals are viewed as actively construing the meaning and significance of the social and physical milieu in which they function.

The third approach emphasizes person-environment relations. The focus is on the dynamic transactions between the individual and environment. Environmental influences depend not on the inherent properties of the stimuli, nor on the perceptions and personal characteristics of individuals. Instead, the impact of environmental demand is understood in terms of interactions between the characteristics of the environment and personal factors such as needs, competence and personality traits.

Much of the current research on environmental influences employs the interactive perspective. The most prominent models from this perspective are based on the concept of person-environment congruence. French, Rodgers and Cobb[14] first outlined the congruence model. The basic idea is that the individual has certain demands, and the environment supplies the resources. Congruence depends on the goodness-of-fit between demand and supply. For example, Kahana, Liang and Felton[15] studied person-environment congruence in the elderly by measuring personal preferences (needs) for environmental attributes and objective indicators of these attributes. This allowed the researchers to measure the congruence or discrepancy between individual needs and environmental supply. Congruence is assumed to be a precondition for subjective well-being; thus deviation from congruence (undersupply or oversupply) can have a negative impact.

Kahana[16] employed the person-environment model to explain and predict the impact of various environmental settings on older adults. Since they frequently experience reduced income, impaired health, and loss of social roles, the elderly have fewer resources to choose or maintain an environment that meets their preferences. Discrepancy can occur when there is a change in the environment (as in relocation) or a change in the individual's resources or preferences. Whenever there is a discrepancy between their needs and living conditions, adaptive strategies are called upon to reduce the mismatch either by changing personal needs or by changing environment. The efficacy of these adaptive strategies determines the level of their well-being.

Kahana has postulated seven dimensions of congruence: (1) The segregate dimension (the homogeneity of the composition of the environment versus the individual's need/preference for homogeneity); (2) the congregate dimension (the extent to which privacy is available versus the individual's need/preference for privacy); (3) the institutional control dimension (the extent to which institutional staff control environmental resources versus the individual's need/preference for autonomy); (4) the structure (the extent to which there is role ambiguity or role clarity versus the individual's tolerance of ambiguity/structure); (5) stimulation-engagement (environmental input versus the individual's need/preference for input); (6) affect (environmental support of affective expression versus the individual's need for affective expression); and (7) impulse control (the extent to which the environment gratifies needs immediately versus the individual's ability to postpone gratification).

According to this model, it is likely that congruity may exist along one or more dimensions while incongruity may exist simultaneously along other dimensions. Rapoport[17] pointed out that one of the shortcomings of this model is that the dimensions are overly focused on institutional settings. Another weakness is that the model does not make specific predictions concerning the results of incongruity.

Gubrium[18-20] proposed a sociological theory to account for the impact of environment on the elderly. He suggests that 'individual' and 'social' contexts interact to affect the level of well-being. The individual context includes three types of resources: good health, financial solvency, and social support. These resources enable the elderly to satisfactorily adjust to normative demands. The 'social' context provides norms that affect behavioral expectations. Adjustment is most likely to be successful when there is congruency between personal expectations and social norms. When individuals with low resources living in an age-heterogenous social context with expectations of high levels of activity, adjustment would be difficult. Similarly, for seniors with high resources living in an age-homogenous environment with low expectations of activities, this lack of congruence should have a negative effect on their well-bcing.

Schooler's[21] theory focuses on the cognitive interpretation of external stimuli with respect to the amount of perceived threat. Whenever a threat is perceived, a stress reaction is activated which leads to individual coping behaviour or a change in environment. If the adaptation is successful in removing the threat, it would result in positive morale, health and life satisfaction. The main focus is on the individual's appraisal of the environment, and the adoption of adaptive coping behaviour to change the environment and reduce stress. According to Schooler, residential environment and environmental change may be appraised as threatening. Thus, stress is a product of the interactions between characteristics of the environment and the cognitive appraisal on the elderly.

Rowles'[22] humanistic, geographical perspective is holistic in approach; it emphasizes a variety of experiences that include action, feeling, reminiscence and fantasy. The main thrust is that the individual strives to maintain consonance between who s/he is and what the environment offers.

Howell[23] is conerned with how people attach personal meanings to their environment. She proposes that a better understanding of person-environment transactions can be gained by taking account of individual life histories and current uses of the physical environment.

Howell[24] notes that one of the problems of an "antecedent-consequent" approach to studying the meaning of places is that it assumes that there is stability over time in the definition of the variables. However, she argues that "a particular consequence may be the result of a different reading of the same antecedent events in place at time 3 rather than at time 2 or than at the original time 1. . . . The problem seems to be that attributes or elements of the built environment which might be classed as antecedents for one or many individuals appear, themselves, to be manipulated and modified *by the individual* over time and thus to vary, in inexplicable ways, in consequences" (p. 98). She therefore proposes that "places are affectively *redefined* in the course of utilizing them in reminiscence, self-concept reviews, problem solving, social rule maintenance, and other operations. Thus, the relative importance of any attribute of place may appear, disappear, and re-emerge in varied form and with associational differences" (p. 99).

The most comprehensive theory of the ecology of aging is developed by Lawton.[3] It incorporates both extra-individual stimuli and intra-individual representations. The main predictive components in this theory are competence and environmental press. Behaviour is assumed to be a function of the competence of the individual and the environmental press of the situation. Competence is considered as a characteristic of the individual, while environmental press as "environmental characteristics with some demand quality for the individual, whether the demand is objective (alpha press) or one construed by the individual (beta press)" (p. 35).[3]

Competence implies varying degrees of capability and efficacy in negotiating the demands of life. The evaluative nature of the term competence makes it different from nonevaluative personal characteristics such as needs and traits. Lawton distinguishes five levels of competence: biological health, sensory perceptual capacity, motor skills, cognitive capacity, and ego strength in ascending order of complexity. "Therefore, competence is defined as the theoretical upper limit of capacity of the individual to function in the areas of biological, health, sensation-perception, motoric behaviour, and cognition"[3] (p. 38). He recognizes that a continuum of different levels of competence is an oversimplification, because an individual may be simultaneously low in biological competence but high in cognitive competence. Therefore, it is best to describe an individual in terms of a competence profile.

Lawton's ecological model includes the environmental docility hypothesis which posits that high competence is associated with relative independence from the effects of environment, whereas low competence is associated with increased vulnerability to environmental press. Thus, either a decrease in competence or an increase in environmental pressure may elicit nonadaptive behaviour or negative affect.

As long as the difference between competence and environmental press does not become too great, outcomes are still positive. However, if demand is much greater than competence, then anxiety, stress, or maladaptive behaviour results. "Clearly, if a person in an unchanging maintenance environment declines in competence, the formerly maintaining environment will become a dysfunctional environment unless personal competence can be reinforced or the now-too-demanding environment can be moderated" (p. 40)[25].

Lawton's model has been criticized by Carp[26] for focusing primarily on environmental demands and vulnerabilities. Carp proposes that a comprehensive model should also consider environmental resources and opportunities which can facilitate personal growth and the gratification of the needs of older persons. In spite of this weakness, Lawton's model is useful in predicting and explaining adjustment outcomes in a wide variety of person-environment interactions.

4. RELEVANT EMPIRICAL FINDINGS

The effects of environmental factors on the well-being of the elderly are, in some cases, self evident. For example, McPherson[9] points out that if transportation is not available, then the elderly become housebound and isolated; such restrictions may result in feelings of alienation and loneliness. He also points out the elderly need to overcome both structural barriers, such as the height of stairs or poor lighting and psychological barriers, such as fear of crowds and crime. However, there has been little systematic study of the relative impact of environmental variables on the aging process.

Faleti[27] observes that there is little evidence of a systematic relationship between physical or psychological changes in older adults and their ability to function in the community. This is because, Faleti argues, that their functioning in daily living is an issue of person-environment interaction. According to Faleti, there are two major bodies of literature that support this view. One involves the development of person-environment models of adaptive behaviour of the elderly.[3,16] The other body of literature focuses on the design of housing and environment for older persons to accommodate the changing competencies of the elderly.

There are certain limitations in prior research. Faleti points our that studies often focus on what the elderly cannot do rather than what they can do. He also argues that person-environment research has been primarily concerned with the general outcomes of transactions (e.g. satisfaction, morale, anxiety) rather than specific characteristics.

In this section, we review a few variables which have been shown to be relevant to the well-being of the elderly. Whenever possible, the results are examined from the perspective of person-environment interactions.

4.1. *Living Arrangements*

Most older people prefer to live in an independent dwelling. In fact, approximately 70 percent of the elderly own their own home.[9] However, for many elderly, this is not an option for financial or health reasons. In short, living arrangements can be affected by social class, health status and various other variables.

Worobey and Angel[28] examined how these variables determine living arrangements among unmarried elderly persons over a two-year period. The results reveal that a decline in functional capacity greatly increases the likelihood that an elderly person will move in with others or become institutionalized. Nonetheless, most single elderly persons continued to live alone two years after the initial interview in spite of a significant decline in health. For those who suffered decline in functional capacity, women were somewhat less likely than men to continue living alone at the follow-up interview; as well, Blacks were more likely than Whites to continue living alone at Time Two.

Magaziner, Cadigan, Hebel, and Parry[29] studied whether living arrangements affect health and whether health problems influence living arrangements. Data were collected during the first wave of a prospective study of 807 women 65 years or older. Living arrangements were coded into one of three categories: living alone, living with a spouse, living with non-spouse others. The results indicate that no particular disadvantage was associated with living alone. Those who lived with non-spouse others had the poorest health. However, when those who chose to live with others because of health problems were removed from the analyses, living arrangement had little relation with health. This study suggests that poor health does not result from living arrangements; rather, health influences living arrangements.

Arling[30] examined the relationship between life-strain (e.g. physical health problems, economic deprivation, and ADL impairment) and distress in old age, while taking into account the effects of age, race, sex education and social resources. The data was drawn from a statewide household survey (N=2,146) of noninstitutionalized older people. Using multiple regression analysis with tests for statistical interactions, the findings revealed that women, whites, those living alone, and those with less education experienced more strain. Individuals with greater sources of strain were also more likely to receive social support. Health status and ADL were the strongest predictors of distress. Social support buffered the effect of ADL impairment and possibly the effects of economic deprivation. However, living arrangement had neither main nor interactive effects on distress symptoms.

The above studies show that generally living arrangements have no direct effects on the elderly. However, it seems plausible that under certain circumstances living alone can be stressful. For example, an elderly individual is likely to experience loneliness if the person is gregarious by nature and needs friends but lives in an isolated rural area. Systematic research is needed to study the effects of congruence or incongruence between living arrangements and personal needs.

4.2. *Neighbourhood*

The community or neighbourhood in which one resides can have a profound impact on one's quality of life. Numerous factors, such as crime rates and accessibility of services, are particularly important for the elderly. In addition to these variables, subjective perceptions of the neighbourhood and personal needs are also important.

Using consensus sketch maps, Regnier[31] has revealed large variations in the city image by residents living in different neighbourhoods. For example, residents of lower-income minority communities use much less detail in describing the city as compared to those with higher incomes and greater physical mobility. On the basis of these findings, Regnier is able to extend Lawton's concept of the resource environment into four resource environments: physical, functional, perceived, and salient.

The physical resource environment consists of all resources available in an area regardless of their relevance or accessibility to the older residents. The functional resource environment consists of those facilities, services, and elements actually utilized by the elderly. The perceived resource environment refers to facilities which are thought of as having relevance to the older person. Finally the salient resource environment consists of those elements which are highly valued for practical or emotional reasons.

Perceived resource environment can be affected by incorrect information, past memories or emotional attachments. For example, changes affecting a park in the neighbourhood can cause distress in the elderly, if they overemphasized the perceived danger of these changes because of their recollection of what the park was like 20 or 30 years ago.

Regnier investigated how 100 elderly residents (average age 71.3 years) perceived their neighbourhood by having them draw a map of their neighbourhood. The hand-drawn consensus map data revealed that salient resources, which represent important community facilities, such as retails, services and recreations, played an important role in their image of the neighbourhood.

Connerly and Marans[32] studied the antecedents of perceived neighbourhood quality. They distinguish between satisfaction and attachment on the basis of cognitive and affective criteria. Neighbourhood satisfaction is viewed as reflecting the cognitive component because it is related to the evaluation of specific neighbourhood attributes. Attachment is viewed as more closely related to the affective component of perceived neighbourhood quality. They predict that social interactions will have a greater effect on attachment, while neighbourhood attributes will be more strongly related to satisfaction.

Data was obtained from interviews of 1194 residents of various age in metropolitan Detroit. Path analyses were conducted to test the predicted relations. Generally support was obtained for the perceived neighbourhood quality model but some modifications to the model were needed. As predicted, evaluation of neighbourhood attributes was found to mediate the effects of general social status, objective neighbourhood attributes and perceived homogeneity of neighbours, but predictions regarding attachment were not supported. Relevance of these findings to the elderly is unclear because it was based on a cross-section of ages. However, the study does suggest that neighbourhood satisfaction is related to perceived neighbourhood attributes.

The determinants of neighbourhood satisfaction in the elderly were studied by La Gory, Ward, and Sherman.[33] They hypothesized that neighbourhood satisfaction is a function of: (1) objective attributes of the neighbourhood, (2) the psychological and physical state of the person making the evaluation, and (3) the person's own subjective definitions of the neighbourhood. The main objective of this research was to weigh the relative significance of these variables. A random sample of 1,185 respondents 60 years and older was interviewed in Albany-Schenectady-Troy, New York.

In general, the sample was functionally able, with 70% indicating no difficulty in going outdoors, walking up stairs, doing household chores, etc. Neighbourhood satisfaction was measured by asking respondents to rate on a four point scale, "their neighbourhood as a place to live." Environmental competence was measured by determining the functional health of the individual and their score on Pearlin & Schooler's mastery scale.

Overall neighbourhood satisfaction was high: 85% gave the neighbourhood a good or excellent rating as a place to live. The central city elderly were less satisfied with their neighbourhood than those in suburban or rural localities; central city elderly also felt less safe.

Multiple regression results suggested that overall neighbourhood satisfaction is primarily the result of perceived rather than objective aspects of the neighbourhood. This finding was interpreted as supporting Rowles'[34] view that the elderly respond to a subjectively constructed rather than an objective environment.

Comparisons of high competent and low competent persons revealed interesting differences. For the low competence group only subjective perceptions of the neighbourhood influence overall satisfaction. For the high competent group, both subjective and objective factors determine neighbourhood satisfaction.

The data suggest that while all three general factors are significant predictors of neighbourhood satisfaction, the older residents' mental portraits of the neighbourhood are the most significant source of neighbourhood satisfaction. They conclude that "this finding indicates support for a synthesis of symbolic interactionist and traditional ecological models of environmental response" (p. 405).

Despite the importance of subjective perception, certain objective factors deserve consideration. Accessibility of services and activities is especially important for the elderly. Such conveniences make it easier for them to maintain their independence. The importance of accessibility has already been substantiated by research. For example, Carp[35] reported that centrality of residence in urban areas was associated with more favourable life-style than peripherality, because it was "related with active, autonomous, and satisfying use of time, space, and social network" (p. 29).

However, centrality of residence must be linked to other factors such as personal safety. Morse[36] studied the experiences of black people in a large northern U.S. industrial city in an oral history project. The most common theme was related to the security of person and property. These older blacks felt imprisoned in their own homes and apartments because of problems of personal security. Another commonly cited problem was transportation. Infrequent bus schedules and long walks to stations made them reluctant to use buses except during business hours.

High crime rates, especially in one's own neighbourhood, can have an averse effect on one's well-being. As people worry about their safety, they are less likely to venture out of their residence for leisure activities. Being a prisoner of fear can greatly reduce one's life satisfaction.

To what extent does crime directly affect the elderly? McPherson[9] states that contrary to popular myth, the elderly are less frequently victims of crimes than other age groups, with the exception of theft, burglary and fraud. Nevertheless, the elderly frequently say that fear of crime is one of their main worries. Both victimization rates and fear of victimization tend to be greater among the economically and socially disadvantaged. McPherson further points out that those who live alone, in public-housing developments or in the central core of large cities are more likely to be victimized.

Cutler[37] used data from two national surveys in the US to examine cohort changes over an eleven year period in fear for one's safety on the streets. Fear of walking alone at night increased among all cohorts, but the greatest increases occurred among the older cohorts. He interpreted these results as consistent with interpretations that identify characteristics of the aging process to account for the elderly's greater fear of vulnerability to criminal victimization. The greatest increase in fear was not among the old-old but among those who were 65–75 at the time of the second survey.

These studies suggest that the elderly, especially those living alone in poor downtown neighbourhoods, have reasons to be worried about their own safety. Therefore, the convenience of living in downtown can be offset by high crime rates and fear of victimization.

4.3. *Effects of Relocation*

Most older people would like to stay in their own homes as long as they can manage. However, changing circumstances and decline in physical health often make it necessary to relocate. In the US, one out of six people over 65 changed his or her place of residence at least once in the previous four years.[9] Relocation is a potential source of stress; however, it depends a great deal on whether the relocation is voluntary and whether the change is an improvement.

Brody[38] studied the effects of voluntary relocation to a community housing project. This housing development was specially designed for the elderly (e.g. rails on outside and inside steps, tamper-proof locks, bathrooms with grab-rails, etc.) and offered a number of services. The study investigated three groups of elderly who had requested help in relocating. The experimental group (N = 24) were randomly selected from those who met eligibility criteria, completed application procedures, and moved to the project. One control group (N = 22) consisted of individuals who moved to other housing, such as other types of apartments, rented rooms, nursing homes; the other control group included individuals who did not move (N=41).

Subjects were all evaluated at baseline (the time of the application) and at six-month follow-up. About half of the applicants wanted to move because they feared the high crime-rate where they lived. As well, many applicants yearned for increased social contact and services promised by the new housing project. The groups did not differ in their satisfaction with their housing at baseline but there were several significant differences at the follow-up: (a) the experimental and mover control groups were significantly more satisfied with their neighbourhood than the nonmover controls, who remained at their baseline levels; (b) only the experimental group showed a significant improvement in their attitudes toward their apartments; (c) both the experimental and control movers showed significant increase in overall satisfaction with the new living arrangements, with the experimental group showing greater gains; (d) significant increases in favour of the experimental group were also obtained in terms of their reported enjoyment of life and number of good friends. These data confirm that relocation can have beneficial effects on the elderly, if it improves various neighbourhood and residential factors.

Voluntary relocation implies choice and control. People typically do not move unless the new residence offers something better than their original residence, such as a better neighbourhood, being closer to their own children, or a more manageable size. Furthermore, people tend to move to a place that allows them to maintain a sense of continuity and identity. Therefore, most studies on voluntary relocation of community dwellers tend to show beneficial effects. Having reviewed a number of studies on the effects of voluntary relocation, Pastalan[39] concludes that the vast majority of elderly who moved voluntarily from one house to another maintained or improved their quality of life; however, there are indications that those who are older and less healthy initially, do not fare as well after the relocation.

Forced relocation, on the other hand, is more likely to be stressful for a number of reasons. According to Kahana & Kahana,[40] forced relocation tends to result in low levels of person-environment congruence. If the new P-E fit does not meet expectations, negative outcomes are likely. It is important to consider not only post-relocation P-E fit but also the relative increase or decrease in fit and the discrepancy between expected and actual fit.

Kahana & Kahana[40] argue that preferences for environmental continuity versus change may be important aspects of P-E fit in relocation studies. They advocate a dynamic approach that takes into account both changes in personal preference and changes in environment. They also propose that consideration of P-E fit requires a different set of environmental dimensions for competent community older persons than for those in institutions.

Pastalan[39] reviewed four studies involving forced relocation from one home to another. Most of these studies reported some negative consequences associated with the move. Those people with health problems were more vulnerable to a move. As well, those forced to relocate tended to score lower on attitude tests following relocation than nonrelocated controls.

What these studies indicate is that we need to take relocation into account in studying the effect of living conditions. The single fact of forced relocation can result in a high level of dissatisfaction in spite of the positive features of the new environment.

Schulz and Brenner[41] postulated the perception of prediction and/or control moderate the negative effects of stress. Therefore, the greater choice an individual has, the less aversive the effects of displacement. The *degree of choice*, in terms of whether the move is *voluntary* or *involuntary*, thus becomes an important factor to be examined separately in studying the effects of environmental change.

If the move is voluntary and the new environment is an improvement, it should have beneficial effects on the elderly. This is exactly what Carp[35] found. He compared differences between movers and a comparison group on health, disability, and mortality. The majority of the respondents were living in small substandard apartments in a rundown area of a city prior to moving to a new public housing project, Victoria Plaza (VP). The comparison group consisted of similar elderly persons who chose not to move in. Individual home interviews were conducted at the end of the first and eighth year of operation of the public housing project.

Health ratings showed significantly less decline among tenants than among members of the comparison group. Although both groups were initially equivalent on this variable, VP tenants, relative to the comparison group, reported fewer specific health problems, spent less time on health care activities per day, showed less increase in restriction in their activity due to health problems, had less medical contact, and had a significantly lower mortality rate. The findings show both short-term and enduring health benefits that come from a voluntary move to an improved living environment.

4.4. *Perceived Environmental and Psychosocial Factors*

Windley & Scheidt[42] investigated the effects of personal variables and perceived ecological/architectural and psychosocial attributes of 18 small towns (2,500 population or less) on the well-being among 989 older residents. The ecological/architectural factors included perceived sensory stimulation, cognitive legibility of neighbourhood and community, residential comfort, privacy, adaptability and control over residence, sociality of settings, community accessibility, population density, the degree of attachment to a setting, and community and residential quality or aesthetic appeal. The multidimensional well-being consisted of housing satisfaction, contact with friends and relatives, mobility, functional health, availability of a confidant(e), feelings of security, and activity participation. Measures of psychopathological symptoms were also included. The three psychosocial dimensions were: (1) satisfaction with opportunities for social relations in the community; (2) satisfaction with community attributes that enhance personal growth and self-esteem; and (3) satisfaction with community resources.

It was hypothesized that well-being would be indirectly affected by ecological/architectural factors but directly by psychosocial factors. When the factor-analyzed well-being domain was regressed on all predictor domains, it was found that five factors were predicted most strongly by demographic variables, five by psychosocial variables, and seven by ecological/architectural variables. A subsequent path analysis revealed that community involvement, community satisfaction, age of the respondent and satisfaction with architectural features of the dwelling unit all had direct predictive effects on well-being. The degree to which environmental barriers hinder older residents from accessing community activities and the degree to which older residents feel they have high control over residence were also found to have direct effects.

One major methodological problem with this study is that the word satisfaction was used in both predictive variables as well as criteria measures. This confound makes it difficult to interpret the results. However, the findings are consistent with other studies which demonstrate the importance of such environmental factors as community accessibility and perceived control over residence.

Kasl & Rosenfeld[43] make three general conclusions on the basis of their review of the impact of the residential environment on the mental health of the elderly: (1) The impact does not appear to be overwhelming. The impact is greatest on housing satisfaction and evaluative perceptions, but less on activities, evaluation of current life circumstances, and least on more traditional indicators of mental health. (2) Social and physical environmental variables interact in their impact. (3) Research on the impact of residential environment on the aged is generally opportunistic and cross-disciplinary, but both of these features are desirable.

5. STRESS, ENVIRONMENT AND WELL-BEING IN THE ELDERLY

There is now a growing body of literature on stress in the elderly. In the present paper, we focus on stress related to living conditions rather than major life events. We are primarily interested in how stress and environment impact on the well-being of the elderly.

In a major study on community elderly, Golant[10] found that the majority of older people (from 65 to 87 percent) were not bothered by common urban nuisances such as the noise, the garbage and litter found on the streets and neighbour's neglect of property. A very small percentage of old people (from 3 to 9 percent) did find them a frequent source of annoyance. Another group (from 6 to 13 percent) were bothered by them more than once or twice in the past few months.

Multiple regression analyses revealed that individuals who rated their health as poorer were more likely to report annoyance with urban problems. This suggests that personal incompetence makes an old person more vulnerable to environmental irritants. Golant also found that those with a high level of perceived internal locus of control were less likely to find caring for their homes too tiring. It appears that self-rated health and perceived control are important predictors of susceptibility to environmental stress.

Golant reported that fear of crimes was a major source of stress. About 61 percent were afraid to go out at night because of the threat of crime; 30 percent were worried about being attacked or robbed while walking, and 35 percent were concerned about break-and-entry into their home. Moreover, he found that the threat of crimes was a greater stress for old people who were less competent physically, socially, and psychologically.

Paulus[44] criticized this research for the lack of objective assessments of the environment. Nevertheless, Goland has provided many insights into the behaviours and feelings of the elderly about a wide variety of issues related to their environment, especially the important role of perceived competence and control. His findings are clearly consistent with Lawton's environmental docility hypothesis.

Langer[45] observes that when individuals believe that they no longer have any control over things important to them, it will have a negative impact on their well-being. The loss of perceived control in institutionalized elderly is stressful at first; but when this perception persists, the individual would feel incompetent, become passive and may display symptoms of reactive depression.

After reviewing previous studies, Fry[46] concluded that the results show the importance of control for the well-being of older people. "The extent to which autonomy and control are encouraged or denied may be an important determinant of improved well-being, life-satisfaction, and physiologic health . . . . The issue of control and generalized perceptions of having control have also come to be associated with positive outcomes such as successful adjustment to old age" (p. 89). Thus, the perception of control is an important variable in studying stress as well as living conditions.

A study by Holahan and Holahan[47] examined the roles of life stress, hassles, and self-efficacy beliefs in predicting adjustment in the elderly. Twenty-six men and 26 women between the ages of 65 and 75 participated in an initial structured interview and a follow-up interview one year later. Measures of negative life change events, daily hassles, and self-efficacy were used to predict depression, psychosomatic symptoms, and negative well-being. It is worth noting that the hassles measure included a substantial number of environmental hassles (e.g. bothered by noise, neighbourhood deterioration, traffic, transportation, home maintenance, etc.).

The results showed that frequency of hassles was the strongest predictor of depression and psychosomatic symptoms both concurrently and one year later, even when initial distress was controlled. Perceived self-efficacy was also predictive of current and subsequent depression, even after initial depression was controlled. However, frequency of negative life events was a weak predictor. These findings suggest that both environmental hassles and personal competence need to be considered to understand mental health in the elderly.

Carp[48] focused on personality determinants of adjustment in the elderly. Two different samples were examined: Sample 1 consisted of 204 elderly in public-housing facilities, 107 of which were later retested; Sample 2 consisted of 212 elderly. Subjects were middle class, generally in good health and with an average WAIS full-scale IQ of 112.

Adjustment was measured by (1) an 8-item happiness scale, (2) peer perceptions as indicated by a 22-item sociometric instrument, and (3) staff ratings of adjustment. Competence variables controlled in the analyses were: (a) cognitive competence (WAIS), (b) composite health index, and (c) sensory-motor status (vision, hearing, movement). Social status indicators were: age, education, marital status, current income, and job level during major working years. For purposes of analysis data were first standardized for all variables. Personality variables were entered in stepwise regression after competence and social status variables.

The results showed that competence and social status variables together account for 4% to 18% of criterion variance, or 3% to 13% when personality traits are partialled out. Personality traits together account for 25% to 65% of criterion variance and for 17% to 63% of the variance over and above that attributable to competence and social status. Results were generally consistent across Sample 1, retest, and Sample 2. The results indicate that personality traits, competence and social status all account for well-being and adjustment in the elderly.

A large body of literature has been reviewed by Lawton[3] as supporting the environmental docility hypothesis. For example, Rosow[49] found that the less competent elderly apartment dwellers were more vulnerable to the press of low age density of the buildings. The research on relocation mortality is also consistent with what might be predicted by the environmental docility hypothesis.

Lawton[3] indicated that institutional relocation mortality data supported the environmental docility hypothesis; those with relatively high competence and who were voluntarily relocating showed no evidence of a relocation mortality effect. However, in most of the studies cited by Lawton, competence was inferred rather than directly measured. In the following section, we provide some new data in which competence was measured by a standardized intelligence test.

6. AN EMPIRICAL STUDY

The present study is part of a larger project on psychosocial factors in successful aging. Among the variables investigated were certain physical and psychological aspects of the environment known to impact on the well-being of the elderly. In this study, we intended to answer three basic questions: (1) What are the relationships between environmental variables and stress? (2) How do stress and environment jointly affect the well-being of the elderly? (3) Does cognitive competence, as measured by an intelligence test, moderate the effects of environmental variables?

The participants were 86 elderly persons (57 females and 29 males) with a mean age of 73.8. They were recruited from two Canadian cities in south-central Ontario. All were community dwellers under a variety of living arrangements. They were either living in their homes or in apartments. Some lived alone. Others lived with their spouse, relatives or friends. About 37% lived in the downtown area, 57% in suburban areas, and the remaining 6% lived in rural settings. The sample was predominantly Caucasian and Protestant; and most were of British descent. In terms of formal education, 23% completed only elementary school, 37% had some secondary school, and 40% had some university education. The sample was slightly above average in intelligence with a mean Full-scale IQ of 108 on the Weschler Adult Intelligence Scale-Revised.

The participants were interviewed several times and were administered a battery of psychological instruments. The environment was assessed by the Living Environment Survey. The physical environment was measured by seven scales, namely, noise level, room temperature, lighting, cleanliness, density, furnishing and facilities. For each scale, there were three levels of responding, with the highest level indicating satisfaction. For example, the three alternatives for noise level were: too noisy, a little bit noisy, just right). The total scores reflect the degree of environmental press. A low total score means a high level of environmental press, because it indicates a very noisy, dirty and uncomfortable living condition.

Psychological atmosphere was measured by five items, namely, pleasantness, friendliness, stimulation, liveliness and supportiveness. Again, each item had three response alternatives. A high total score would indicate a very positive, warm and inviting psychological climate.

In addition to the overall atmosphere, we also measured controllability and predictability, because such perceptions have important implications for one's well-being. To measure personal control, we asked the respondents: "To what extent do you have control over what is going on at the place where you live? A five-point response format was utilized, with the five indicating "complete control" and one indicating "no control whatsoever." Similarly, predictability was tapped by the question: "To what extent can you predict what happens at the place where you live?" Again, a five-point rating scale was provided.

A Multifaceted Stress Questionnaire was used to assess levels of stress. Developed in cooperation with G. Reker, this questionnaire measures various sources of stress. For the present study, we considered Societal Stress and Everyday Stress. The former refers to potential stressors that affect the whole community, such as crime rates, pollution and inflation. Thus, it reflects the quality of the community. The latter refers to the hassles and frustrations of everyday life, such as interruptions, transportation problems and relationship difficulties.

For each source of stress, we measured both the frequency of stressful events and the degree of negative impact. Respondents were asked to check off any of the events in the Stress Questionnaire that applied to them, and include any additional ones. This frequency score is a relatively objective count of events that are applicable to respondents.

Having checked off the relevant events, the respondents were asked to indicate which of these events were stressful to them and to what extent each of these events had a negative impact on their lives. Thus, the greater the frequency count, the more eventful their lives; the greater the negative rating scores, the more stress they experience.

With respect to outcome measures, we focused on mental health rather than psychopathology. Two self-report instruments were used. One was the Memorial University of Newfoundland Scale of Happiness (MUNSH) developed by Kozma & Stones.[50] This has been demonstrated to be a psychometrically reliable and valid measure of happiness. The second instrument was the Perceived Well-being Scale (PWB) developed by Reker & Wong.[51] It assesses both psychological and physical well-being; six items are related to psychological well-being (e.g. "It is exciting to be alive.") and eight items are related to physical well-being (e.g. "I am in good shape physically"). The response format for all items was a seven-point Likert-type scale. The reliability and validity of this instrument with the elderly have been demonstrated and the two components of well-being have been found to load on two relatively independent factors.

6.1. *Results and Discussion*

Since environment is assumed to be a major source of stress, we wanted to explore the relationships between environmental variables and stress measures. The zero-order correlations between these two sets of variables are shown in Table I.

A number of significant relationships were found. Controllability of the environment was negatively related to three of the four stress indicators. The largest correlation was between controllability and the frequency of societal stressors; thus, as the frequency of societal stressors increases, the less controllable the

Table I Correlations between types of stress and environmental attributes

Stress	Physical environment	Controllability	Predictability	Atmosphere
Everyday No.	−.23*	−.16	−.23*	−.27*
Everyday Rating	−.12	−.22*	−.27*	−.34*
Societal No.	−.14	−.49**	.09	−.10
Societal Rating	−.09	−.29**	−.02	−.01

* $p < .05$ $p < .01$**

elderly tend to view their environment. One interpretation of this finding is that societal stress, which is generally beyond personal control, tends to sensitize the elderly to the uncontrollable aspects of their immediate living conditions. It is also possible that when the environment is perceived as controllable, the elderly become more sensitized to the uncontrollable societal stressors.

Both the atmosphere of the living environment and its predictability were negatively correlated with everyday stress; increased predictability and a more positive atmosphere in the living environment were associated with lower levels of everyday stress but were unrelated to levels of societal stress. These findings suggest that the more everyday hassles experienced by the elderly the more likely they are to perceive their living conditions as unpredictable and unpleasant.

It is worth noting that physical environment was only significantly related to the frequency of everyday stress. It is risky to draw any causal inference from correlational data. However, it seems highly plausible that when there is a high level of environmental press (e.g. low satisfaction with physical environment) because of noise and crowding, one is more likely to experience such hassles as interruptions and frustrations.

Multiple regression analyses were then performed to determine which environmental variables predict the experience of life stress. Such analyses would give us a clearer picture of the relationship between environment and stress. To control for possible confounds with demographic variables (age, income, sex, education, marital status), these variables were forced into the regression equation on the initial step and then the four environmental variables were allowed to enter in a stepwise manner. The results of these analyses are summarized in Tables II and III.

For the prediction of everyday stress, highly similar results were obtained for the frequency measure and the negative impact ratings. In both cases, the demographic variables did not emerge as significant predictors of stress and atmosphere was the only environmental variable that independently predicted everyday stress. Predictability and controllability did not contribute to any additional variance beyond what was predicted by atmosphere; it appears that these perceptions make up part of the overall psychological evaluation of the living condition. Thus, a desirable place to live is also an orderly, predictable place where one has a high sense of control.

When societal stress was the criterion, the demographic variables again did not emerge as significant predictors. Controllability was a significant predictor for both the frequency of societal stress and the negative impact ratings of these

Table II Summary of multiple regression analyses predicting everyday stressful events from environmental variables

Step	Predictors	Mult*R*	*R*sq	*R*sqCh	*F*Ch
Frequency					
1	Demographics	.260	.067	.067	1.11
2	Atmosphere	.359	.129	.062	5.36*
Negative Rating					
1	Demographics	.220	.048	.048	.78
2	Atmosphere	.386	.149	.101	8.99**

* $p < .05$, ** $p < .01$, *** $p < .001$

Table III Summary of multiple regression analyses predicting societal stressors from environmental variables

Step	Predictors	Mult*R*	*R*sq	*R*sqCh	*F*Ch
Frequency					
1	Demographics	.321	.103	.103	1.77
2	Controllability	.611	.373	.270	32.75***
3	Predictability	.638	.407	.034	4.35*
Negative Rating					
1	Demographics	.257	.066	.066	1.08
2	Controllability	.392	.154	.088	7.93**

*$p < .05$, **$p < .01$, ***$p < .001$

stressors. It can be seen from Table III that controllability accounts for 27% of the variance in predicting frequency, and 9% of the variance in the impact ratings. The demographic variables, controllability, and predictability of the environment were able to account for approximately 41% of the variance in the frequency of societal stressors.

While atmosphere was important in predicting everyday stress, control perceptions were important in predicting societal stress. This pattern of results makes good sense, because the psychological climate of the living conditions should be closely linked to the level of perceived stress in everyday life. A pleasant environment not only means a low level of environmental press but also implies a life that is relatively free from hassles and upsets. Controllability, on the other hand, should be more related to perceived societal stress, because a low level of perceived control tends to heighten one's anxiety about societal stress, such as crime rates.

To investigate how stress and environment affect well-being, a series of multiple regression analyses were performed. Each of the three measures of well-being (happiness, psychological well-being, and physical well-being) were considered as criterion measures and were regressed on the environmental and stress variables. For these regression analyses, demographic variables were again entered first and partialled out as an initial block; the four living environment variables were then allowed to enter in a stepwise manner and finally the four stress variables were allowed to enter in stepwise fashion. The result of these analyses are summarized in Tables IV, V, and VI.

For both happiness and psychological well-being, atmosphere emerged as a significant predictor, accounting for 23% of the variance in happiness and 21% of the variance in psychological well-being. No other environmental variable

Table IV Summary of multiple regression analyses predicting happiness from environmental and stress variables

Step	Predictors	Mult*R*	*R*sq	*R*sqCh	*F*Ch
1	Demographics	.330	.109	.109	1.88
2	Atmosphere	.582	.339	.230	26.37***
3	Everyday-frequency	.640	.409	.071	9.00**

**$p < .01$, ***$p < .001$

Table V Summary of multiple regression analyses predicting psychological well-being from environmental and stress variables

Step	Predictors	Mult*R*	*R*sq	*R*sqCh	*F*Ch
1	Demographics	.353	.124	.124	2.19
2	Atmosphere	.580	.336	.212	24.23***
3	Societal-frequency	.612	.375	.039	4.62*

*$p < .05$, **$p < .01$, ***$p < .001$

Table VI Summary of multiple regression analyses predicting physical well-being from environmental and stress variables

Step	Predictors	Mult*R*	*R*Sq	*R*sqCh	*F*Ch
1	Demographics	.191	.036	.036	.58
2	Everyday-frequency	.522	.272*	.236	24.64***

**$p < .01$, ***$p < .001$

independently contributed to variance in happiness and psychological well-being. Frequency of everyday stress was a significant predictor of happiness, resulting in an additional 7% increment in predicted variance. In predicting psychological well-being, frequency of societal stress was significant, yielding a 4% increment in variance.

These findings have clearly established that atmosphere matters. Intangible as it may be, the psychological climate of the residence is an important factor in one's mental health. Further research is needed to determine what contributes to the perception of a positive atmosphere. The present survey indicates that friendliness, supportiveness, liveliness are all ingredients of the atmosphere measure. It means that the people you live with are important. The kinds of activities that take place where you live are important. It is also possible that physical attributes, such as colour, decor, decorations, furnishings, and the landscape all play a part in creating an enjoyable atmosphere.

One problem with self-report measures is the possibility of confounds. It is likely that both perceived atmosphere and perceived well-being overlap. Thus, a happy person is more likely to perceive the place as being pleasant, while a depressed person is more likely to perceive the same place as gloomy. It is also likely that both measures may reflect the same personality traits, such as optimism. This problem can be overcome if one obtains atmosphere ratings by objective observers or by other individuals living in the same place.

It is not clear why everyday stress predicts happiness, while societal stress predicts psychological well-being. One plausible interpretation is that happiness is an affective state which is intimately related to everyday events, while perceived well-being is more cognitive and evaluative, and therefore more sensitive to the perception of societal stress. This is an interesting new finding. Prior research on stress has focused on major life events and everyday hassles. This study suggests that societal stress is relevant to the well-being of the elderly.

With respect to perceived physical well-being, no environmental variable emerged as a significant predictor; however, the frequency of everyday stress was an important predictor, accounting for approximately 24% of the variance. This finding suggests that an increase in the number of everyday hassles can make the elderly feel physically worn out and tired, resulting in a low level of perceived physical well-being. It is also plausible that those who have poor health are more likely to encounter difficulties in everyday functioning.

The main message from this set of findings is that both environment and stress can impact on the well-being of the elderly and that psychological aspects of the environment deserve more attention. Of all the environmental variables investigated in this study, atmosphere is the only significant predictor. This suggests that in designing any living environment for the elderly, it is important to consider what kind of psychological atmosphere is being created.

The ultimate test of any building, be it residential or institutional, is whether it exudes a pleasant and uplifting atmosphere for those who live or work there. When a living environment passes this test, it is likely to contribute to the happiness and well-being of its occupants. What we have in mind is far less mysterious than the popular Chinese concept of "Fungshui," because the proper blend of the right ingredients that produce a positive psychological environment can be discovered through systematic research.

6.2. *A Test of Lawton's Model*

Lawton's model has posited that environmental press interacts with individual competence to determine the adjustment outcomes. The present study allows us to test Lawton's model. To investigate whether the elderly's competence level moderates the impact of environmental press on well-being, a series of hierarchical multiple regression analyses were performed to determine whether significant environment by competence interactions could be obtained. Cognitive competence was measured by WAIS-R. The three outcome measures were happiness, psychological well-being and physical well-being. The four environmental measures were: physical environment, psychological atmosphere, controllability and predictability. For each of these environmental measures, a low score indicates a high level of environmental press.

A separate analysis was performed for each of the four environmental variables for each of the three outcome measures. Following standard procedures for detecting interaction effects, the environment and intelligence variables were first entered into the regression equation followed by the interaction term.

Significant interactions were found in five of the twelve analyses. Atmosphere and intelligence interacted significantly in the prediction of happiness, F (1, 82) = 10.73, $p < .01$, (R^2 Change = .07) and psychological well-being, F (1, 82) = 16.64, $p < .001$, (R^2 Change = .11). Controllability interacted significantly with intelligence in the prediction of physical well-being, F (1, 82) = 4.50, $p < .05$, (R^2 Change = .05). Predictability and intelligence interacted in the prediction of happiness, F (1, 82) = 5.09, $p < .05$, (R^2 Change = .05) and in the prediction of psychological well-being, F (1, 82) = 4.82, $p < .05$, (R^2 Change = .05). There were no significant interactions involving the physical environment variable.

The nature of each of the significant interactions was examined by plotting the regression lines for high (+1 SD) and low (−1 SD) intelligence scores. For atmosphere and predictability, the results indicated a classical buffering effect of intelligence for both the happiness and psychological well-being measures. Figures 1 and 2 illustrate the nature of these interactions.

These figures show that when environmental press was low (i.e. high scores in terms of atmosphere and predictability), the well-being scores of high and low

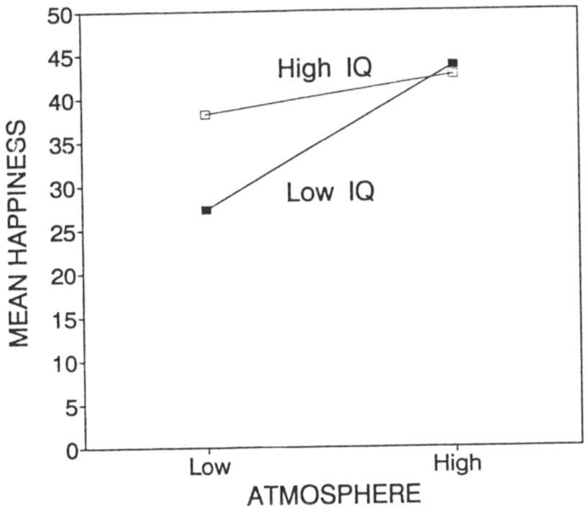

Figure 1 The moderating effect of intellectual competence on the relation between atmosphere and happiness.

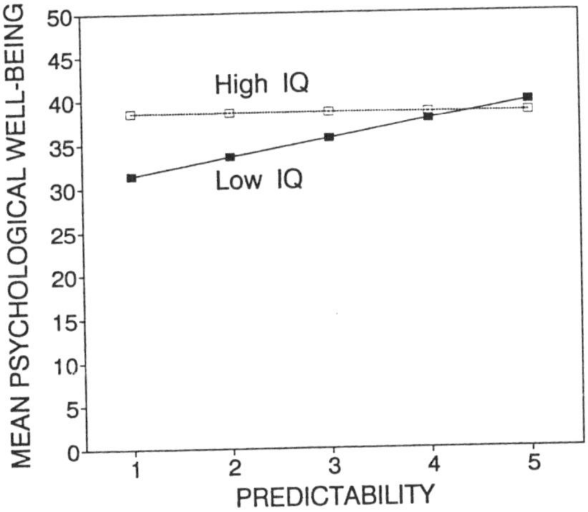

Figure 2 The moderating effect of intellectual competence on the relation between predictability and psychological well-being.

intelligence individuals were very similar; however, when environmental press was high, individuals with high intelligence had substantially higher well-being scores than low intelligence individuals. In other words, low competent elderly were more vulnerable to the adverse effects of a negative psychological atmosphere and a chaotic living place. Competent individuals, on the other hand, were far more resistant to unfavourable environments as hypothesized by Lawton. In fact, the high IQ elderly were able to maintain the same level of psychological well-being even when the environment was perceived as low in predictability.

The other interactions followed the same pattern except for the interaction between controllability and intelligence as depicted in Figure 3. It shows that when the environment was low in controllability (i.e., high environmental press),

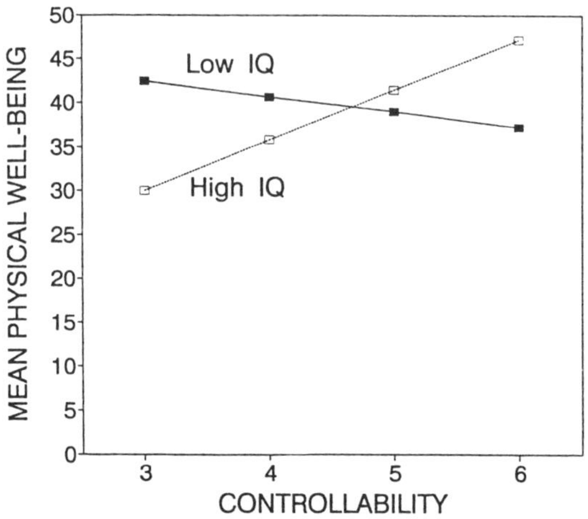

Figure 3 The moderating effect of intellectual competence on the relation between controllability and physical well-being.

low IQ individuals reported greater physical well-being than high IQ individuals, but when the living environment was high in controllability, high IQ individuals reported better physical well-being.

This pattern of interaction is difficult to interpret because it is opposite to the traditional pattern of moderator effects, and suggests that low competent seniors actually fare better when the living environment is perceived as uncontrollable. The only way one can make sense of the present finding is to propose that perhaps for the low IQ elderly, a high degree of control may mean a high level of personal responsibility, hence a high level of stress. It is plausible that low competent elderly may prefer a low controllable environment, where other people take care of many of the demands of life; they may find a highly controllable environment too demanding and burdensome. This line of reasoning suggests that we cannot always know what constitutes environmental press. Consistent with the person-environment congruence perspective, we should take into account individual's needs. When the environment meets personal needs, there is little or no environmental press. Thus, when incompetent individuals need others to take care of their chores, then a low controllable environment may indeed mean low environmental press.

The present findings on the mediating role of cognitive competence are generally in agreement with Lawton's model, but we need to be cautious in defining environmental press. Perhaps, the best way to test Lawton's model is to directly measure perceived environmental press.

CONCLUSIONS

The present study has shown that physical aspects of the environment have very little or no effect on the well-being of the elderly. This is perhaps because most of the people interviewed in this study lived in rather good quality environments. Clearly, if the living quarters were filthy and without basic facilities such as running water and plumbing, the well-being of the occupants would be adversely affected. However, when basic amenities do exist, psychological aspects of the environment have a greater impact on one's well-being.

It is helpful to remember that environment is much more than a place or a building where we can find refuge from the elements and conduct the business of daily living. The environment may be conceived as an extension of ourselves—our personalities, needs, aspirations and memories. Where we live is, indeed, a major part of our lives; it becomes an intimate, inseparable part of our psyche. That is why the study of subjective perceptions of the environment is so essential if we want to understand how environment and self influence and shape each other.

We began by considering the need for an environmental taxonomy. Perhaps, this task becomes more tractable, if separate taxonomies are developed for the psychological and physical aspects of the environment. The Living Environment Survey employed here is a limited attempt to quantify both aspects of the environment. We have also briefly reviewed the various person-environment congruence models. The interactionist perspective seems most promising, but we still do not have an over-arching theory to integrate the various findings of environmental influences.

The promise of environmental psychology is that ultimately it will inform us as to how we can design different types of environments that would meet our needs at various levels of human development. In view of the greying of our population, it is incumbent on the researchers to find out what is the ideal place for people to grow old and spend their remaining days. Perhaps, there are different "ideal" places for different types of older adults. Maybe the starting point is to identify the common elements that are necessary to produce an enjoyable and uplifting psychological atmosphere for the elderly.

References

1. P. T. P. Wong and G. T. Reker, "Pitfalls in stress research: Resolving the hassle" *Stress Medicine* **2**, 1–4 (1986).
2. L. A. Pervin, "Definitions, measurements, and classifications of stimuli, situations, and environments" *Human Ecology* **6**, 71–105 (1978).
3. M. P. Lawton, "Competence, environmental press, and the adaptation of older people" **In**: M. P. Lawton, P. G. Windley, and T. O. Byerts (eds.) *Aging and the Environment: Theoretical Approaches* (Springer, New York, 1982) pp. 33–59.
4. B. R. Little, "Personality and the environment" **In**: D. Stokols and I. Altman (eds.), *Handbook of Environmental Psychology* (Wiley, New York, 1987) pp. 205–244.
5. K. H. Craik, "The prospects for an environmental psychology" *IPAR Research Bulletin* (University of California, Berkeley, 1966).
6. K. H. Craik, "The comprehension of the everyday physical environment" *J. Amer. Instit. Planners* **34**, 646–658 (1968).
7. G. E. McKechnie, "The Environmental Response Inventory in application" *Environ. and Behav.* **9**, 255–276 (1977).
8. G. E. McKechnie, "Environmental dispositions: Concepts and measures" **In**: P. McReynolds (ed.), *Advances in Psychological Assessment* (*Vol. 4*) (Jossey-Bass, San Francisco, 1978).
9. B. McPherson, *Aging as a Social Process: An introduction to individual and population aging* (Butterworths, Toronto, 1983).
10. S. M. Golant, *A Place to Grow Old: The meaning of environment in old age* (Columbia University Press, New York, 1984).
11. K. Koffka, *Principles of Gestalt Psychology* (Harcourt, Brace, New York, 1935).
12. K. Lewin, *Dynamic Theory of Personality* (McGraw-Hill, New York, 1935).
13. H. A. Murray, *Explorations in Personality* (Oxford, New York, 1938).

14. J. P. R. French, W. Rodgers and S. Cobb, "Adjustment as person-environment fit" **In**: G. V. Coelho, D. A., Hamburg and J. E. Adams (eds.), *Coping and Adaptation*. (Basic Books, New York, 1974).
15. E. F. Kahana, J. Liang and B. J. Felton, "Alternative models of person-environment fit: Prediction of morale in three homes for the aged" *J. Geront.* **35**, 584–595 (1980).
16. E. F. Kahana, "A congruence model of person-environment interaction" **In**: *Aging and the Environment: Theoretical perspectives* M. P. Lawton, P. G. Windley and T. O. Byerts (eds.) (Springer, New York, 1982) pp. 97–121.
17. A. Rapoport, "Aging-environment theory: A summary" **In**: *Aging and the Environment*, M. P. Lawton, P. G. Windley, & T. O. Byerts (eds.) (Springer, New York 1982) pp. 132–149.
18. J. Gubrium, "Toward a socio-environmental theory of aging" *Gerontologist* **12**, 281–284 (1972).
19. J. Gubrium, *The Myth of the Golden Years: A socio-environmental theory of aging* (C. C. Thomas, Springfield, Ill., 1973).
20. J. Gubrium, *Living and Dying at Murray Manor* (St. Martin's Press, New York, 1975).
21. K. Schooler, "Response of the elderly to environment: A stress theoretic perspective" **In**: P Windley and G. Ernst (eds.), *Theory Development in Environment and Aging* (Gerontological Society, Washington, D.C., 1975).
22. G. Rowles, *Prisoners of Space? Exploring the geographical experience of older people* (Westview Press, Boulder, 1978)
23. S. Howell, "Environments as hypotheses in human aging research" **In**: L. Poon (ed.), *Aging in the 1980s* (American Psychological Association, Washington, D.C., 1980).
24. S. Howell, "The meaning of place in old age" **In**: G. D. Rowles and R. J. Ohta (eds.), *Aging and Milieu* (Academic Press, New York, 1983).
25. M. P. Lawton, "Three functions of the residential environment" *J. of Housing for the Elderly* **5**, 35–50 (1978).
26. F. M. Carp, "Effects of the living environment on activity and use of time" *Int. J. Aging Hum. Devel.* **9**, 75–91 (1978–79).
27. M. V. Faletti, "Human factors research and functional environments for the aged" **In**: I. Altman, M. P. Lawton & J. F. Wohlwill (eds.), *Elderly People and the Environment* (Plenum Press, New York, 1984).
28. J. L. Worobey and R. J. Angel, "Functional capacity and living arrangements of unmarried elderly persons" *J. Geront.* **45**, S95–101 (1990).
29. J. Magaziner, D. A. Cadigan, J. R. Hebel and R. E. Parry, "Health and living arrangements among older women: Does living alone increase the risk of illness?", *J. Geront.* **43**, 127–133 (1988).
30. G. Arling, "Strain, social support, and distress in old age" *J. Geront.* **42**, 107–113 (1987).
31. V. Regnier, "Urban neighborhood cognition: Relationships between functional and symbolic community elements" **In**: G. D. Rowles and R. J. Ohta (eds.), *Aging and Milieu* (Academic Press, New York, 1983).
32. J. E. Connerly and R. W. Marans, "Comparing two global measures of perceived neighborhood quality" *Soc. Indicators Res.* **17**, 29–47 (1985).
33. M. La Gory, R. Ward and S. Sherman, "The ecology of aging: Neighborhood satisfaction in an older population" *Soc. Quart.* **26**, 405–418 (1985).
34. G. Rowles, *Prisoners of Space? Exploring the geographical experience of older people* (Westview Press, Boulder, 1978).
35. F. M. Carp, "Life-style and location within the city" **In**: T. O. Byerts, S. C. Howell and L. A. Pastalan (eds.), *Environmental Context of Aging: Life-styles, environmental quality, and living arrangements* (Garland STPM Press, New York, 1979).
36. D. W. Morse, "Aging in the ghetto: Themes expressed by older black men and women living in a northern industrialized city" *Indust. Geront* 1–10 (Winter, 1976).
37. S. J. Cutler, "Safety on the streets: Cohort changes in fear" *Int'l J. Aging and Human Devel.* **10**, 373–384 (1979–80).
38. E. M. Brody, "Service-supported independent living in an urban setting: The Philadelphia Geriatric Center's community housing for the elderly" **In**: T. O. Byerts, S. C. Howell and L. A. Pastalan (eds.), *Environmental Context of Aging: Life-styles, environmental quality, and living arrangements* (Garland STPM Press, New York, 1979).
39. L. A. Pastalan, "Environmental displacement: A literature reflecting old-person–environment transactions" **In**: G. D. Rowles and R. J. Ohta (eds.), *Aging and Milieu: Environmental perspectives on growing old* (Academic Press, New York, 1983).
40. E. Kahana and B. Kahana, "Environmental continuity, futurity, and adaption of the aged" **In**: G. D. Rowles and R. J. Ohta (eds.), *Aging and Milieu: Environmental perspectives on growing old* (Academic Press, New York, 1983).

41. R. Schultz and G. Brenner, "Relocation of the aged: A review and theoretical analysis" *J. Geront.* **32**, 323–333 (1977).
42. R. J. Scheidt and P. G. Windley, "The ecology of aging" **In**: J. E. Birren and K. W. Schaie (eds.), *Handbook of the Psychology of Aging* (Van Nostrand, New York, 1985).
43. S. Kasl and S. Rosenfeld, "The residential environment and its impact on the mental health of the aged" **In**: J. E. Birren and R. B. Sloane, *Handbook of Mental Health and Aging* (Prentice-Hall, Englewood Cliffs, 1980).
44. R. B. Paulus, "An old approach to a new problem" *Contemporary Psych.* **31**, 225–226 (1986).
45. E. J. Langer, *The Psychology of Control* (Sage, Beverly Hills, 1983).
46. P. S. Fry, "The person-environment congruence model: Implications and applications for adjustment counselling with older adults" *Int'l J. Adv. Counselling* **13**, 87–106 (1990).
47. C. K. Holahan and C. J. Holahan, "Life stress, hassles, and self-efficacy in aging: A replication and extension" *J. Applied Soc. Psych.* **17**, 574–592 (1987).
48. F. M. Carp, "Relevance of personality traits to adjustment in group living situations" *J. Geront.* **40**, 544–551 (1985).
49. I. Rosow, *Social Integration of the Aged* (Free Press, New York, 1967).
50. A. Kozma and M. J. Stones, "The measurement of happiness: Development of the Memorial University of Newfoundland Scale of Happiness (MUNSH)" *J. Geront.* **35**, 906–912 (1980).
51. G. T. Reker and P. T. P. Wong, "Psychological and physical well-being in the elderly: The Perceived Well-Being Scale (PWB)" *Canadian J. Aging* **3**, 23–32 (1984).

4. SEASONAL FACTORS AND SUICIDE

K. EGASHIRA* and K. ABE*

Seasonal distribution of suicide and its secular trends in various countries are demonstrated. It was noticed that, in spite of geographical differences, there is a common process for suicide deaths to change their seasonal pattern with time. Many industrialized countries seem to have accomplished a greater change in their seasonal pattern and now have a smaller suicide seasonality. The correlation between suicide deaths and environmental stress factors, such as hours of sunshine, was examined. Based on annual, monthly and daily statistical data, a significant correlation between suicide deaths and hours of sunshine was found in cooler periods, but not in warmer ones.

1. INTRODUCTION

Suicidal behavior is associated with mood, especially with depression.[1] Depression is induced not only by psycho-social stresses but also by physico-environmental ones. From the point of view of environmental stresses, meteorological factors like sunshine, temperature, atmospheric pressure and humidity may be associated with mood.[2–6] Therefore, suicidal behavior may also be affected by such meteorological factors.

These meteorological factors are, on the other hand, associated with seasons. For example, sunshine hours are short in the rainy season. In the rainy month of July in Japan, a remarkable trough was observed in suicide rates, as will be mentioned later in this chapter. It is well known that suicide deaths show seasonal variations[7–43] and typically they are most frequent during the months of April-July, and least frequent during December-February in the northern hemisphere.

Over the years, two schools of thought have evolved[19] to explain the mechanisms of the seasonal variations of suicide deaths. One attributes the seasonal pattern to socio-economic factors,[28,32] and the other to meteorological factors,[20,31,40] or to both factors.[29, 33, 34]

As regards the influence of environmental stresses on suicides, we have investigated the relationship between meteorological factors and suicides, and some of the findings will be presented below in the following order: first, the seasonal distribution of suicide deaths in several countries and its secular trends are examined; secondly, the correlation is noted of monthly suicide deaths with monthly hours of sunshine in eight districts of Japan; thirdly, the more direct connection of suicide deaths with sunshine duration is examined by studying correlations between the daily number of suicide deaths and daily hours of sunshine in a southern Japanese city; lastly, the effect of meteorological factors, especially of sunshine, on suicidal behavior will be discussed with reference to seasonal affec-

*Department of Psychiatry, University of Occupational and Environmental Health, Iseigaoka, Yahatanishi, Kitakyushu, 807 Japan.

tive disorder and its phototherapy,[44–48] biochemical findings in suicidal patients[49] and seasonal variation of biochemical markers.[50–57]

2. SEASONAL DISTRIBUTION OF SUICIDE DEATHS

Traditionally, it is presumed that autumn is a melancholic season and that suicidal behavior would be most common during this time. Hence, early investigators were surprised to find that suicide deaths increased during the late spring and early summer and decreased during the winter.[19] Further studies revealed that the seasonal patterns of suicidal behavior caused secular changes.

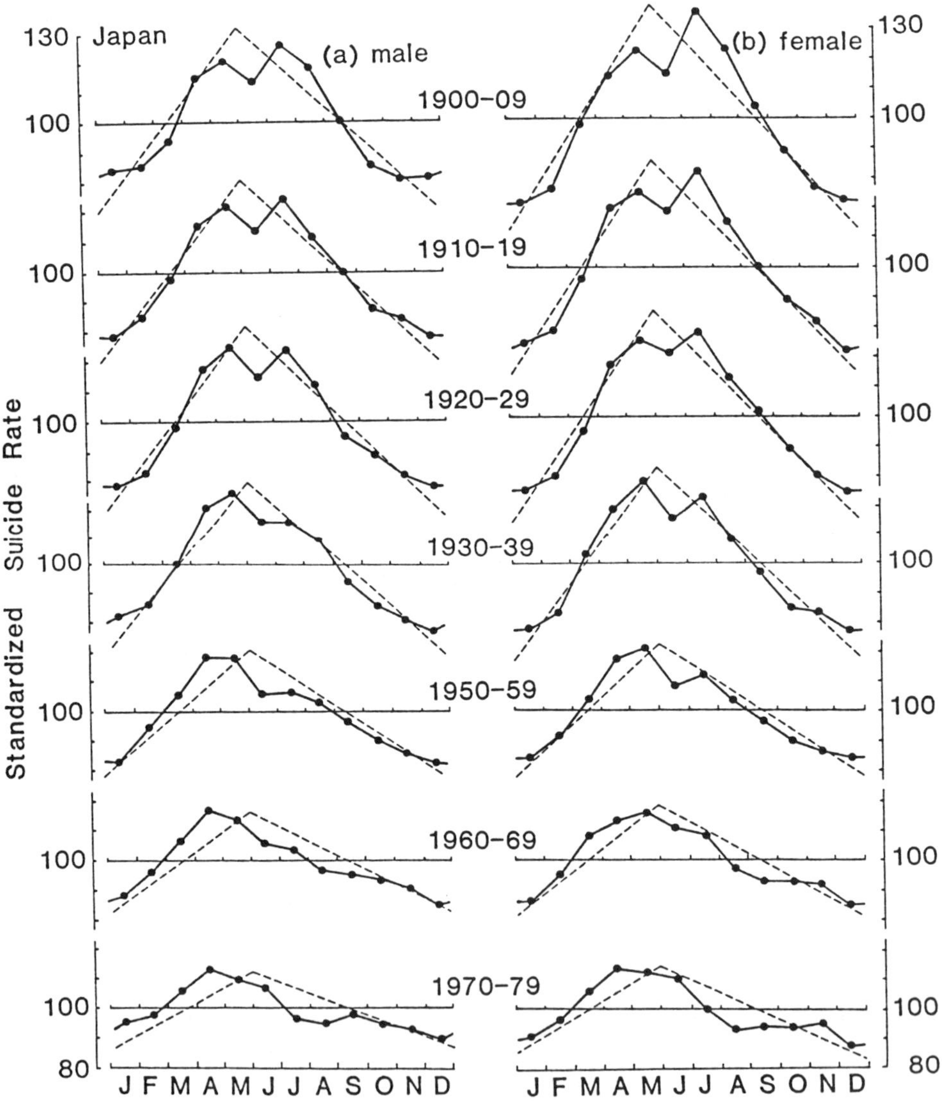

Figure 1 Standardized suicide rate of Japanese (a) males and (b) females from 1900–09 to 1970–79. Broken lines are hypothetical curves for the seasonal distribution of the standardized suicide rate. (After Fig. 3, Egashira and Abe[34])

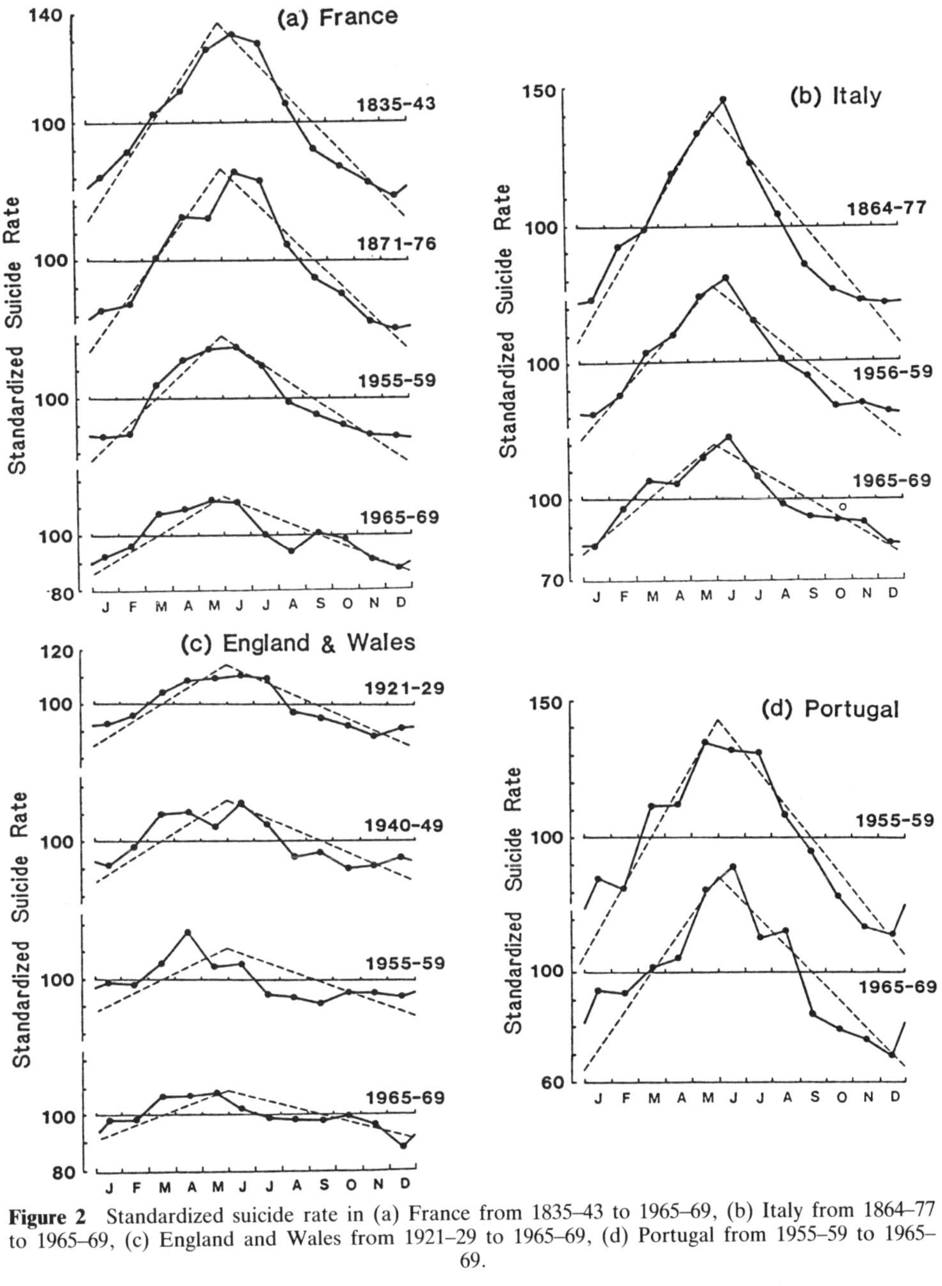

Figure 2 Standardized suicide rate in (a) France from 1835–43 to 1965–69, (b) Italy from 1864–77 to 1965–69, (c) England and Wales from 1921–29 to 1965–69, (d) Portugal from 1955–59 to 1965–69.

Seasonal patterns of suicide deaths in several developed countries and their secular changes[34] are shown in Fig. 1 and Fig. 2. These statistical data on suicide deaths in each country were collected from different organizations and investigators such as the World Health Organization, Japan Social Welfare Ministry, Morselli,[7] Swinscow[11] and McKinlay.[10] In these figures the "standardized suicide rate" has been defined as follows: the raw data were first normalized to an

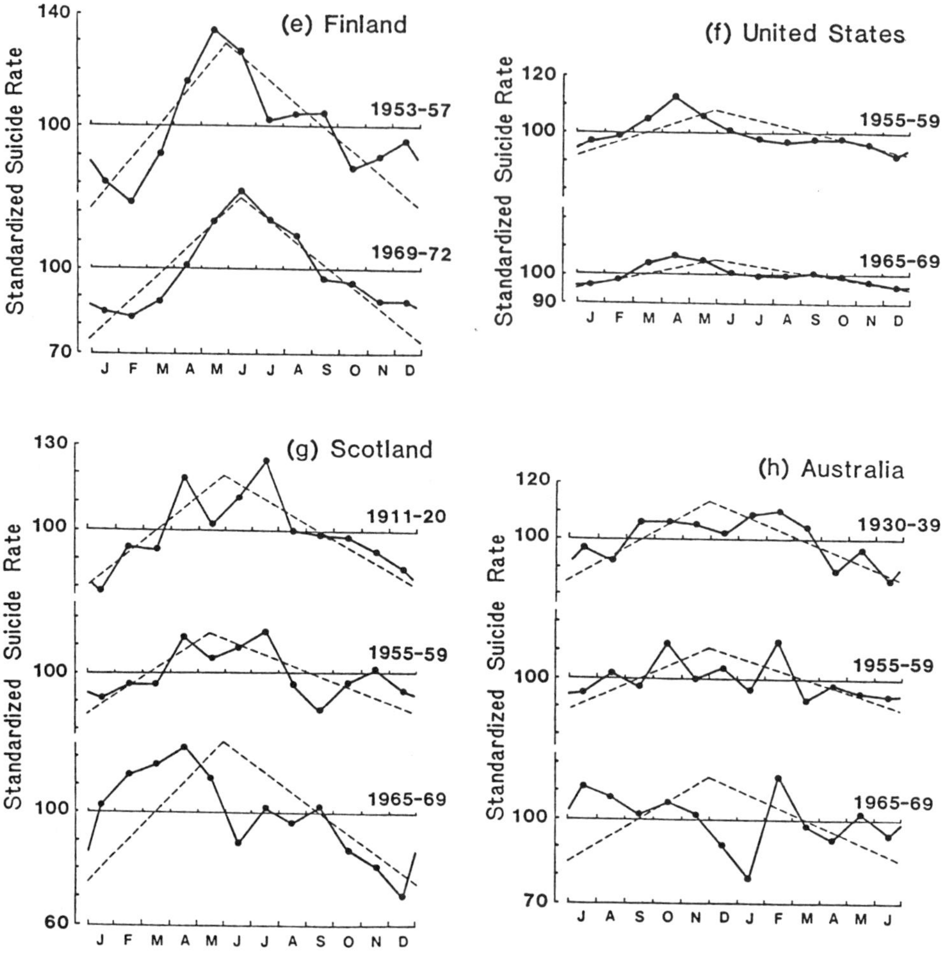

Figure 2 *(continued)* (e) Finland from 1953–57 to 1969–72, (f) USA from 1955–59 to 1965–69, (g) Scotland from 1911–20 to 1965–69 and (h) Australia from 1930–39 to 1965–69. (Egashira and Abe[34])

equivalent 30-day month length; then the normalized monthly suicide data were averaged out throughout a whole year; thirdly, the annual average was put as 100, and each set of normalized monthly data was standardized to the same proportion, leading to the standardized suicide rate.

The statistical data in Japan in the earlier decades from 1900–09 to 1920–29 showed that the maximum peak of suicide deaths was in July, followed by the next highest peak in May (Fig. 1). Only in June did the Japanese data show a trough. This finding suggests an influence of the monsoon on Japan. It has been noted that the peak of suicides in July decreased gradually and no peak was found in the 1960s. In the 1970s there also appeared a small nadir during July-August, thereby producing the next highest peak in autumn. The amplitudes of the curves of the seasonal distribution of suicides have decreased over the decades.

In several European countries and the United States, one can see a large maximum peak around July in former times (Fig. 2). This peak seems to have

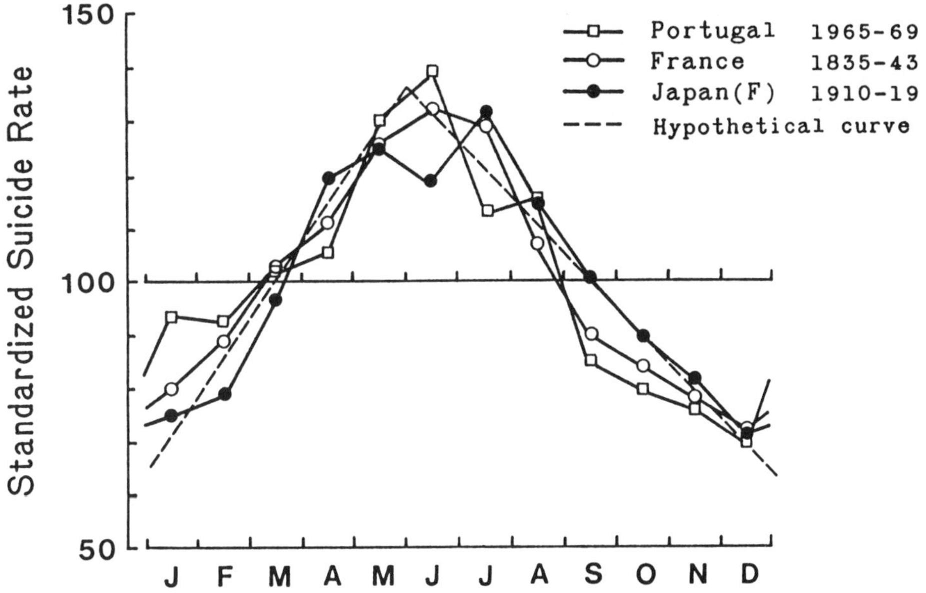

Figure 3 Standardized suicide rate in Portugal 1965–69, France 1835–43 and Japan (female) 1910–19. The broken line shows a hypothetical curve with a peak in May–June and a trough in December–January. (Egashira and Abe[34])

moved toward the earlier months as time went on. The secular trend of the seasonal suicide distribution in these countries is almost similar to that of Japan. The amplitude of the seasonal curves were the least in the United States among the countries studied. In Australia the seasonal pattern in 1930–39 is almost similar to that observed in the northern hemisphere, but in recent years a typical seasonal pattern has not been observed.

In Fig. 2, the secular decrease of the amplitude of the seasonal curves seems to be associated with the economic growth in each country. In spite of the amplitude decrease, the main peak which has shifted to spring remains until today. As Näyhä points out, the new small peak in autumn is likely to be a product of the modern age.[23]

In Fig. 3 one can see the similarities of the different curves regarding the seasonal distribution of suicide deaths in different countries. Three curves are shown in one setting for France 1835–43, Portugal 1965–69 and Japan (female) 1910–19. These curves resemble each other as regards the differences of their location and the time observed. The broken line which has a peak at midpoint in the months of May-July and a trough at midpoint in the months of December-January is a hypothetical one. This hypothetical curve was used for the calculation of an equation regarding "suicide seasonality" (S), where S is defined as the sum of twelve values of the deviation of the monthly standardized suicide rate relative to the annual average of 100. In that equation S has been expressed as a function of Δ, where Δ stands for the ratio of suicide deaths in the cooler months of November-February to the total annual suicide deaths. A geometrical calculation has lead to the following equation:[34]

$$S = -\frac{630}{23}\left(\Delta - \frac{100}{3}\right).$$

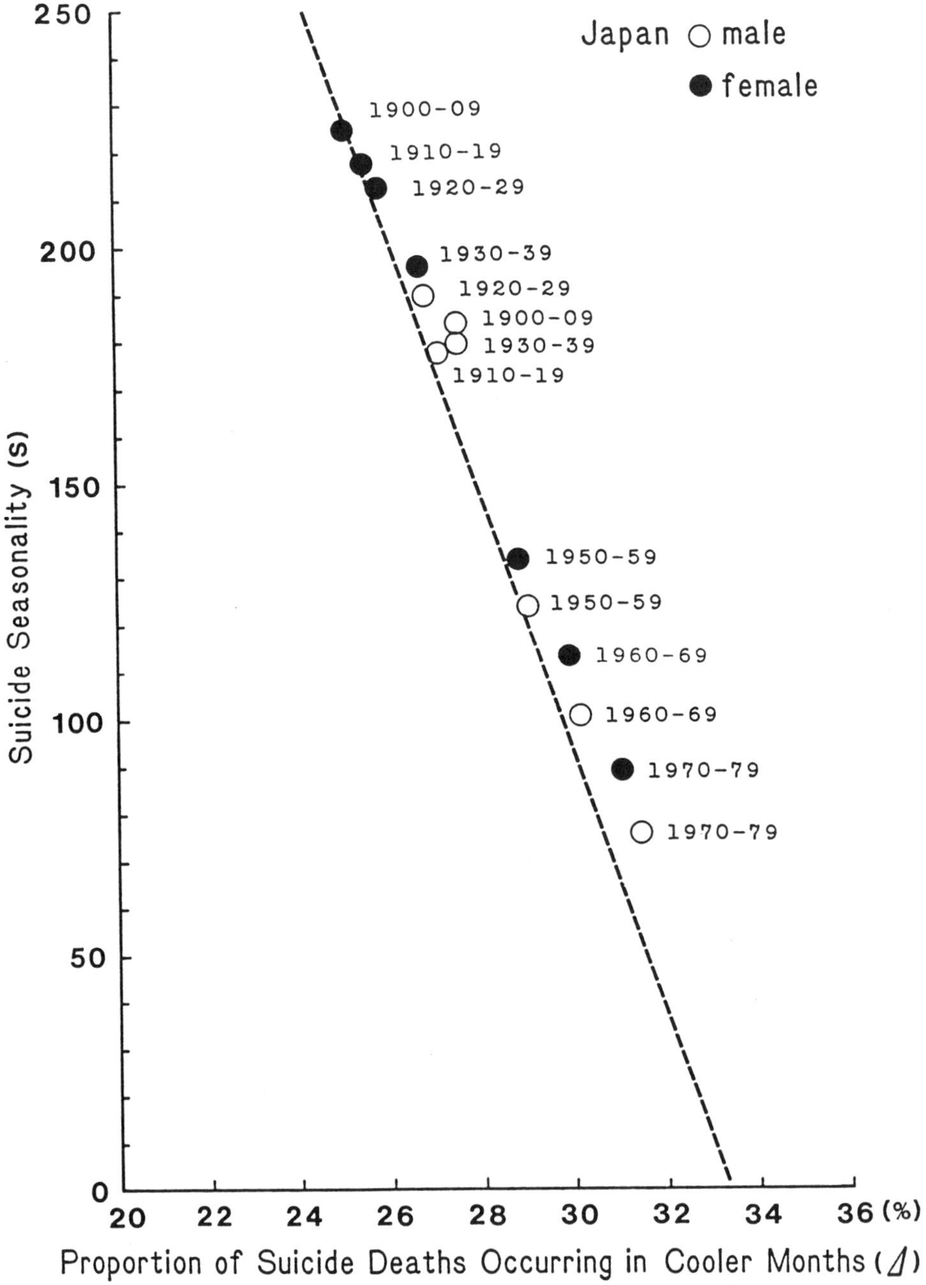

Figure 4 Suicide seasonality (*S*) vs. proportion of suicide deaths occurring in the cooler months from November to February to the total annual deaths (Δ) in Japan. (Egashira and Abe[34])

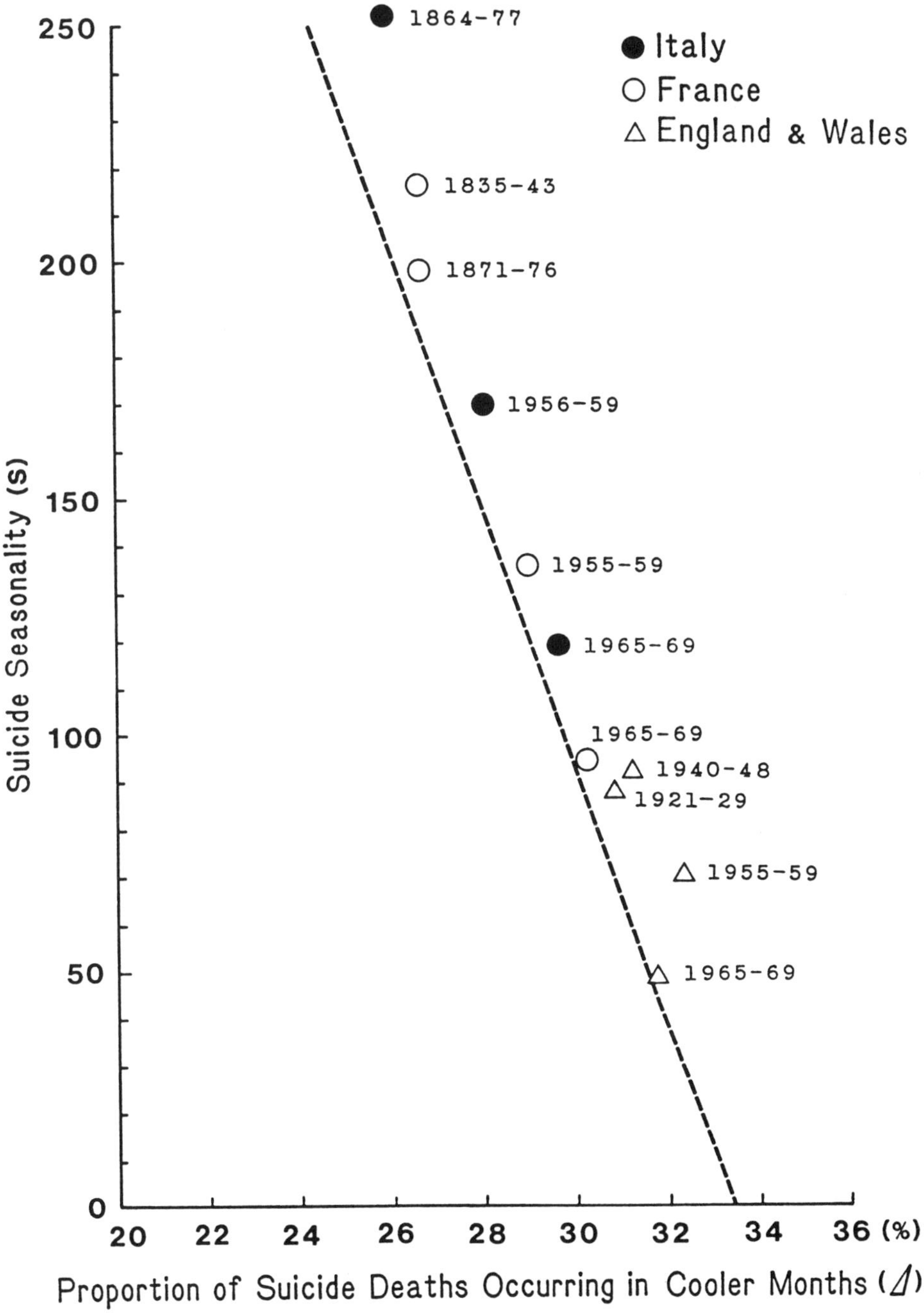

Figure 5 Suicide seasonality (S) vs. proportion of suicide deaths occurring in the cooler months (Δ) in three European countries. (Egashira and Abe[34])

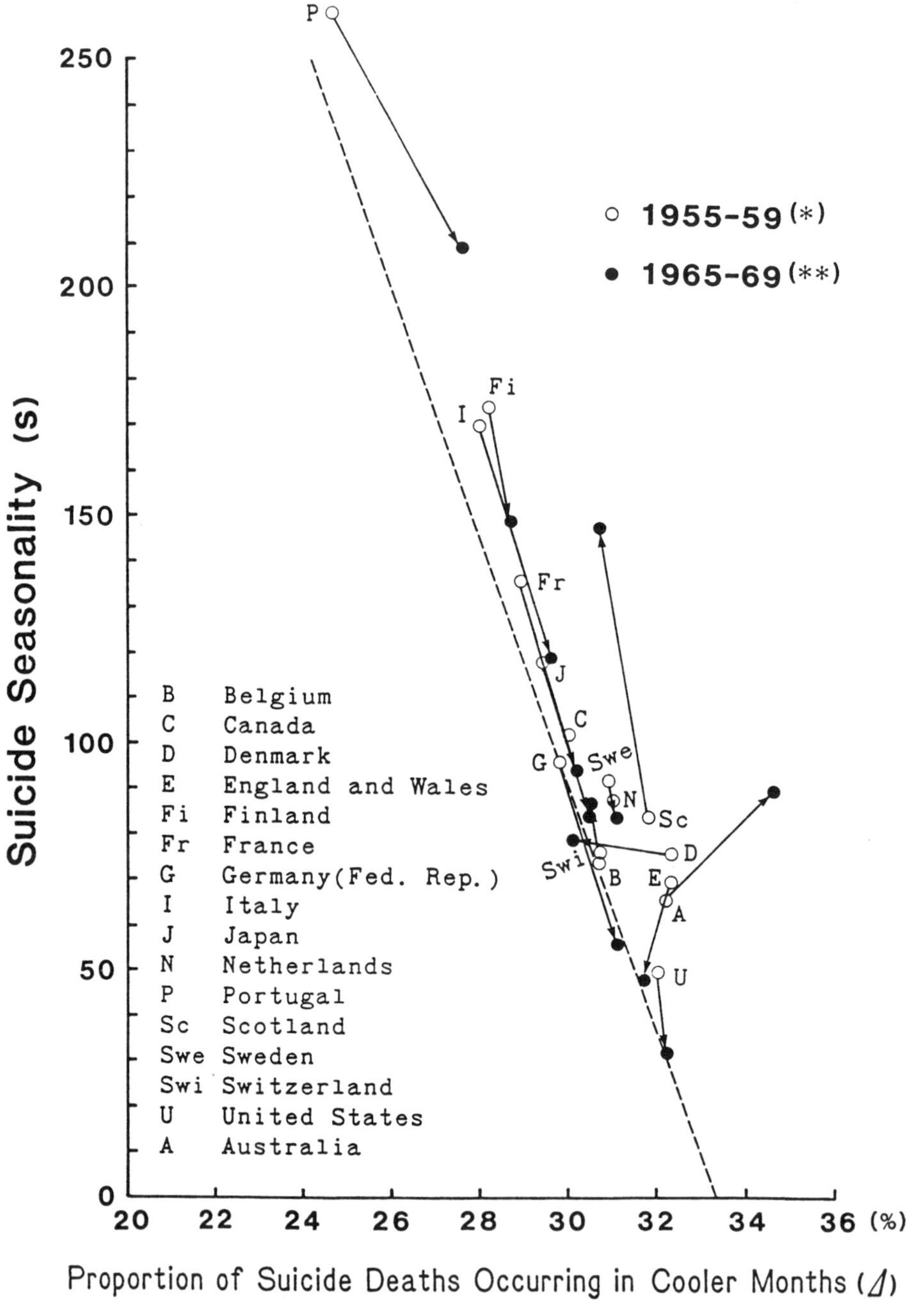

Figure 6 Suicide seasonality (S) vs. proportion of suicide deaths in the cooler months (Δ) in various European countries, USA, Canada, Japan and Australia, where changes from 1955–59 to 1965–69 are shown by arrows. (Egashira and Abe[34])

Fig. 4 shows the suicide seasonality (S) vs. the proportion of suicide deaths in the cooler months (Δ) in Japan from the years 1900–09 to 1970–79. There is a linear relationship between observed values of S and Δ. The magnitude of S decreased gradually with time along the linear line which was calculated from the above equation.

A similar relationship regarding S and Δ was observed in England and Wales, and in France and Italy, as shown in Fig. 5. We found the same linear relationship in more than ten European countries and in North American countries in the 1950s and the 1960s, as shown in Fig. 6. We also noticed that more industrialized countries tend to have a smaller suicide seasonality.

All the facts mentioned hitherto in this chapter suggest that there may be a common process in different countries for the pattern of seasonal variation of suicides to change as time goes on. In short, three facts are found: (1) in the past, a high peak in early summer was dominant; (2) in recent years, the decrease in the height and shifting of the position of the dominant peak resulted in a peak in the spring; and (3), also in recent years, a new second small peak appeared in the autumn.

The questions thus arise: what had caused the main peak in summer in the past? What made the summer peak diminish? In relation to this, we will now describe the influence of sunshine on suicide deaths.

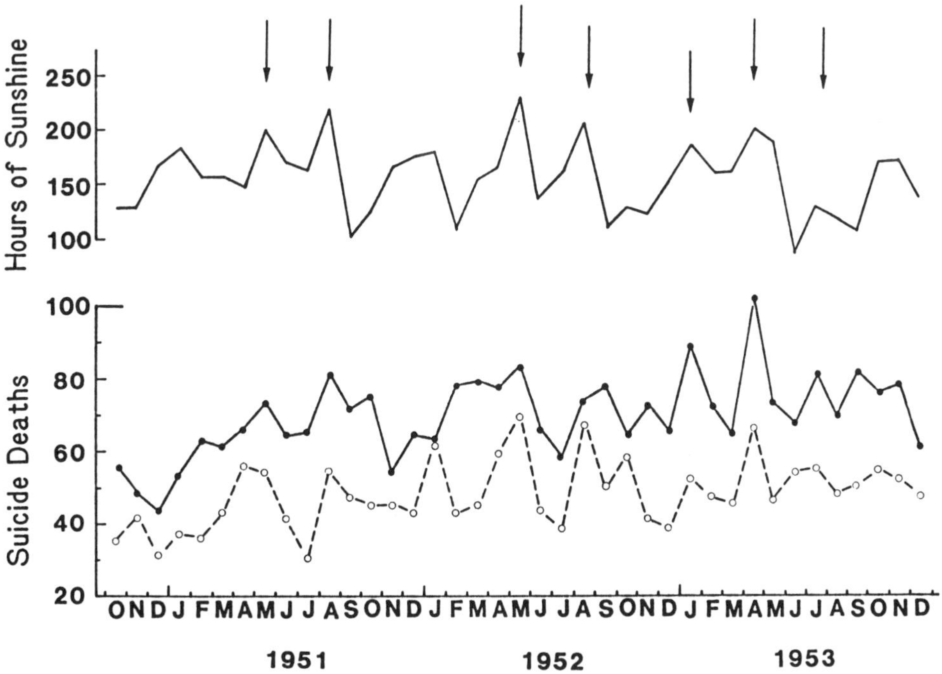

Figure 7 Monthly hours of sunshine (top) and suicide deaths in males (●) and females (○) in Tokyo for each month from October 1950 to December 1953. Arrows indicate where peaks of suicide deaths coincided with those of sunshine hours. (Egashira and Abe[42])

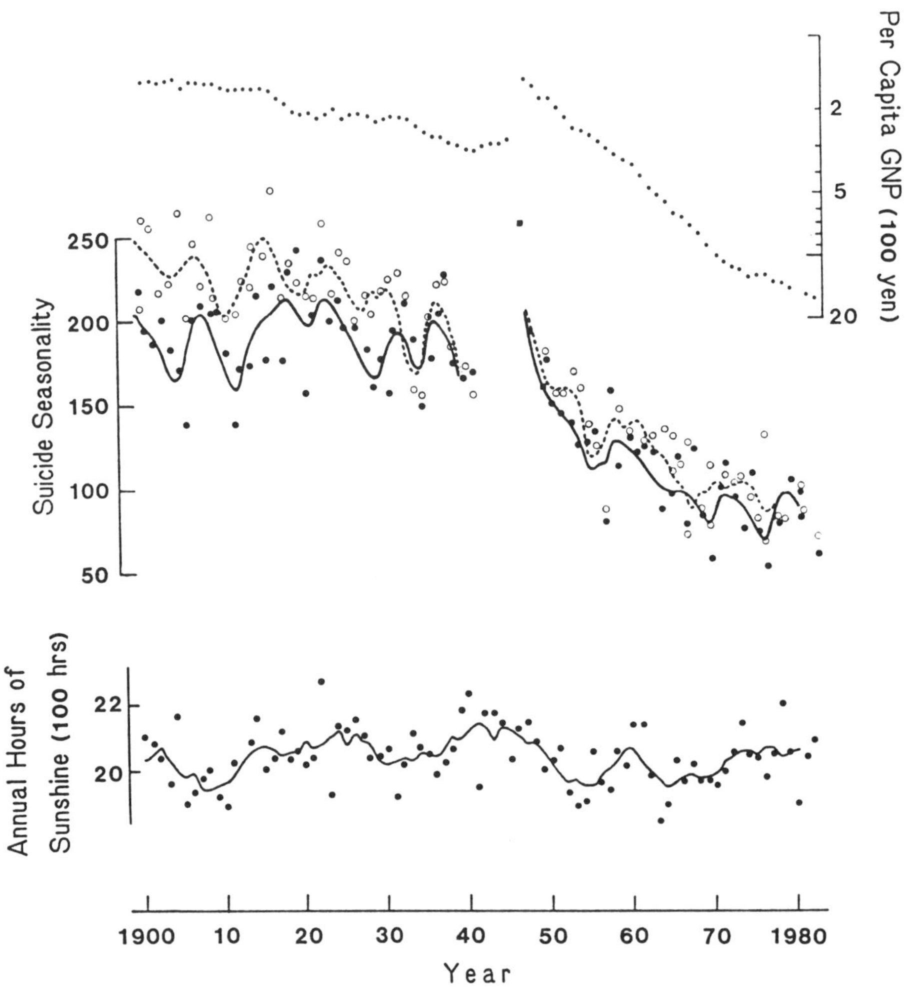

Figure 8 Suicide seasonality (●:men, ○:women, ■:both sexes) and its five-year moving average (---:men, ⋯:women) compared with per capita GNP at constant prices (1934–35) in Japan and annual hours of sunshine and its five-year moving average. (Egashira and Abe[34])

3. ANNUAL AND MONTHLY HOURS OF SUNSHINE AND SUICIDE

It is well known that the photoperiod or daytime plays an important role in the hibernation and reproduction of animals. What about the effect of sunshine on human beings?

Fig. 7 shows sunshine hours and suicide deaths in Tokyo for each month from October 1950 to December 1953.[42] As may be seen from this figure, many of the peaks of sunshine hours coincide with those of suicide deaths both in males and females. Thus the direct influence of sunshine on the suicidal behavior of human beings is evident.

In Fig. 8 we show the relationship of suicide seasonality with annual hours of sunshine in Japan from 1900 to 1982.[29] In this figure, three variables, viz. the suicide seasonality, annual hours of sunshine and per capita GNP, are shown.

The curves represent 5-year moving averages for the suicide seasonality and the sunshine hours. The 5-year moving averages were introduced in order to smooth out the temporally fluctuating raw data of the suicide seasonality and sunshine hours. The suicide seasonality and per capita GNP show opposite trends over these years, and the decreasing trend of the suicide seasonality is very similar to the increasing trend of the per capita GNP. One also notices that many of the peak positions of the suicide seasonality and the annual sunshine hours seem to coincide. Is there any correlation between the two variables? A correction was made on the suicide seasonality values in relation to its decreasing trend with the per capita GNP, which was called "corrected" suicide seasonality. The correlation coefficient between the 5-year moving averages of the "corrected" suicide seasonality and the annual hours of sunshine was 0.69 for men and 0.64 for women. Thus we were able to determine the significant positive correlation between the suicide seasonality and the annual hours of sunshine.

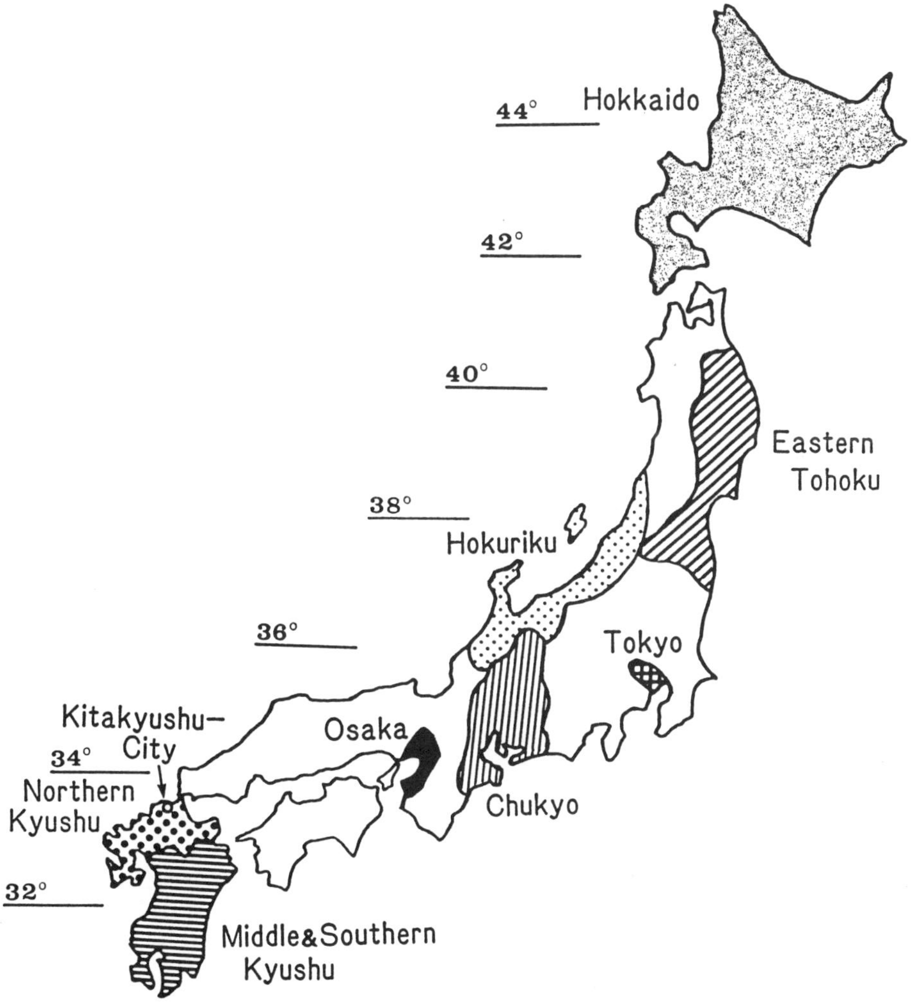

Figure 9 Location of eight districts and Kitakyushu City, which were investigated in this study.

What about the influence of sunlight on monthly suicide rates in various districts of Japan? We investigated the monthly suicide rates in a relationship to monthly hours of sunshine in eight districts in Japan (Fig. 9) from 1950 to 1957, whose climatic conditions differ distinctly from each other.[31] The results are shown in Fig. 10, where the monthly suicide rates, monthly hours of sunshine and monthly hours of sunshine in the cooler months from November to February vs. monthly suicide rates of the next month from December to March are presented.

As is obvious from this figure, the pattern of the seasonal distribution of the sunshine hours differs from district to district, reflecting differences in their climatic conditions. However, there is a remarkable similarity between the curve of suicide rates and that of sunshine hours in the cooler months for each district. For example, in Hokkaido, which is the northernmost island of Japan, the sunshine hours are at a minimum in December-January and reach a maximum in May, while the suicide rate is smallest in January and increases greatly in May. In Tokyo, which is located in the central part of Japan facing the Pacific Ocean, the sunshine hours in winter are not short and the suicide rate shows little decrease in this season. In Hokuriku, which is located in central Japan on the Japan Sea coast, the sunshine hours are extremely short and the suicide rate is lowest in winter. In Hokkaido, Eastern Tohoku, Hokuriku and Northern Kyushu, both the suicide rates and the sunshine hours show a trough in the winter, whereas in Tokyo, Chukyo, Osaka and Middle and Southern Kyushu, both show a rather flat variation in this season. As a whole, in the cooler months there is a remarkable similarity between the seasonal distribution of monthly suicide rates and that of monthly sunshine hours in each district.

In Hokkaido, a trough of the monthly hours of sunshine in winter occurs in December-January, while a trough of the suicide rate appears later in January. In Eastern Tohoku, a trough of the monthly hours of sunshine and that of the monthly suicide rate in winter are both in January. In Tokyo, however, a small peak of sunshine hours appears in January and a small suicide rate peak appears a month later in February. In Chukyo, a small trough of sunshine hours is seen in December-January, and a suicide rate trough was noticed in January. In Osaka and Northern Kyushu, a trough of sunshine hours and of suicide rates both appear in January. In Middle and Southern Kyushu, a trough appears in January and that of the suicide rate in January-February. In spite of a remarkable similarity between the curve of the monthly sunshine hours and that of the suicide rate, the latter shows a maximum phase delay of one month.

On the other hand, in warmer months no evidence of correlation between the monthly hours of sunshine and the suicide rates was found. In six districts other than Hokkaido and Eastern Tohoku, the monthly sunshine hours increased during the months of July-August but not the suicide rates.

In Hokkaido, the monthly sunshine hours from November to February were significantly correlated with the monthly suicide rates of the next month from December to March (Fig. 10a). Correlation coefficients for Hokkaido and the other seven districts are summarized in Table I. In this table correlation coefficients by the month-shift are shown, that is, correlation coefficients of the monthly sunshine hours during the cooler months with the suicide rate of the same month (0-month shift), of the following month (one-month shift) and of the succeeding month (two-month shift). As may be seen in the second column of the Table (1-month shift), significant correlations of moderate intensity were obtained with the suicide rate of the following month for five districts, Hokkaido,

Eastern Tohoku, Tokyo, Chukyo and Osaka, but not for Hokuriku, Northern Kyushu, and Middle and Southern Kyushu. These coefficients are larger than for the same month (0-month shift) and the succeeding month (2-month shift). For example, in Tokyo the correlation coefficient of the monthly sunshine hours with the suicide rate of the following month was 0.40 ($p < .02$), whereas that of the same month was −0.00. On the other hand, in Hokuriku, the correlation coefficients were largest for the suicide rate in the same month ($r = .50$, $p < .01$). In both Northern Kyushu as well as Middle and Southern Kyushu, which is the southern island of Japan, no significant correlations were observed between monthly sunshine hours and suicide rates.

As for other meteorological factors than sunshine hours, we studied temperature, but no significant correlation with suicide deaths was recognized. For example in Hokkaido 1950–57, no significant correlation was found between the suicide seasonality and the average temperature in the cooler months from November to February.[41]

4. MORE IMMEDIATE EFFECT OF SUNSHINE ON SUICIDE

We described long-term effects of sunshine hours on suicide deaths in the previous section. In this section we will discuss more immediate influences of sunshine on suicidal behavior.

In order to investigate a more direct relationship between daily hours of sunshine and daily suicide deaths, statistical data for Kitakyushu City and its suburbs were analyzed. This city and its suburbs are in northern Kyushu and have a population of about 1.4 million. A daily number of inspections of suicide deaths during 1986 to 1988 were provided by the Regional Police Headquarters. Actual dates of suicides may precede the date of inspections, but this discrepancy was

Table I Correlation coefficients between monthly hours of sunshine and monthly suicide rates in the cooler months in eight districts of Japan 1950–57, examined for three shift patterns of months. (Egashira *et al.*[31])

Shift	0-month	1-month	2-month
Hokkaido	.34+	.40*	.37+
Eastern Tohoku	.25	.49**	.42*
Tokyo	−.00	.40*	.35+
Chukyo	.33+	.42*	.23
Osaka	.13	.49**	.21
Hokuriku	.50**	.22	−.29+
Northern Kyushu	.15	.18	.01
Middle and Southern Kyushu	.16	.25	.11

Combination of month-shifts are as follows:

	Sunshine				Suicide			
0-month	Dec.	Jan.	Feb.	Mar.	Dec.	Jan.	Feb.	Mar.
1-month	Nov.	Dec.	Jan.	Feb.	Dec.	Jan.	Feb.	Mar.
2-month	Nov.	Dec.	Jan.	Feb.	Jan.	Feb.	Mar.	Apr.

**p<.01; *p<.02; +p≤.05

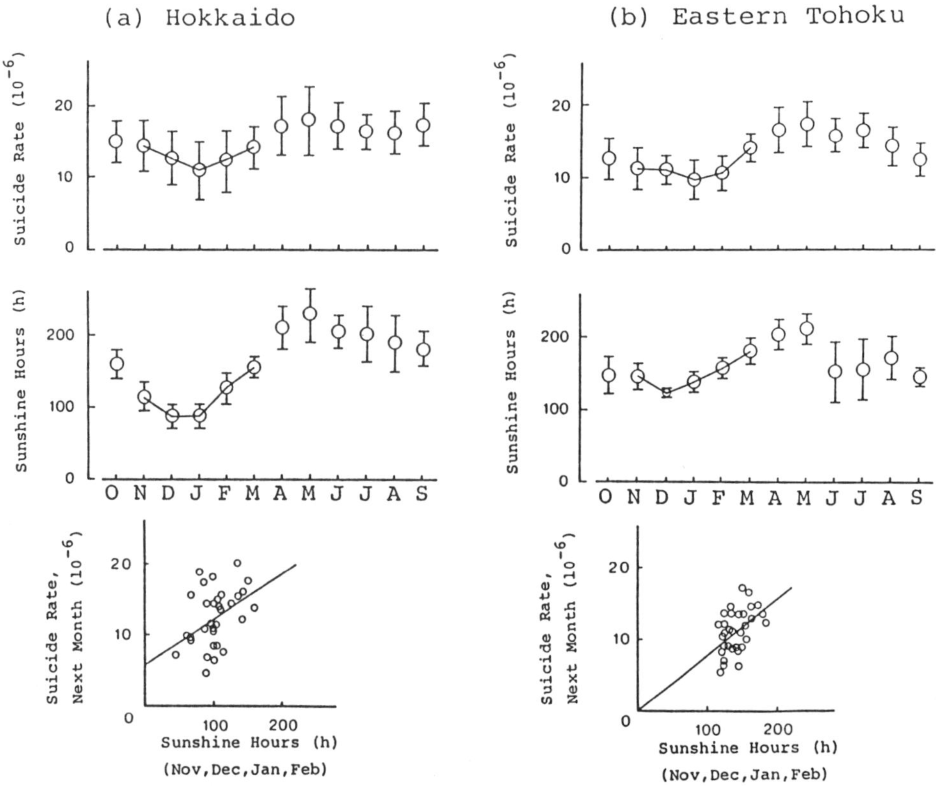

Figure 10 Seasonal variation of monthly suicide rate (top), and monthly hours of sunshine (middle) and monthly hours of sunshine in the cooler months from November to February vs. the monthly suicide rates of the following months in eight districts in Japan during 1950–57. (Egashira *et al.*[31])

neglected in this study. Data on daily hours of sunshine were provided by monthly weather reports from the Meteorological Observatory.

In the cooler months from November to March during those three years, suicide deaths and the sunshine duration in a 10-day unit were significantly correlated to each other ($r=0.44$, $p<.01$). No significant correlation was found in the warmer months from May to September.

In March 1986, when fine weather lasted for many successive days, more suicide deaths were observed than in March 1987 and 1988. On that occasion, as seen in Fig. 11, in the peaks of suicide deaths and sunshine hours, there was only a few days delay, which suggests an immediate influence of sunshine on suicide deaths. A more detailed analysis is now in progress and will be presented elsewhere.

5. DISCUSSION

The investigation of the effects of meteorological factors on suicides was approached in two ways: one is to study the direct influences of weather conditions on each suicide attempt; the other is to study the effects of meteorological factors on the seasonality of suicide deaths.

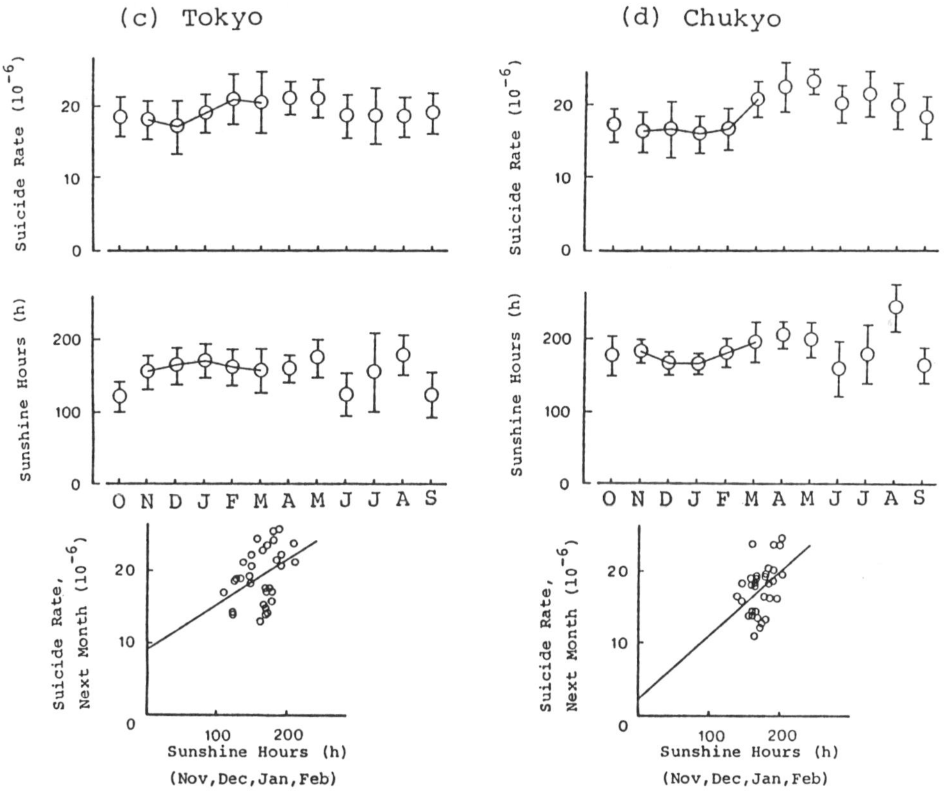

Figure 10 *Continued*

In spite of many investigations no clear answer has been obtained regarding the relationship between suicide and weather conditions. Faust[2] paid special attention to changes in weather and demonstrated that foehn and a warm front influence suicide attempts. Breuer *et al.*[4] took dynamic parameters into consideration and found significant correlations between the time of attempted suicide and the weather parameters such as upslides, fog, thunderstorm and warm air. Thus it may be deduced that the exogenous factors, like weather conditions, may play an important role in the timing of suicide attempts. These results, however, do not explain the seasonal variation of suicide deaths.

Another way to study the meteolorogical effects is to search for the causes of the seasonal variation of suicides. To accomplish this purpose we first have to clarify common features of the seasonal variation of suicide deaths. Many investigators have demonstrated seasonal patterns with various timings of the seasonal peaks and troughs.[19] The inconsistencies in their findings may be attributed to the differences of the statistical methods used or the differences of data sources temporally or geographically.[26] Some data are based on a single year, some on several years and some on a local area. The use of a single year's data introduces the possibility that the year chosen is not representative of the true seasonal pattern. On the other hand, using local data may lead to erroneous conclusions, because of the small sample sizes or possible local preventive interventions. Besides, the sea-

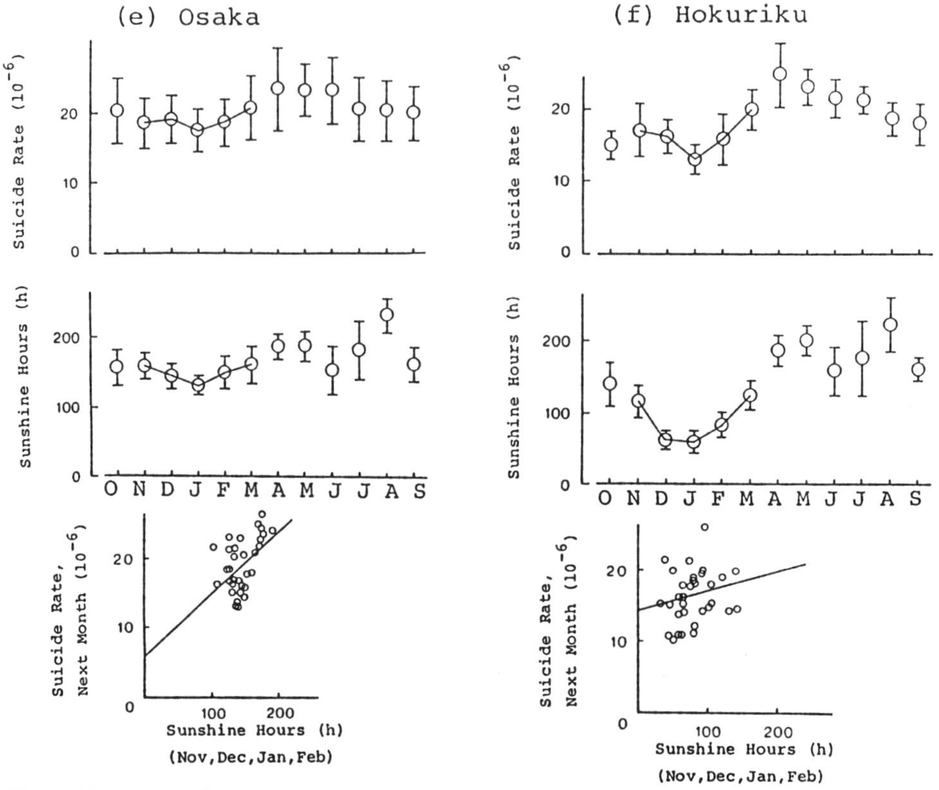

Figure 10 *Continued*

sonal distribution of suicides differs also by age and sex.[21–23,30,36,38] Of course, it is important to obtain more detailed information, such as age or sex, but it is more necessary to clarify the more essential features of the seasonal variation of suicide.

In order to avoid the errors mentioned above, we used data with sufficient sample sizes and over a period of several years. This made it possible to compare the seasonal patterns of suicide and their secular trend in different decades and countries. We found that the seasonal variation of suicides is common to various parts of the world, i.e. (1) the dominant peak was observed in early summer in the beginning of this century, (2) it has decreased its height and shifted its position to spring in recent years, and (3) a second small peak has appeared in autumn also in recent years. (Fig. 1–Fig. 3). The degree of suicide seasonality, which is defined and called S, decreased as time passed (Fig. 4–Fig. 6), and in the 1960s such countries as England and Wales, Germany and the United States showed the smallest seasonality values. Thus, more economically developed countries showed a smaller seasonality. This finding is also supported by another study, [28] which demonstrated a correlation between the decrement of the suicide seasonality and the annual growth of the gross domestic product in several European countries, the United States and Australia. On the other hand, Portugal lately shows a larger seasonality (Fig. 6). This may be related to the stages of economical development of this country. The suicide seasonality of Portugal is

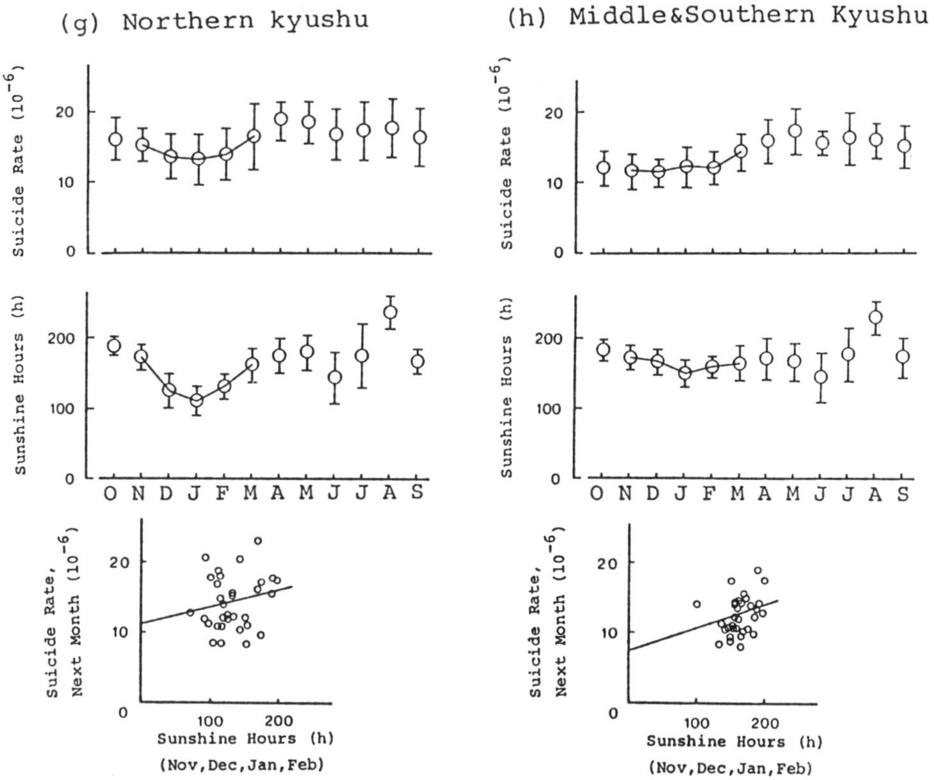

Figure 10 *Continued*

also decreasing with time with a linear relationship to Δ (Fig. 6), where Δ is the proportion of suicide deaths in the cooler months to the total annual deaths. The speed of the temporal change of the suicide seasonality of each country is likely to be related to the degree of its economical development. Thus we emphasize the importance of studying both the temporal and geographical aspects of the seasonal variation of suicide deaths.

We demonstrated that the suicide seasonality correlates with the total annual sunshine hours in Japan 1900–82 (Fig. 8). We speculated that the cause of this correlation was the following: the annual hours of sunshine may correlate positively with the sunshine hours in summer, the latter may correlate positively with the peak height of seasonal suicidal curves, and hence with the suicide seasonality. This speculation may be applicable to the earlier decades of this century. In accordance with this speculation, there was a decrease in the summer peak of the seasonal suicide distribution in 1931, when summer was cooler and sunshine hours shorter than in the adjacent years.[34] These findings do not contradict those of Aschoff[20], who studied suicide data of many countries in 1920–70. He stated that the amplitudes of the seasonal variation of suicides are maximal at about 40° N, where the amplitude of monthly hours of sunshine reaches its maximum.

At this latitude of 40° N there is Portugal, which shows a marked suicide seasonality. As mentioned above, suicide seasonality depends also on the state

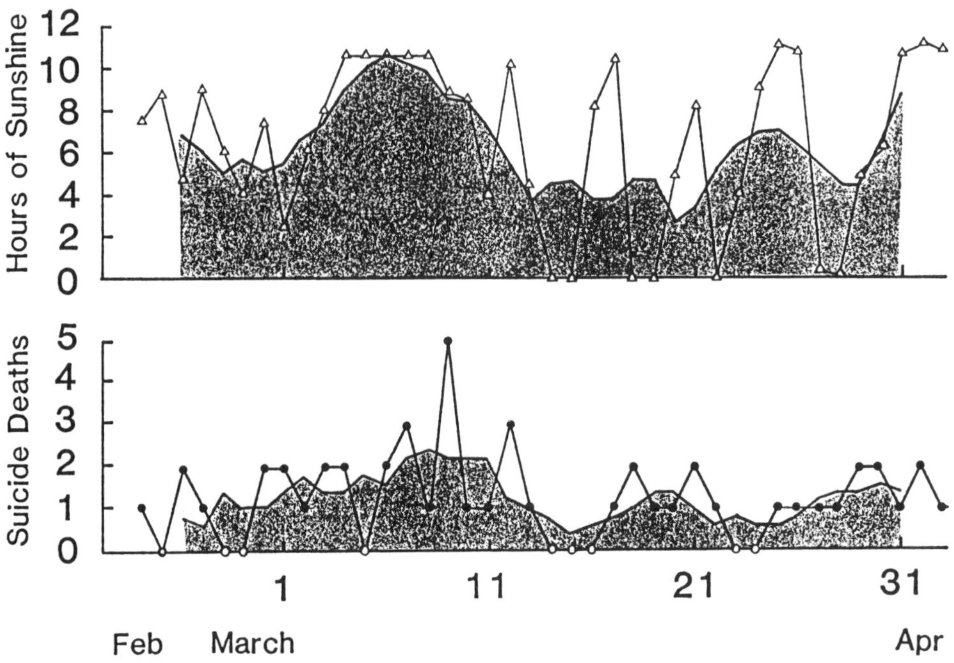

Figure 11 Daily hours of sunshine and suicide deaths in Kitakyushu City and its suburbs in March 1986, where five-day moving averages of the two variables are shown by dark areas.

of economical development, i.e. less developed countries tend to show a larger suicide seasonality.[28,29] Therefore, in order to interpret the geographical difference of suicide seasonality, one should take not only meteorological factors but also economic ones into consideration.

In recent years, the summer peak in the seasonal distribution of suicides has diminished in many industrialized countries. The cause of this diminishment will be discussed later in association with the apparent change of influences of sunshine due to industrialization.

We found a significant correlation in the monthly hours of sunshine in the cooler months with the monthly suicide rates (Table I). Correlation coefficients of the monthly sunshine hours with the monthly suicide rate of the following month were larger. In warmer months, however, no correlation was found between the monthly sunshine hours and the monthly suicide rate. Souêtre *et al.*[40] reported that the rhythm of suicides was highly correlated with daytime and the mean temperature shifted one month backward in France 1978–1982. Differing from our present analysis, they used the 12-month data for analyzing. One should wonder, however, whether or not a meteorological factor brings about the same effects throughout the whole year. The present results suggest that the effect of sunshine is more enhanced in cooler months than in warmer months. If this is the case, the rapid increase of suicide rates from later winter to early spring is likely to be due to rapid prolongation of sunshine hours during that period.

As mentioned above, there is a question regarding the decrease of the summer peak which was predominant in former times. Does sunshine have something to do with this decrease? One speculative answer is concerned with the industrialization of our society which introduces changes in our life style. Today people spend more time indoors than in earlier times. Therefore, the intense summer sunlight could not today affect human beings or their suicidal behavior in the same way as before and has resulted in the diminishment of the summer peak. This assumption is consistent with the findings on urban/rural differences; the seasonal variation is greater in the rural areas than in the urban areas.[43] In the rural areas, where outdoor-life is more predominant than in the urban areas, greater influence by sunlight is expected.

We also suggested an immediate influence of sunshine on suicides. This influence seemed to be induced in cooler months within several days, especially in March (Fig. 11). It is our impression that this direct influence is noticeable when very fine weather lasts for several days. Further analyses of these observations are now in progress.

It is well known that suicide is strongly associated with depression.[1] Recently, seasonal affective disorder (SAD), a type of depression, has been identified. SAD shows a depression in autumn/winter and a remission or hypomania in spring/summer.[44–47] SAD is likely to be associated with a seasonal change of the photoperiod, and it can be treated successfully with artificial bright light.

Terman[47] reported bimodal seasonal mood changes in a general population with a larger peak in May and a smaller one in September, i.e. most people feel better in late spring and in early autumn. Interestingly enough this seasonality of mood resembles the suicide seasonality observed in recent years.

In contrast, the seasonal mood changes in SAD patients are unimodal, with a summer peak and a winter trough,[47] i.e. most SAD patients feel best in summer and worst in winter. Besides, Carney[6] showed that admission rates of mania were higher in the sunnier months and correlated significantly with the total monthly hours of sunshine. Such unimodal seasonality of mood or manic symptoms resembles markedly the seasonal distribution of suicide observed in the earlier decades.

The facts mentioned above indicate that the seasonality of suicides corresponds inversely to the seasonality of depressive mood felt by healthy people or SAD patients. If suicides are committed while one is in a depressive state, these facts are contradictory. Actually, suicide attempts occur most frequently when patients are in a phase of recovery from a severe depression. At this phase of depression, patients are so unstable that they are in a state of high risk for suicide attempts. Depression is a pre-suicidal state, which is likely to lead to suicide attempts triggered by social or environmental precipitating factors. Prolongation of sunshine hours in late winter through early spring may be one of the important environmental factors.

As regards biochemical markers of suicide and depression, many investigations have been carried out.[49–57] Serotonin, one of the chemical transmitters in the human brain, is known to be associated with depression. In suicidal patients the metabolite of serotonin 5-HIAA is decreased.[49] Therefore, 5-HIAA is regarded as one of the biochemical markers of suicidal risk. There are also reports on seasonal changes of such biochemical factors.[50–57] Whitaker *et al.*[55] reported that ^{3}H-imiplamine binding increased in winter and decreased in summer. But contradictory results are also reported by other investigators.[56] Thus reports on the seasonality of biochemical factors associated with suicide still remain inconsistent and controversial.

CONCLUSIONS

First, seasonal distribution of suicide and its secular trends in various countries are demonstrated. We noticed that, in spite of geographical differences, there is a common process for suicide deaths to change their seasonal pattern with time. Many industrialized countries seem to have accomplished a greater change in their seasonal pattern and now have a smaller suicide seasonality.

Secondly, the effects of meteorological factors on suicidal behavior are discussed. The correlation between suicide deaths and hours of sunshine was examined using annual, monthly and daily statistical data. A significant correlation between the current month's suicide rates and the previous month's sunshine hours was demonstrated in cooler months, but not in warmer months. An immediate influence of sunshine on suicide, which occurs within several days, is also suggested. Thus, it was found that sunshine, an environmental and meteorological factor, is one of the stress factors which affect suicidal behavior.

Lastly, the cause of the seasonal variation of suicides was considered in reference to previous studies.

We did not take into consideration such subclass variables as age, sex, social status, etc., and further studies are required on these points.

References

1. K. Achté, "Depression and Suicide" *Psychopathorogy* **19**(S2) 210–214 (1986).
2. V. Faust, "Suizid und Wetter" *Fortschr. Med.* **91** 154–158 (1973).
3. A. Larcan *et al.*, "Conduites suicidaires et conditions météorologiques." *Ann. Medico-psycholog.* **134** 51–61 (1976).
4. M. Breuer *et al.*, "Social, toxicological and meteorological data on suicide attempts" *Eur. Arch. Psychiat. Neurol. Sci.* **235** 367–370 (1986).
5. D. Lester, "Suicide and homicide rates: Their relationship to latitude and longitude and to the weather" *Suicide Life-threat. Behav.* **16** 356–359 (1986).
6. P. A. Carney *et al.*, "Influence of climate on the prevalence of mania" *Brit. J. Psychiat.* **152** 820–823 (1988).
7. H. Morselli, *Der Selbstmord* (Brockhaus, Leipzig, 1881).
8. J. R. Miner, "Suicide and its relation to climatic and other factors" *Am. J. Hyg. Mongr.* **2** 1–140. (1922).
9. H. W. Gruhle, *Selbstmord* (Georg Thieme, Leipzig, 1940) pp. 83–93.
10. P. L. McKinlay, "Mortality from suicide in Scotland" *Health Bull.* **6** 29–35 (1948).
11. D. Swinscow, "Some suicide statistics" *Brit. Med. J.* **II** 1417–1423. (1951).
12. K. Dreyer, "Comparative suicide statistics" *Danish Med. Bull.* **6** 75–81 (1959).
13. E. Takahashi, "Seasonal variation of conception and suicide" *Tohoku J. Exp. Med.* **84** 215–227 (1964).
14. J. E. Edwards and F. A. Whitlock, "Suicide and attempted suicide in Brisbane: 1" *Med. J. Australia* **1** 932–938 (1968).
15. W. W. K. Zung and R. L. Green, "Seasonal variation of suicide and depression" *Arch. Gen. Psychiat.* **30** 89–91 (1974).
16. M. R. Eastwood and J. Peacocke, "Seasonal patterns of suicide, depression and electroconvulsive therapy" *Brit. J. Psychiat.* **129** 472–475 (1976).
17. M. R. Eastwood and S. Stiasny, "Psychiatric disorder, hospital admission and season" *Arch. Gen. Psychiat.* **35** 769–771 (1978).
18. D. Lester, "Temporal variation in suicide and homicide" *Am. J. Epidem.* **109** 517–520. (1979).
19. S. M. Kevan, "Perspectives on season of suicide: Review" *Soc. Sci. & Med.* **14D** 369–378 (1980).
20. J. Aschoff, "Annual rhythms in man" **In:** (J. Aschoff, ed.) *Handbook of Behavioral Neurology 4*, Biological Rhythms (Plenum, New York, 1981) pp. 475–487.
21. R. Meares *et al.*, "A sex difference in the seasonal variation of suicide rate: A single cycle for men, two cycles for women" *Brit. J. Psychiat.* **138** 321–325 (1981).

22. G. M. Skutsch, "Sex difference in seasonal variation in suicide rate" *Brit. J. Psychiat.* **139** 80–81 (1981).
23. S. Näyhä, "Autumn incidence of suicides re-examined: Data from Finland by sex, age and occupation" *Brit. J. Psychiat.* **141** 512–517 (1982).
24. G. Parker and S. Walter, "Seasonal variation in depressive disorders and suicidal deaths in New South Wales" *Brit. J. Psychiat.* **140** 626–632 (1982).
25. S. Näyhä, "The bi-seasonal incidence of some suicides," *Acta Psychiat. Scand.*, **67** 32–42 (1983).
26. C. W. Warren *et al.*, "Seasonal variation in suicide and homicide: a question of consistency" *J. Biosoc. Sci.* **15** 349–356 (1983).
27. K. MacMahon, "Short-term temporal cycles in the frequency of suicides, United States, 1972–1978" *Am. J. Epidem.* **117** 744–750 (1983).
28. K. Abe *et al.*, "Suicide seasonality and economic growth," *Stress Med.*, **2** 79–81 (1986).
29. K. Egashira *et al.*, "Suicide seasonality and sunshine," *Jpn. J. Biometeor.*, **23** 65–70 (1986).
30. S. Araki *et al.*, "Seasonal variation in suicide rates by cause and sex," *J. Biosoc. Sci.*, **18** 471–478 (1986).
31. K. Egashira *et al.*, "Monthly suicide rates and sunshine hours in eight districts in Japan," (in Japanese) *Seishin-igaku*, **29** 735–740 (1987).
32. S. Araki and K. Murata, "Suicide in Japan: Socioeconomic effects on its secular and seasonal trends," *Suicide Life-threat. Behav.*, **17** 64–71 (1987).
33. E. Souetre *et al.*, "Seasonality of suicides: Environmental, sociological and biological covariations," *J. Affect. Disord.*, **13** 215–225 (1987).
34. K. Egashira and K. Abe, "Secular trends of suicide seasonality: A comparative study of various countries" *Jpn. J. Biometeor.* **25** 97–109 (1988).
35. R. Micciolo *et al.*, "Geographical variation in the seasonality of suicide" *J. Affect. Disord.* **15** 163–168 (1988).
36. D. Lester and M. L. Frank, "Sex differences in the seasonal distribution of suicides" *Brit. J. Psychiat.* **153** 115–117 (1988).
37. D. Lester and M. L. Frank, "Seasonal variation in suicide rates in the United States" *J. Clin. Psychiat.* **49** 371 (1988).
38. R. Micciolo *et al.*, "Seasonal variation in suicide: Is there a sex difference?" *Psychol. Med.* **19** 199–203 (1989).
39. D. Lester and M. L. Frank, "Geophysical variables and behavior: LIX. The seasonal variation in suicide rates: Three studies" *Percept. Motor Skills* **70** 57–58 (1990).
40. E. Souêtre *et al.*, "Influence of environmental factors on suicide behavior" *Psychat. Res.* **32** 253–263 (1990).
41. K. Egashira and K. Abe, "Correlation of suicide seasonality to sunshine hours in cooler months in Hokkaido (1950–57)" *Jpn. J. Biometeor.* **27** 3–7 (1990).
42. K. Egashira and K. Abe, "Season, sunshine duration and suicide" **In**: (S. Takahashi, K. Takahashi and K. Honma, eds.) *Rinsho Jikan-Seibutsugaku (Clinical Chronobiology)* (in Japanese), Asakura, Tokyo, 1990, pp. 191–203.
43. R. Micciolo *et al.*, "Geographical and urban–rural variation in the seasonality of suicide: Some further evidence" *J. Affect. Disord.* **21** 39–43 (1991).
44. A. J. Levy *et al.*, "Bright artificial light treatment of manic depressive patient with a seasonal mood cycle" *Am. J. Psychiat.*, **139** 1496–1498 (1982).
45. N. E. Rosenthal *et al.*, "Seasonal affective disorder: A description of the syndrome and preliminary findings with light therapy" *Arch. Gen. Psychiat.* **41** 72–80 (1984).
46. C. Thompson and T. Silverstone (eds.) *Seasonal Affective Disorder* (CNS, London, 1989).
47. M. Terman, "On the question of mechanism in phototherapy for seasonal affective disorder: Considerations of clinical efficiency and epidemiology" *J. Biol. Rhythms* **3** 155–172 (1988).
48. W. E. Stumpf and H. P. Privette, "Light, vitamin D and psychiatry: Role of 1,25 dihydroxyvitamin D_3 (soltriol) in etiology and therapy of seasonal affective disorder and other mental processes" *Psychopharmacol.* **97** 285–294 (1989).
49. L. Traskman-Bendz *et al.*, "Biochemical aspects of suicidal behavior" *Prog. Neuro-Psychopharmacol. Biol. Psychiat.* **13** S35–S44 (1989).
50. A. Wirz-Justice *et al.*, "Diurnal and seasonal variations in human platelet serotonin in man" *J. Neural Transmission* **41** 7–15 (1977).
51. A. Wirz-Justice and R. Richter, "Seasonality in biochemical determinants: A source of variance and a clue to the temporal incidence of affective illness" *Psychiat. Res.* **1** 53–60 (1979).
52. C. Swade and A. Coppen, "Seasonal variations in biochemical factors related to depressive illness" *J. Affect. Disord.* **2** 249–255 (1980).
53. A. Carlson *et al.*, "Seasonal and circadian monoamine variations in human brains examined post-mortem" *Acta Psychiat. Scand. (suppl.)* **280** 75–85 (1980).

54. R. C. Arora *et al.*, "Seasonal variation of serotonin uptake in normal controls and depressed patients" *Biol. Psychiat.* **19** 795–804 (1984).
55. P. M. Whitaker *et al.*, "Seasonal variation in platelet ^{3}H-imipramine binding: Comparable values in control and depressed populations" *Psychiat. Res.* **11** 127–131 (1984).
56. A. Goziotis and S. W. Tang, "A longitudinal study of intact platelet ^{3}H-imipramine binding in 12 normal human subjects" *Psychiat. Res.* **26** 157–162 (1988).
57. D. Marazziti *et al.*, "Lack of specificity of reduced platelet imipramine binding in different conditions" *Psychiat. Res.* **30** 21–29 (1989).

5. STRESS, MENTAL HEALTH AND THE ENVIRONMENT

HUGH FREEMAN*

The paper examines the impact of environmental stressors on people from a point of view of controllability and importance of setting in which the stressors occur; social stress and social class factors are then analysed in some detail. This is followed by a study of environmental stress, i.e. imbalance between the demands of the environment and the individual's ability to adjust to them. The relationship of bodily health to the structural environment is a complex matter, since psychological and social pathologies which seem to accompany mass urbanization are not yet fully understood. Various aspects of this problem are presented, e.g. mental health and planning in cities, indicators of the psycho-social environment, psycho-social stress in urban areas, and the mental health of young people in such environments. Professor Freeman concludes that current urban programmes are not conducive to mental well-being and that it is necessary to carry out much research to investigate the dangers of urban sprawl and obliteration of communities' landmarks. He advocates an alliance between the mental health professions and city planners and architects to "humanise" the environment on the basis of an objective and sound scientific enquiry.

1. STRESS

In the search for a general explanation of the mediating process between environment and individual reactions, the concept of stress has emerged as particularly useful heuristically. Research on this subject has been dominated by two models:[1] the physiological, derived from Selye, implies both a homeostatic concept of bodily function and negative effects from the processes of coping with stress, independently of the impact of the stressor itself. The psychological model, on the other hand, emphasises the cognitive interpretation of environmental conditions, influenced by characteristics of the individual and the context; it is not clear, though, which model is more crucial in determining human reactions. Biological approaches tend to overemphasise pathogenic outcomes of a physiological kind, but psychological effects, e.g. on mood or interpersonal relationships, are also very important, both as outcome measures in themselves and through affecting the susceptibility of the host organism to disease.

The impact of environmental stressors on people is likely to depend on both the importance and controllability of the setting in which the stressors occur.[2] It can wrongly be concluded, however, that certain stresses have no negative effects because the individual has made successful efforts to maintain normal performance through adaptation, though these efforts themselves may well be harmful

*21 Montagu Square, London W1H 1RE, UK.

to him or her in the long run.[1] If various kinds of changes in environmental conditions can all be conceptualised as "stressful", they would need to share some common mechanism, or else the construct of stress fails to retain credibility.

However, the processes that make certain environmental conditions stressful can also be seen in various terms: an arousal model proposes that they affect the reticular activating system, and that stress will result if arousal levels thereby move outside an optimum range; an overload model assumes that an excess of environmental stimuli has harmful effects; and a systems model sees stress as resulting from individual needs being thwarted by environmental conditions.[1] Cassel[3] pointed out that because of the tendency to think in mono-aetiological terms, much work on the social and psychological antecedents to disease has tried to identify particular forms of stress that would have a specific causal relationship to clinical entities. Yet animal data show that a wide variety of pathological conditions may emerge following changes in the social milieu, and in humans also, variations in group relationships could be expected to enhance susceptibility to disease in general, rather than having a specific aetiological role. In different individuals, the specific manifestations of disease following stress would depend both on their genetic predisposition, metabolic state, etc., as well as on any adverse physico-chemical or infective factors to which they might have been exposed. Unfortunately, research on stress up to now has nearly all been restricted to short time-spans; results are not available from studies carried out over the lengths of time that are necessary to answer any of the fundamental questions.

In some societies at least, an association between crowding and physical disease may derive from the particular kind of stress that arises from accompanying disordered social relationships;[4,5] adverse effects of living at high density might be mainly confined to periods following migration, and once settled, people may generally adapt to these conditions. Levy & Herzog[6] concluded from a review of the literature that crowding is unlikely to be a serious problem in the absence of abject poverty, lack of sanitation, or poor nutrition; otherwise, it is more of an irritant and social strain, which may precipitate psychoneurosis or other mild forms of maladjustment, but is probably irrelevant in the main to physical disease and major mental disorder.

1.1. *Social Stress*

"Social stress" is a concept which grew partly out of wartime experience, such as that of Dunkirk survivors.[7] Epidemiological psychiatric studies like the Midtown Manhattan,[8] the Stirling county work, and Hollingshead & Redlich's[9] investigation of social class and mental illness agreed in showing a consistent relationship between indicators of social stress and rates of psychiatric morbidity. Though there is little agreement in the literature as to which factors should be included, this concept assumes that such phenomena as stigmatisation, effects of poverty, role disturbance (e.g. one-parent families), or high parity are of aetiological significance in psychiatric disorder. In Mechanic's view,[10] evidence of some relationship between social stress and psychiatric disorders, as well as with other diseases, has been inconsistent, but since such stress appears to be a factor contributing to the rate of medical consultation, it is essential to separate the role of stress in the aetiology of a condition from its influence on seeking professional help. In spite of some strong statistical associations, the aetiological significance of social stress in psychiatric disorder remains to be clearly established, but this work does provide a possible basis for focussing preventive and therapeutic inventions more effectively within defined areas.[11]

Most studies which have attempted to investigate social causes of mental illness have used social stress as a frame of reference; Murphy[12] conceptualised this as including social forces or events which threaten psychological well-being through the failure to satisfy needs. These needs may be operationalised as including creature comforts, affiliation with others, coherence in values, social role training, self-expression, and recognition of the individual as a unique and valued person. Stress theories used in psychiatric epidemiology reflect the view that mental health is promoted by social situations in which there are "adequate" resources for such needs, and though these theories remain controversial, they have not yet been replaced by others which are more solidly founded. Since research on cultural differences has produced very little evidence of a relationship with differences in levels of psychiatric morbidity (though the expression of this morbidity may vary), Murphy concluded that if social stress has anything to do with mental illness, it must be of a type which is ubiquitous in human populations. It appears, in fact, that no group of people large enough to be self-perpetuating as a culture is free from psychiatric disorder or lacks a view of mental illness.

If social stress is a useful concept, it is almost certainly relevant to the situation of these living in the inner city areas of Britain (about four million people) and of other older industrialised countries. Such areas are characterised by:

> falling population, increasing concentration of the poor and those of lower status in society, a predominantly poor housing stock and neglected environment, subjected to large-scale physical change, and virtual abandonment by economic activity and productive investment.[13]

These processes are part of a cycle which began with the Industrial Revolution, 200 years ago, and which now involves the collapse of the economic base which led to the establishment of these cities as places where people came to work. The long-standing residents of inner cities undoubtedly feel powerless to influence such overwhelming changes in the economic pattern of their society, as well as "the uncertain and deteriorating future of their area."[14]

Whether a useful distinction can be made in these circumstances between social stress and social disintegration is uncertain; furthermore similar processes can be seen in rural areas which are affected by migration, economic decline, etc. Ridley[15] regarded young people in inner city areas as subject to passive alienation—"A generation idle and frustrated because unemployed; rejected by employers, thus alienated; concentrated in certain districts where the environment itself is grim"; they must inevitably be at high risk of involvement in crime, extremist politics, or simply random violence.

Social change, especially if rapid, has often been considered a form of social stress, though Ruesch's[16] opinion still has validity, that "the tolerance limits for social change show great individual variation, and what causative role social change plays in the aetiology of mental disease has not yet been clearly established." Selye[17] had proposed that the phenomenon of change itself, beyond the capacities of the organism to adapt—irrespective of the results of the change—is the critical precipitant of stress-related pathology, whether physical or psychiatric. However, Brenner[18] added that deleterious life changes should be regarded as more harmful because they are capable of producing stresses which in turn lead to other life changes and stresses ("acceleration of stress"). From analysis of macroeconomic data at the national level, he concluded that abrupt economic changes, regardless of direction, are stress-provoking, but that undesirable changes such as unemployment and income loss are substantially more pathogenic. This appeared to confirm Selye's view that extremely high stimulation has an overall stressful effect, though the whole issue remains a controversial one, and Brenner's economic modelling has been strongly criticised.

There is also the likelihood that social changes may become cumulative and self-accelerating; Myrdal[19] states that it is usual for supporting processes to emerge, rather than countervailing ones, thus moving the change further in the same direction. Involuntary moves of residence may represent an important example of social change which is likely to have adverse mental health consequences. Social-cultural change may or may not disrupt the integration of a society, depending on the relevance of this change to the self-esteem of its individuals; if demoralisation is caused, e.g. through damage to the symbols representing that society's dominant sentiments, then distress is likely to be widespread. This may be relevant to the disintegration of the former Soviet Union.

Reviewing the social stress concept, Dohrenwend[20] emphasised highly consistent relationships between the estimated prevalence of various psychiatric disorders and three sociocultural factors—sex, urban/rural residence, and social class. These relationships might be explained if the factors caused people to be exposed, but there were great gaps in the evidence linking variation in these factors to variations in the stress to which people are in fact exposed, and in the evidence on how types of stress are related to disorders in individuals. Some support for this view of the effects of environmentally induced stress can be seen in the sequelae of extreme situations (e.g. disaster victims) and in the relationship of stressful life events to psychiatric disorder. Also, evidence for genetic effects in the genesis of schizophrenia leaves much of the variability still to be explained, but this argument by subtraction does not prove that environmentally-induced stress is important aetiologically in its own right. Dohrenwend concluded that existing evidence failed to demonstrate that sociocultural and social-pyschological factors are important in the aetiology of major psychiatric disorders, but in view of the complexity of the issue and the often insidious onset and long duration of the conditions involved, this was not surprising. Much more detailed investigations would be needed, for instance to separate the influence of biological from social differences in variations of morbidity between the sexes.

This impressive analysis is rather flawed, however, by the fact that one of the three main props of the argument—differences between overall urban and rural psychiatric morbidity—no longer seems valid. Cassel[3] suggested that those involved in rapid social change become unfamiliar with the cues and expectations of their society, find that their actions are unlikely to lead to anticipated consequences, and experience stress, which leads to neuroendocrine disturbances, and so to increased susceptibility to disease. This model remains to be tested.

1.2. *Social Class*

Closely related to social stress is the possible influence of social class. From their survey of the literature, the Dohrendwends[21] concluded that a higher rate of psychiatric symptoms was to be found in the lower socio-economic classes, and that it resulted from the response of their members to relatively frequent and severe stresses, which are an inevitable part of life at that level; but this was mainly an urban phenomenon. Working-class people are particularly subject to stresses of a developmental nature, e.g. disruptive effects from high rates of births, family deaths, and marital disruption. In addition, they have to adjust to stress without the resources and social support available to the middle class; for instance, the fact that the family unit is often broken or discordant may make it a stressor rather than a resource. In fact, any excess in rates of psychiatric morbidity is concentrated in the very lowest socio-economic class, and it is there also that a higher prevalence of weak and unstable social network ties is found.[22] In Leighton's view,[7] life in the lower social classes is associated with stress, which is essentially that of living in a socioculturally disintegrated environment, though once again, this would only apply to the bottom of the socio-economic scale.

Although it has become a form of conventional wisdom that low socio-economic status is associated with a higher prevalence of psychiatric disorder—Brenner[18] describes this as "the single most consistent empirical regularity in the field of psychiatric epidemiology"—there is no agreement as to whether social stress or selection would provide the primary explanation for this. In other words, do genetically vulnerable individuals accumulate in the lower socio-economic groups, or do the effects of adverse environmental and social factors correlate with low socio-economic status? The Midtown Manhattan study concluded that both social stress (childhood deprivation) and social selection (inter-generational mobility) contribute to the strong inverse relationship between psychopathology and social class, but it would be just as plausible to substitute genetic predisposition for childhood deprivation.[20] There are also diagnostic differences; schizophrenia and personality disorder are consistently found to be at their highest rate in the lowest socio-economic class, but this is not the case for neurosis or for manic-depressive psychosis. For personality disorder, it is not clear how important is social selection, relative to environmentally-induced stress, in determining the relationship with social class.

Whilst one may assume that crowding occurs when the personal space of the individual is invaded, the nature of that stress is yet to be determined. The body-buffer zone of the individual is a mobile territory—the equivalent of the individual distance in ethology—though this use of the term "territory" is by analogy, and may tend to reify it to a greater extent than is justified. The invasion of this personal space, however, is likely to provoke a more intense response than that caused by invasion of the core area, as typified by the home. Personal space is indicated by decreasing eye contact as proximity increases,[23] followed by an increased sensation of discomfort and probability of flight. Explanations for the effect of crowding include: sensory overload (leading to dominance hierarchies); a reduction in choice; humiliation from the lack of a fulfilling role (as there is competition for the few tasks available); and anxiety caused by invasion of instinctive boundaries. Schmidt & Keating[24] regarded crowding as an attribution of settings in which there is a loss or lack of personal control. Implicit in most of these theories is the view that crowding in itself does not cause the stress, but provides an environment where these stresses are more likely to impinge; however, this view is not universally shared.[25] In an extreme situation, if a prison houses one man to a cell and communal activities are forbidden, the stresses of crowding are likely to be replaced by those of isolation. Baron[26] states that field studies provide no indication of a link between crowding and crimes of violence, but the failure to find this may result from the complex nature of environment-behavior interactions.

Stokols[27] divided crowding experiences firstly into personal, or neutral, where the stress is seen as an inevitable consequence of restraints within the environment, and secondly into those taking place in the primary environment, like the home or place of work, or in the secondary environment, such as the train, street, or theatre. This would separate four stress areas, e.g. marital conflict (occurring in the personal primary area); conflicts over the sharing of rooms, baths, etc. (in the neutral primary area); mugging (in the personal secondary area); and queuing (in the neutral secondary area). High density would exacerbate the latter two conditions, whilst crowding within the home would do so with the former. It is likely that such stresses would have different effects on different individuals, eliciting emotions ranging from fear to anger, or even resignation. At this point, however, the huge variation between individual personalities and coping skills makes questionable the attempt to relate such specific stress to individual effect.

2. ENVIRONMENTAL STRESS

Virtually every type of stress can be said to be environmentally engendered, i.e. resulting from an imbalance between the demands of the environment and the individual's ability to adjust to them. Assuming, on the basis of biological theory, that an organism strives to maintain a state of homeostasis with the demands of his natural, social, and cultural environments, should these demands exceed his ability to adjust to them, he will experience a stress reaction or disequilibrium. This can take many forms, including anxiety, fear, defensiveness, and avoidance behaviour. Disequilibrium implies a threat situation, which the individual will endeavour to reduce to the point where the environmental field is seen as "problem-free", and these adaptive reactions of the organism, in response to a stimulus overload from the environment, may embody stress. As Howard & Scott[28] suggested, "Stress occurs if the individual does not have available to him the tools and knowledge to either successfully deal with or avert the challenges which arise in particular situations," but unfortunately, the stress concept has seldom been rigorously defined in a scientific sense.[29] Rather, it operates as a heuristic device to provide a framework within which to study emotional disturbance (and other behaviour) from diverse perspectives.[30]

Demands or threats, and the stress to which they give rise, derive from the interaction of the individual and environment, but whether a particular situation is conducive to stress depends on the individual's emotion health, vulnerability, and personality structure. Continued exposure to the wear and tear of even mild stressors can result in development of one of the emotional, behavioural, or physiological "diseases of adaption" because of the natural reaction of the human organism to prepare for a fight or flight when faced with a threatening stressor. If he flees or fights, the physiological system will afterwards return to normal, and the stressful situation has been coped with. This allows for a natural cycle of reaction followed by readjustment to any unusual demands. The problem in contemporary society, however, is that very few stressful situations allow for either fight or flight; the stress is not dissipated and the physiological reactions accumulate, to the detriment of the organism.

Both life-threatening and more chronic psychosomatic disorders have been linked to stress, and in particular to occupational stressors. Though it would be ideal to be able to associate each stressor with a specific problem or condition, the phenomenon appears at present to be largely non-specific, in both its determinants and its consequences. The syndrome can be activated by any number of factors, and the reactions to stress vary over a wide range of social and psychophysical problems and diseases.[31] Rabkin & Struening[32] stressed the degree of evidence for the aetiological relationship of excessive stress to pathology, and considered that the stress itself is the product of socially structured situations and environmental factors. They state that the aetiological model of the onset of illness is generally associated with a number of potential factors, which include "stressful environmental conditions, perception by the individual that such conditions are stressful, the relative ability to cope with or adjust to these conditions, genetic predisposition to a disease, and the presence of a disease agent." They regard stress as a general concept, describing the organism's reaction to environmental demands, and its utility as deriving from its role in identifying productive lines of research on the aetiology of disease, which encompass external events that influence individuals and populations. In Detroit, Kasl & Harburg[33] devised an Ecological Stress Index, based on socio-economic level and measures of instability-disorganization, for each census tract. Differences in stress levels were not found to be related to differences in indicators of mental health and well-being, though these latter did have some relationship to variations in perception of the urban environment. However, limitations in the experimental design reduced the significance of these findings.

It is widely recognised that chronic stress is normally engendered, not by a single traumatic event, but rather by a series, any one of which may be temporarily annoying, but which together combine to induce a state of morbid anxiety or depression. Thus it would be advisable to concentrate on those potentially stressful factors which endure over time, and from which there is little possibility of escape. Amongst these, environmental conditions may be an incessant source of irritation, are often difficult to avoid, and concomitantly less amenable to improvement.

An example is rapid social change; Brown *et al.*[34] concluded from their comparative studies of Inner London and Hebridean Islands that the main psychiatric impact of industrialisation and urbanisation has been in causing a very high rate of depression among working-class city women. Middle-class women there have largely avoided this, tough they experience more depression than women in a traditionally structured community. Within the Hebrides, there is a higher rate of depression among certain groups of women, who have moved further from a traditional way of life, and this seems to be linked to the same kind of provoking agents that are found in the inner city, particularly among working-class women.[35] However, it was concluded that "a simple urban/rural dichotomy does not give an accurate picture of the differing rates of depression within the two populations." Another illustration of the effects of rapid social change came from a longitudinal study of drinking practices in two communities of the Shetland Islands since the development of the oil industry there.[36] Increases in mean alcohol consumption were found to be twice as large in an area directly affected by the oil developments as in a central area, which was in a protected zone, where any such developments were banned. However, in both communities, increases in drinking were largely concentrated among those aged under 30.

Another important aspect is in relation to population density. Miller[37] concluded that for each individual, there is an optimum range beyond which, in either direction, he begins to experience stress. This range varies according to such factors as personal characteristics of the individual, cultural aspects of his group, and sociological characteristics of his environment. The disruptive effects of extreme density occur within a narrower range for some psychological-behavioural dimensions (e.g. affective interpersonal behaviour) than for others (e.g. task performance), while for a given individual or sub-population, the range of non-stressful densities is more restrictive than that for a large population. Moving an individual or sub-population along the density dimension to a point outside the usual experience creates stress, which is relieved by adaptation over time. Miller concluded that the exact relationship between stress and feeling crowded, or between feeling crowded and pathological behaviour, is unresolved.

A dramatic illustration of the effects of stress on physical and mental health came with the Athens earthquake of 1981.[38] Though several lines of evidence had linked fatal heart attacks, and particularly sudden deaths, with acute psychological stress, there had not previously been an adequately controlled epidemiological study demonstrating the association in a human population. In this case, a very large population, with adequate health and registration services, was suddenly and simultaneously exposed to events that invoked the danger of death or serious injury. It was found that earthquake-related stress increased the short-term probability of a fatal cardiac event by about 50 per cent, and the short-term probability of death from atherosclerotic heart disease as underlying cause by about 100 per cent. The peaking of the excess of cardiac deaths during the third, rather than the first day after the major earthquake, was thought to be related to lesser quakes having occurred on the second and third days. The symbolic danger of these may have generated stress as effectively as the real danger, because conditioned reflex pathways had become established. The facts that deaths from quake-related cardiac events were more common among men than women, and among subject aged 69 or less than among the older, were also thought to confirm the general interpretation of the findings.

Though there has been a general tendency to regard stress as primarily an urban phenomenon, Keller & Murray[39] point out that:

> "Rural people have always been stressed by the environment in which they live. Lack of control over (it) is keenly felt . . . (they are) ultimately dependent on the weather for their success or failure. Even without bad weather, isolation may lead to loneliness, depression, a sense of helplessness, and loss of interpersonal skills."

They added that further emotional stress may result from the challenges which modern communication media bring to rural values and life-styles, particularly causing conflict in the young, and that "rural life is complicated by many significant social, economic, and environmental problems that take a heavy toll on those who live there." It would therefore be simplistic and misleading to regard environmental stress as particularly characteristic of either urban or rural life alone. De Vos[40] stated that one of the main difficulties of theories which regard the transition to urban life as stressful is not so much an incorrect view of city life as "a mistaken and romanticized image of what village, rural, or tribal life is typically like."

Physiological changes produced by noise consist of non-specific responses, often associated with stress reactions.[17,41] These include changes in electrodermal activity, cardiovascular activity, and hormone secretions; they can affect respiratory and digestive systems. An extensive review of relevant studies was carried out by McLean & Tarnopolsky;[42] it should be borne in mind that the majority of these studies have involved short-term exposure to relatively high sound levels in the laboratory, so that extrapolation to environmental noise effects must be tentative. "Noise annoyance" consists of the disturbance of normal activities such as speech and sleep,[43] but reactions to environmental noise depend as much on the person as on the type and level of noise. Noise worsens the performance of complex and intellectual tasks, increases anxiety and aggression, and reduces the frequency of helpful interpersonal behaviour. Sensitivity to noise appears to act as a non-specific marker for increased vulnerability to other stresses in the environment.[44]

Miller[37] also used a stress model to analyse the psychiatric implications of geographical mobility in terms of four main stress points: antecedent events, which have not been significantly studied; separation, which is unlikely to have serious psycho-social sequelae if the migration is unforced, and if the lost social organisation can be replaced; the process of moving, which is relatively tolerable intranationally, but may be associated with greater stress internationally; and re-establishment, when the degree of stress depends on many factors, both in the individual, his family, and the host community. Miller concluded that the relationship between psychiatric morbidity and mobility behaviour is a complex one, and that the two main explanatory hypotheses are both much too simplistic; these are "social causation" (that migration is more stressful than non-migration) and "social selection" (that the mentally ill migrate in greater numbers than the non-mentally ill).

It could be argued that in many cases, the stresses generated in new environments encountered after migration merely trigger off disorders in these susceptible populations.[45] Certainly, some groups of migrants have a pre-migration history of psychiatric illness.[46] Although the majority of migrants move voluntarily, usually motivated primarily by economic and vocational considerations, many have done so unwillingly; studies in Australia, for instance, have shown that wartime and political refugees from Eastern European countries were particularly vulnerable to many severe forms of psychiatric disorder.[47]

For many migrants, the stresses generated by actually settling in the new environment of the receiving countries are capable of precipitating psychiatric disorders. Such stresses are clearly greatest for those migrants who differ most markedly from the host populations in terms of such important variables as colour, language, and religion. However, the often traumatic and stressful experience of settling in the new country generally dissipates after a few years, resulting in lower levels of illness,[48] though for some groups, the problems appear to persist. Thus in Victoria State, Australia, Krupinski[49] found that high rates had persisted among middle-aged housewives from southern European countries; although they had been resident in Australia for more than seven years, they were housebound and had not learned English, whereas their husbands and children had become assimilated. A second high-risk group identified by Krupinski comprised well qualified professional immigrants, who had experienced a fall in both social and occupational status because their qualifications had not been recognised by Australian professional bodies.

Although the conceptual and methodological frameworks of these behavioural studies are quite diverse, most authors emphasise the importance of the role of stress in inducing residential moves, even among voluntary movers. For some social groups, however, the levels of stress are clearly much greater than for others. Thus the residents of older, rented, inner-city housing are particularly affected by the precipitants which generate involuntary and forced moves. Several authors in the US have attributed the higher frequency of local, inner city moves among blacks than among whites to the additional constraints induced by living in ghettoes.[50] These moves have been interpreted as largely vain attempts by black residents to find satisfactory neighbourhoods and housing milieux within the predominantly poor quality rented housing stock found inside the ghetto. The stresses induced by such constraints are particularly marked among those blacks who have high ambitions to improve their educational and socio-econmic status. The subject clearly deserves further research, for even those few blacks who have managed to move upwards socially and outwards spatially from the ghetto apparently experience even greater levels of stress (as indexed by mental hospital admission rates[51]) than the majority of their brethren, who remain trapped in the ghetto.

In Thamesmead, a very large new housing development in South London, Higgins[52] calculated that if it can be assumed that those who moved out within three years were twice as likely as the rest to have been affected by environmental stress, then of the 747 families who moved in during 1969 and 1970, there were 75 in which at least one adult had an emotional disturbance related to environmental stress. This is likely to be the minimum figure, and includes only those who were distressed enough to seek help from a doctor. The most serious stress appears to be financial, due to high rents; even those not disturbed are recorded as troubled by this. The physical environment was occasionally held to blame, but in most cases the stress was due not to the problem itself, but to frustration; either the people responsible could not be identified, or they could be identified but not induced to respond.

Thus, the main causes of stress lay in the social environment. In some instances, the adverse social circumstances could not have been prevented. No-one could really be held to account for the effects of a disturbed family on its neighbours. The GLC was not responsible for the system of support for those in need; its component parts had been developed by different authorities, in different circumstances, over the years. The system was—and still is—distinguished by its complexity, its lack of coordination and by the burdens it puts on those most needing help.

3. THE STRUCTURAL ENVIRONMENT

The relationship of bodily health to the structural environment is now reasonably well understood in terms of water supply, sewage disposal, protection from weather, etc., though unpleasant surprises such as Legionnaire's disease may still emerge from time to time. But the complex and interrelated causes of those psychological and social pathologies which seem increasingly to accompany mass urbanization have yet to be fully identified.

The social environment has been defined by Burke[53] as consisting of "the norms, values, customs, fashions, habits (which might include work), prejudices and beliefs of a society." Though these vary enormously from one society to another, "in each one, their profile will be more or less supportive of health. They are modulated through the mass media . . . and institutionalised in the family, the community (which may be defined ethnically as well as geographically) and the nation."

"Mental health" can be defined in innumerable ways—most of them too vague or idealized to be of much value. Practical considerations, though, require it to be used primarily in terms of the levels of emotional symptoms such as anxiety or depression which people report; if environmental changes in an area resulted in significant reductions in the levels of these complaints, then it would be reasonable to say that the mental health of the population had improved.

"Social pathology" is usually taken to mean significant deviation from prevailing norms of behaviour, expressed in such forms as delinquency and crime, disordered or broken family relationships, and addictions. Most people in any population who behave badly are not psychiatrically disordered, though the exact proportion who are is unknown; but it appears that communities with high rates of mental illness do also tend to show high levels of behavioural deviance and of psychiatric sequelae to crime and social disturbance.

On the other hand, it would not be true to say that urban living as a whole is less healthy in psychiatric terms than life in the country,[54] because as units of analysis, "urban" and "rural" are now too large and heterogeneous to be meaningful. The characteristics of particular environments therefore need to be specified more clearly, if their possible effects on mental health are to be analysed.

There may, however, an optimum size beyond which the costs of cities start to outweigh their benefits, resulting in an escalation of levels of social pathology, particularly crime.[55] The sheer size of today's urban settlements may involve what Cappon[56] has called "the destruction of whatever natural order of territoriality was established by historical and social usage"; for example, Mexico City has already passed the size of any previously known metropolitan area. From the anthropological point of view, Fox[57] has pointed out that "We are certainly evolved to be gregarious, but not in nations in 500 million or in cities of 15 million."

In many industrial cities, "inner-city malaise" has been described in recent decades. A British Government policy statement of 1977[58] identified the common factors underlying this phenomenon in different places as: decayed infrastructure, bad physical conditions, high levels of unemployment, limited job opportunities, and a concentration of people with social difficulties. However, attention to inner cities should not cause the problems of peripheral public housing areas to be ignored: these one-class ghettoes often suffer from shoddy buildings, lack of social or shopping facilities, and isolation from employment and civic amenities. Demoralisation, substance abuse, family breakdown, crime, and alienation would not be surprising in such circumstances—and these are often described there.

3.1. *The Inter-Relationship Between Mental Health and Planning*

Those responsible for the building of large structures, transport developments, and changes in the pattern of industry and commerce have had a general lack of interest in the human consequences of their actions. Economic considerations have been overwhelming, irrespective of the political ideology with which any society operates.

One of the most important urban processes that is relevant to mental health is redevelopment and residential relocation. In the UK, this process began on a large scale after 1918, and resulted in four million homes being built within 20 years, which housed over eight million people, mainly of upper working- or lower middle-class social status. The density of this new environment was on average about one-third that of the old, so that the urban areas exploded in size. In this process, the standard of domestic facilities such as bathrooms and hot water systems became generally much higher.

However, there was a social price to be paid, in that these new residential areas lacked the established kinship and mutual-help networks of older communities, while most new occupants tended to place greater emphasis on their domestic privacy, in place of social interactions with those living around. At the same time, local building styles and materials were abandoned, so that traditions and landmarks tended to become submerged; identical streets of small houses, with their nearby shopping areas, could now be found from one end of the country to the other.

In 1938, the description of "suburban neurosis" by Taylor[59] recorded the stresses which were commonly occurring after removal from more central areas, e.g. higher expenses, social isolation, distance from employment, and loss of familiar surroundings. Similar findings emerged in 1957 from a comparison by Martin *et al.*[60] of a housing estate on the edge of London with an older working-class area: it was suspected that this was partly due to the shock of rehousing and partly to the poor social facilities on the estate, which tended to provoke loneliness and social isolation. The way in which a new housing estate was built in Liverpool without any essential services, but with design faults which made life almost intolerable for so many of its residents, has been described by a family doctor who worked there.[61]

Hansen & Hillier[62] point out that such developments were designed on the assumption that the undesirable features of the old communities could be planned away, and that life would be better in an environment consisting only of adequate housing, together with (eventually) a few essential shops. Here, the planning cliché of separation of land uses had been fully worked out. But in fact, virtually none of the anticipated benefits of this change have been shown to occur: these housing estates have no meaningful integration into the wider community, and even within them, the layout seems to have provoked more suspicion than sociability.

Although three million people were relocated from old urban areas in the UK in the period 1955–75 (half a million leaving London alone within ten years), hardly any research has been carried out into possible adverse psychological and emotional effects of this process on individuals, or of social effects on the areas they left behind. That such effects might have occurred is very likely, considering that many of the cleared areas were long-established communities, with a characteristic culture of their own and important social networks, which could never be reproduced artificially. It has become increasingly recognized in recent years that the social support derived from these networks is of fundamental importance in the maintenance of mental health.

Thus, the demolition of a neighbourhood is not just the destruction of buildings, but also that of a functioning social system. While "slum clearance" was a powerful slogan in achieving the removal of often unsatisfactory old environments, it ignored the fact that most of the social problems found in slums were not directly related to their physical structure. Furthermore, the redevelopment process—from first plans to resettlement of all the residents—is usually a long one, during which time people are worried about their future, watch the decay of the familiar neighbourhood around them, and may finally be moved against their will to a home they do not want.[63] In this process, the boundaries and spatial arrangements which different ethnic or social groups have established over time are wiped out, and with them often their *modus vivendi* with society in general—having predictably adverse effects on the social order of cities.

While inner city areas in the UK and other Western countries have been largely emptied of their previously established working-class residents, much of the accommodation that remained has been reoccupied, predominantly by members of minority races or by people with socially marginal characteristics, such as alcoholics, drug addicts, or those with severely disordered personalities. During this process, the nature of large urban communities has changed fundamentally, but in such a way as to foster the development in many places of major social problems, particularly as it has coincided with their economic decline and deindustrialization.

Dispersal of urban communities was promoted first, by an assumed desirability of low density (though it had never been shown that high density *per se* is harmful); secondly, by a belief in "shortage of land" (though all cities contained enormous unused or partly used areas), and thirdly, by the assumption that old houses should be demolished rather than rehabilitated (though no evidence existed that this was in fact preferable).

Grandiose road developments were also ruthlessly imposed—and continue to be imposed—on (mainly poor) communities, bringing them few if any benefits, but enormous social costs. In tearing apart compact and human-scale environments, highways caused the sub-division of urban areas into artificial and now segregated cells, between which movement became very difficult, particularly for pedestrians or cyclists. Whereas it is the conventional wisdom that there is never enough public investment to undertake the changes that would be desirable, this was in fact a case not of too little being spent, but far too much. Not for the first time, such policies tended to have the opposite effects to those claimed for them.

The UK also carried out probably the world's largest programme of planned new town construction, following a tradition begun early in the century at Letchworth, Welwyn, and Port Sunlight. The most significant features of these new communities were that they embodied a conscious element of social engineering, and were designed to become balanced and self-sufficient, rather than merely dormitory areas. Through their choice of residents, they were protected from the worst social problems of the inner cities and from the burdens imposed by the elderly and chronic sick—though such cases must steadily increase in number with the passage of time.

In general, the British new towns were designed as multiples of neighbourhood units, and it was assumed that these would establish local social values and a positive feeling among residents that they had a relationship towards their unit. However, we have no means of knowing whether or not this formula achieved what was expected of it, because the matter was never investigated.[64] In fact, this huge national investment was accompanied by hardly any interdisciplinary scientific effort to study its human consequences, so that no confident statement can be made as to whether or not it was generally worthwhile. Without such information, there is no way of assessing the "opportunity costs", i.e. the possible benefits that might have been achieved by spending the same amount of money in different ways. (For instance, was it right to have built the new city of Milton Keynes so near to London, rather than making the same investment in the rehabilitation of inner cities or the north-east of England?)

Many factors have to be included in the balance sheet of new towns, expanded towns, or peripheral housing estates, but it is very likely that a large proportion of the people who migrated to them did so reluctantly and out of economic necessity—i.e. they could not find a more central home at a price they could afford. If they had not had to migrate, kinship and other social networks would have been preserved, many established urban communities that no longer exist would have retained a healthy social and economic life, and much stressful daily travel would not have become necessary.

In the UK and some other countries, one of the foremost urban policies of the post-war period was the construction of high-rise residential blocks as a "quick technological fix" for the housing shortage. This was carried out in the absence of any systematic enquiry as to what kind of accommodation people wanted,[65] and in combination with the use of untested industrialized building methods, nearly all of which have proved to be disastrous. There has been little systematic research into the effects on mental health of high-rise living for the working class, but the balance of evidence suggests that this has been adverse for families with young children, and possibly in some other cases.[66] Possible explanations for this effect include poor construction, lack of sound insulation, poor maintenance, difficulty in monitoring the surrounding space, difficulties in informal interaction between residents, inability to make physical changes to the home, lack of opportunity to participate in management of the block, and unpleasant appearances, both inside and out.[67]

These environments are also lacking in shelter (which is hardly appropriate for the UK climate) and subject to the gusting of high-speed winds, causing both psychological and physical stress, particularly for those in poor health and for the old and very young. Was any thought given by architects or planners to how a woman pushing a pram, and with another small child, was to cross a footbridge which has stairs up and down to it, or get up to a 20th-storey flat when the lifts were not working?

Many cities lack effective planning authorities or the means to implement their decisions; in England, the authorities carrying out these functions at a strategic level in metropolitan areas have been abolished. Yet, as Sherlock[68] points out, without such planning, "the problems of urban transport and land use will never be solved. As a result, urban living is likely to become even more unattractive and, if this drives people out of urban areas, it will take many of the urban problems, including traffic congestion, to the surrounding countryside."

3.2. *Indicators of the Psycho-Social Environment*

Planning authorities and local government have had an overwhelming preoccupation for many years with reduction of urban population density. Choldin[69] has pointed out, though, that "It is more important to recognise that high density is a basic feature of the urban community and to discover ways in which to organise community life and to build satisfying environments."

It is even more important to recognize that relatively high density is actually essential for the positive qualities of towns and cities, such as cultural life and specialized professional services; the low densities of suburbs make these facilities difficult to provide without a degree of car travel which is environmentally damaging and unjust to the more disadvantaged sections of society. Newman[70] has drawn attention to the way in which suburban communities endlessly reproduce the same relatively low level of collective amenities, while leaving behind "the decaying remnants of a once rich urban environment," in their flight from the pathology of the inner city. A principal casualty of this endless process is "the disappearance of the urban life-style."

Density-pathology research has mostly assumed that dense living conditions have "pathological" effects on the inhabitants of urban areas, but Magaziner[71] points out that no consistent or linear relationship has been demonstrated between the two. Living alone, which is "the extreme of in-unit density, must be viewed as separate from the continuum of low- to high-density environments." His own data from Chicago (where the greatest amount of research into this subject has been done) show that density outside the dwelling units interacts with the in-unit density to promote or prevent psychopathology. Thus, if opportunities for social contacts are readily available in the neighbourhood, living alone need not cause social isolation. This joint effect was more apparent in the older age-groups, presumably because these are generally less mobile, whereas younger people can usually obtain their social contacts more widely, unless prevented by poverty or lack of public transport. Mitchell[72] has pointed out that density (number of persons per dwelling unit) is different from crowding (excessive demands on available space), which is strongly influenced by psycho-social factors. Except in extreme cases, the amount of space per person is less important for mental health than lack of personal control over it, with the accompanying disordered social relationships.

Housing may be regarded as that part of the physical environment which is most directly relevant to mental health; since it usually contains the immediate family, it also relates to the most significant aspect of the social environment. However, relatively little is known in scientific terms about the direct effects of any specific aspects of housing on psychiatric morbidity. One fact which is clearly established, though, is that there is no simple dichotomy whereby unsatisfactory physical conditions impair only physical health, whilst poor psychological or social conditions impair mental health. In fact, the two constantly interact. Much of the research that has been undertaken in this field had focussed on residential movement, but this is far from ideal, since moving home tends to be accompanied by changes which were the reasons for the move, rather than its effects. Examples of these factors include changes in economic status and moving into a different age-range.

The most important investigation up to now into the psychiatric sequelae of relocation is that of Fried[73] on people who had been compulsorily moved from the West End of Boston—a long-established, mainly Italian, working-class community. About 25% of the total sample still felt significantly depressed about this move two years after it occurred, another 20% had such feelings for at least six months, while less severe responses were reported by many others. The two most important components of this grief reaction were fragmentation of people's sense of spatial identity and dependence of group identity on stable social networks. In such a working-class community, the sense of continuity—which is believed to be an important factor for mental health—appears to be strongly related to the external stability and familiarity of places and people. It seems likely that an integrated sense of spatial identity is fundamental to human functioning and that, in the working-class at least, this is generally tied to a specific place.

However, the bereavement-type reactions which have been shown to follow such enforced removal from a well-established community would presumably not apply to the areas which are undergoing rapid social change of an adverse king, or where there has been a breakdown of law and order. Similarly, in the case of people who have marked neurotic or personality problems, a new housing environment may represent little more than a matrix for the projection of long-standing difficulties.[74] Unless adverse practical problems become overwhelming, most families eventually adapt to most new residential environments, but this still leaves a substantial minority who do suffer distress from such changes, and the distress should be largely avoidable.

Another aspect of the environment which needs to be examined is noise. The conclusion of the Fifth International Congress on Noise as a Public Health Problem was that this was not to be considered merely as a nuisance, but as posing a threat to human health—mental health is the aspect likely to be affected first. However, there are great cultural differences in the tolerance of noise, and threshold levels beyond which noise would have various pathogenic effects are difficult to define. The city of Barcelona has succeeded in reducing average traffic noises by 6 dB(A), but it is impossible to assess whether this has improved mental health, particularly in the short term (Information Note EH6).

3.3. *Promoting Mental Health in Cities Through the Social Environment*

The Ottawa Charter states that health promotion actions include the building-up of a social and physical environment which is supportive to health; to achieve this, intersectoral collaboration is needed. "Supportive environment" includes both physical and social environments, but with particular emphasis on the pre-requisites of ecological stability and resource sustainability.

Up to now, any suggestions for the reform of society by the application of medical (and specifically psychiatric) knowledge have had unimpressive results, probably because they have been focussed at too general and ambitious a level. Yet more modest efforts, on the basis of existing knowledge and resources, to reduce psychiatric or social pathology could be extremely worthwhile for urban communities. Certainly, cities do tend to house concentrations of multiply dis-advantaged people, who may show poor mental or physical health along with other unfavourable attributes. However, it would be quite wrong to extrapolate this negative picture to whole city populations.[75]

When prevention of psychiatric disorder is discussed, it is often assumed that this must in some way deal with "basic causes"; yet this involves a naive and simplistic view of aetiology, most of which still remains unknown. The result is a tendency to wait for the millenium, mostly because of entirely unproved beliefs that psychiatric and social pathologies are direct expressions of the harmful structure of society. There is an important social contribution to the causes of these disorders—which includes the influence of the physical environment—but the inter-relationships of the various aetiological factors are highly complex and often subtle.

Although the relationship between physical surroundings in general and mental health lacks reliable data, it is not unreasonable to believe that an attractive environment will tend to make people generally happier than they would be in an unattractive one. There is evidence, though, that dimly lit and hostile-looking constructions which lack the human scale of more traditional environments are likely to make many people feel anxious or afraid. This is not without reason, since the danger of harassment or attack is far greater in areas with footbridges, subways, multi-storey garages, and empty spaces between high-rise blocks.

The direct influence of particular structural forms of urban development on mental health has yet to be established, but it seems likely that there are human advantages in such traditional forms as arcaded streets or enclosed markets. For instance, many agoraphobics feel less anxious if they go round the edge of a large open space, while people often experience anxiety on pedestrian bridges over main roads, or when a lift opens onto a dimly-lit balcony. "Architecturally over-determined" new environments have generally broken away from established patterns of movement, even abolishing such a fundamental scene of human life as the street. In these environmental changes, the formerly rich pattern of interaction that occurred around homes, local shops, and cafés or bars—which were an important source of social support—have been largely destroyed. That kind of social support is widely believed to be an important underlying factor in mental health. Faria[76] has suggested that mental health promotion should include the reinforcement of social networks that might help individuals to cope with the more distressing situations.

We await more systematic information, in an area that has so far been subject to little scientific enquiry. However, it would not be unreasonable to maintain that mental health is unlikely to be promoted by incomprehensible urban sprawl, severed by dangerous motorways, and full of monotonous blocks, with unwelcoming spaces between them. There are also the likely ill-effects of population dispersal, such as the time and energy wasted by millions of people every day in commuting, together with the stress that must come from this frustrating activity. As a consequence, city centres are deserted at night and weekends and suburbs empty during the weekdays, causing further undesirable social and psychological effects.

More recently, the construction of orbital motorways around the outer suburbs of cities has produced a new generation of suburb-to-suburb traffic, pulling industry, business and even professional activity out of the central areas. Promoted as a solution to urban traffic congestion, these developments tend to create a "suburban glue-pot" of new traffic movements, while accelerating the decline of city centres, and making suburb-to-suburb public transport uneconomical through loss of passengers—again harming those with least resources. Thus, those who are multiply disadvantaged become ecologically trapped, unable to work because employment has moved to distant suburbs or become strung along motorways.

This dispersal and areal differentiation of human activities also makes the control of deviant behaviour more difficult, since people are constantly on the move (except for the poor and the handicapped), and the unofficial monitoring that neighbours and workmates carry out in less complex societies is no longer possible. Instead, a concern for people's mental health would suggest the need to restore to large human settlements the benefits of urbanity—of a social matrix that would provide the environmental structure in which a worthwhile quality of life and work could grow. For instance, much of the old commercial centres of Manchester and Liverpool are empty, yet still consist of soundly constructed buildings; to provide the resources to convert these to residential use would be little problem for institutional investors. The social benefits would be immense, while in the long run even the financial returns might be hardly less favourable than some of those obtained from present investment policies. Regrettably, there is so far little sign that such considerations carry any weight.

The International Hearing on Achieving Better Living Conditions in Towns, held in Strasbourg in October 1988, concluded that a mixed-use policy of combining housing, commerce, and business could play a significant role in reducing levels of urban delinquency and vandalism. It condemned housing policies based on demolition and rehousing elsewhere of the inhabitants, which tend to create a climate of tension and insecurity and to provoke a breakdown of collective life. Architects were considered to have a responsibility to produce a stimulating environment, where individuality can be expressed, where there is a sense of belonging, and where places are created in which the residents are proud to live.

Many cities have experienced an increasing separation of housing areas along the lines of social class, and this has a variety of adverse effects. People operating the services of central areas (transport, catering, cleaning, mail, etc.) who generally receive low earnings, have the greatest difficulty finding affordable accommodation within reach of their work. At the other end of the scale, those professionals and owners of small businesses who formerly lived within mainly working-class areas now rarely do so, so that the community's most effective leaders and spokesmen have mostly been removed. A better social mix, e.g. through the return of middle-class housing and small enterprises to city centres, could be expected to improve the morale, services, and living standards of these areas.

A number of structural features in public housing have been found to have a direct relationship with aspects of mental health. In London, Coleman[77] found that indicators of social malaise all had a direct relationship (of worsening) against an upward trend of four design variables—dwellings per entrance, dwellings per block, number of storeys, and number of overhead walkways. In New York, Newman[78] showed that the number of dwellings per entrance was the most important factor affecting the rate of crime in the block, i.e. the more dwellings, the more crime. Though crime levels are not a direct proxy for mental ill-health, fear of crime and the psychological and social consequences of having been exposed to it should represent two important elements in any overall measure of such morbidity. Coleman also found that design features which are bad from the human point of view tend to cluster together in the same scheme, and so reinforce each other. Newman's concept of "defensible space" has been of great heuristic value in considering these questions: personal control over one's home environment and its immediate surroundings is clearly important to the morale of both individuals and communities, as well as influencing their behaviour favourably.

3.4. *Psycho-Social Stress in Urban Areas*

Other than the study of Cohen *et al.*[9] in California, there have been few attempts to examine either the interplay between multiple environmental stresses themselves or the influence that these may have on health. In fact, the direct effects of such stresses on either behaviour or bodily physiology may be less harmful than the consequences of individuals' coping responses to these forces.[1] It appears that the combination of multiple stresses results in a very much greater combined adverse effect than the mere summation of the effects of the separate stresses.[80]

Though city living seems to be associated with decline of social support from neighbours and of participation in the local community life, this pattern is not homogeneous across urban populations.[81] It might be especially characteristic of areas such as the inner city or "overspill" housing estates, where there tends to be a high turnover of residents and changing ethnic composition. Within such areas, the effects of reduced supportive ties would be greater in those not in contact with people outside the immediate locality, e.g. housewives (particularly with young children at home or not working) and the elderly. These groups might not find adequate supportive relationships among neighbours, especially if neither had lived there for long. Greater urban prevalence of depression has been found not only in working-class women, but also in the elderly.[82]

Health workers have become increasingly aware of the human ill-effects of comprehensive urban redevelopment, high-rise housing, and the widespread dispersal of communities. In high blocks particularly, people complain of feeling remote and cut off from any community life, with no real neighbourly contacts, yet at the same time lacking privacy, e.g. through the absence of sound insulation. Such disadvantaged groups as the elderly and physically handicapped often find these situations hazardous and stressful. This is because of their dependence on lifts and rubbish chutes which often fail to work, because of their isolation (compared with conventional housing), and because they are often harassed by groups of adolescents who hang around public areas and are remote from any parental observation or control. The common pollution of public areas by litter, vandalism, urine or faeces, and graffiti is also a source of much distress to residents, not only because it is offensive, but also because it demonstrates the breakdown of the moral order of society—a situation out of control.[77] In such circumstances, it is surely impossible to feel any sense of comfort or familiarity about the surroundings of one's home.

A factor generally ignored in the plans of public housing authorities is that high-technology environments such as large housing blocks involve large maintenance costs; they can only function successfully when there is money available permanently to ensure that all the mechanical services are working and the communal areas clean and free from crime. In contrast, these costs are relatively small in areas of traditional housing, such as terraced streets. Furthermore, while the faults of new buildings have often remained incurable, the tenants themselves can do nothing about them; the rigidly bureaucratic structures of many housing authorities have also been incapable of responding to the volume of justified complaints from tenants, quite apart from financial restrictions on repairs. With economic recession, a vicious circle of environmental decline often begins: respectable working-class families do everything possible to find accommodation elsewhere, and tend to be replaced by families with multiple social problems. In this process, the chief sufferers are those who start by being already disadvantaged; they now find themselves trapped in an escalation of vandalism, crime, and despoiling of their surrounding environment.

3.5. *The Mental Health of Children and Adolescents in Large Cities*

The connection between high rates of social deviance and psychiatric disorder, which tends to be found in certain areas of cities, is most marked in the case of children. Rutter[83] points out that inner-city life is associated with a wide range of problems—not only crime and delinquency, but depression, emotional disturbance, educational difficulties, and family breakdown. However, he adds that it is not possible to say what are the mechanisms by which city life has adverse effects on families, and through families on children, when these are compared with children in a more rural setting. The critical factors do not appear to be overall population density, migration, urbanisation, or industrialisation in themselves: most psychosocial problems in British children are not due to lack of money or low social status, and social isolation is not common in those urban areas. In fact, the inner-city excess of childhood psychiatric disorder could be accounted for by the combined adverse effects of certain schools and certain families; no broader ecological influence needed to be invoked.[84]

Yet even in a deprived neighbourhood, more than a quarter of children suffering the combined disadvantages of parental criminality, bad child-rearing, poverty, low intelligence, and large family size showed no evidence themselves of anti-social behaviour or delinquency.[85] There is considerable individual variation, probably of genetic origin in the main, in response to environmental stresses, though of course, such stresses should still be removed whenever possible. However, it may be useful to focus help on the most vulnerable, when these can be identified. Children exposed to a single risk factor were found to be no more likely to have psychiatric disorder than children with no such factors, but when two or more stresses were present together, the risk of disorder increased very greatly, due to interaction effects.[84]

Children and adolescents have suffered particularly from the consequences of contemporary environmental changes in many cities—destruction of established neighbourhoods, high-rise housing, urban motorways, loss of public transport, etc. Sherlock[68] points out that they "can no longer play spontaneously in their street or cycle to school. They may now live in more hygienic houses, but their peer community has been destroyed. If they want to visit a friend, they have to be escorted by an adult." He adds that this sacrifice of children's freedom should have some gain to show for it, but in fact, unrestricted motor traffic has actually made the movement around of both people and freight more difficult, while public transport has also suffered greatly. Since streets have been taken over by traffic, the usual response is to provide fenced-off play areas, to which children mostly have to be escorted by adults. In fact, these tend to be either unused or become centres for vandalism, delinquency, or drug-taking.

Jane Jacobs[86] first pointed out that established neighbourhoods are self-policing, in that children outside their homes are constantly seen by passing neighbours or relatives, as well as being observed by "the eyes in the street"—people who watch the passing scene from their homes. Contact with many adults then provides a variety of role models for the process of socialisation into the society's mores. However, in redeveloped areas, particularly with high-rise buildings, this informal supervision can no longer take place, while the lack of distinction between public and private space means that learning the limits of privacy becomes largely impossible. Young people also tend to remain separate from adults in these environments, e.g. in the gang territory of empty spaces.[77] Coleman also points out that all the options available to a mother of young children in a high-rise flat are unpalatable—keeping them indoors, leaving them to play on a balcony, leaving them on a corridor, or allowing them outside, where they cannot be supervised.

For adolescents, redeveloped environments—particularly those with heavy traffic—are unfriendly and fail to provide opportunities for the kinds of social interactions and physical activity that they need. In these circumstances, vandalism, drug-taking, and the defacement of every accessible surface with graffiti are likely to be on a massive scale, compared with older communities. Such "incivilities" are especially frightening to older people, leading to a deterioration in the community's morale and so to poorer mental health.

Faria[76] has proposed that young people should constitute the priority target group for mental health promotion through the reinforcement of social networks: this action might take place particularly in schools, where attempts could also be made to strengthen the psychological, emotional, and social abilities of individuals so that they could cope better with stressful situations.

4. CONCLUSION

To promote endless urban sprawl, while obliterating communities' landmarks and replacing them with the tasteless impersonality of most present-day buildings, seems like a recipe for social disaster in the long run. However, it remains the task of scientific enquiry to demonstrate these dangers in a way that can influence both the political process and the practice of architecture and planning.

Progress in the humanising of environments can only come through scrupulous, painstaking research, based on clear concepts, and through action which is directly related to what has been scientifically demonstrated. In this process, the mental health professions could form a useful alliance with those architects and planners who have a genuine concern with the human consequences of their activities.

Note

Some material in this chapter previously appeared in *Mental Health and the Environment* (H. L. Freeman, editor), Churchill Livingstone, London, 1985.

References

1. G. W. Evans, "General Introduction" **In**: (G. W. Evans, ed.), *Environmental Stress* (Cambridge University Press, Cambridge, 1982).
2. D. Stokols, "Environmental psychology" *Annual Review of Psychology* **29**, 253–95 (1978).
3. J. C. Cassel, "Psychiatric epidemiology" **In**: (G. Caplan, ed.), *American Handbook of Psychiatry, Vol. 2*, Second Edition, (Basic Books, New York, 1974).
4. J. C. Cassel, "Physical illness in response to stress" **In**: (S Levine and N. A. Scotch, eds.) *Social Stress* (Aldine, Chicago, 1970).
5. J. C. Cassel, "Health consequences of population density and crowding" **In**: National Academy of Science (ed.) *Rapid Population Growth: consequences and policy implications, Vol. 1* (John Hopkins University Press, Baltimore, 1971).
6. L. Levy and A. N. Herzog, "Effects of population density and crowding on health and social adaptation in the Netherlands" *J. Health and Social Behaviour* **15**, 228–240 (1974).
7. A. H. Leighton, *Caring for Mentally Ill People* (Cambridge University Press, Cambridge, 1982).
8. L. Srole, T. S. Langner, S. T. Michael, M. K. Opler and T. A. C. Rennie, *Mental Health in the Metropolis* (McGraw-Hill, New York, 1962).
9. A. B. Hollingshead and F. C. Redlich, *Social Class and Mental Illness* (Wiley, New York, 1958).
10. D. Mechanic, "The contribution of sociology to psychiatry" *Psychological Medicine* **3**, 1–4 (1973).
11. D. A. Regier, "Research progress 1955–1980" **In**: (O. Wagenfeld, P. V. Lemkau and B. Justice, eds.) *Public Mental Health: Perspectives and Prospects* (Sage, Beverly Hills, 1982).
12. J. E. Murphy, "Social causes: the independent variable" **In**: (B. H. Kaplan, R. N. Wilson and A. H. Leighton, eds.) *Further Explorations in Social Psychiatry* (Basic Books, New York, 1976).
13. Liverpool Inner Area Study, *Third Study Review: Issues and Policies* (HMSO, London, 1975).
14. Exit Photography Group, *Survival Programmes in Britain's Inner Cities* (Open University Press, Milton Keynes, 1982).
15. R. F. Ridley, "Unemployment in Liverpool" *Political Quarterly* **52**, 1–22 (1981).
16. J. Ruesch, "Social psychiatry: an overview" *Archives of General Psychiatry* **12**, 501–509 (1965).
17. H. Selye, *The Stress of Life* (McGraw-Hill, New York, 1956).
18. M. H. Brenner, "Influence of the social environment on psychopathology: the historic perspective" **In**: (J. E. Basset, ed.) *Stress and Mental Disorder* (Raven Press, New York, 1979).
19. G. Myrdal, *Economic Theory and Underdeveloped Regions* (Methuen, London, 1957).
20. B. P. Dohrenwend, "Sociocultural and sociopsychological factors in the genesis of mental disorders" *J. Health and Social Behavior* **16**, 365–393(1975).
21. B. P. Dohrenwend and B. S. Dohrenwend, "Psychiatric disorders in urban settings" **In**: (G. Caplan, ed.) *American Handbook of Psychiatry, Vol. 2*, Second Edition, (Basic Books, New York, 1974).
22. D. P. Mueller, "Social networks" *Social Science and Medicine* **14A**, 147–161 (1980).
23. M. Argyle and J. Dean, "Eye contact, distance and affiliation" *Sociometry* **28**, 289–304 (1965).
24. D. E. Schmidt and J. P. Keating, "Human crowding and personal control: an integration of the research" *Psychological Bulletin* **86**, 680–700 (1979).
25. P. B. Paulus, G. McCain and V. C. Cox, "Death rates, psychiatric commitments, blood pressure

and perceived crowding as a function of institutional crowding" *Environmental Psychology and Non-Verbal Behaviour* **3**, 107–116 (1978).
26. R. A. Baron, *Human Aggression: Perspectives in Social Psychology* (Plenum, New York, 1977).
27. D. Stokols, "The experience of crowding in primary and secondary environments" *Environment and Behaviour* **8**, 49–86 (1976).
28. A. Howard and R. A. Scott, "A proposed framework for the analysis of stress in the human organism" *Behavioural Science* **10**, 141–160 (April, 1965).
29. E. Hinkle Jr., "The concept of stress in the biological and social sciences" *Science, Medicine and Man* **1**, 31–48 (1973).
30. E. C. Lehman, "An empirical note on the transactional model of psychological stress" *The Sociological Quarterly* **13**, 484–495 (1972).
31. H. Selye, *Stress in Health and Disease* (Butterworths, Boston, 1973).
32. J. E. Rabkin and E. L. Struening, "Life events, stress and illness" *Science* **194**, 1013–1020 (1974).
33. S. V. Kasl and E. Harburg, "Mental health and the environment: some doubts and second thoughts" *Health and Social Behaviour* **16**, 268–282 (1975).
34. G. W. Brown, S. Davidson, T. Harris, U. Maclean, S. Pollock and R. Prudo, "Psychiatric disorders in London and North Uist" *Social Science and Medicine* **11**, 367–377 (1977).
35. G. W. Brown and R. Prudo, "Psychiatric disorder in a rural and an urban population: 1. Aetiology of depression" *Psychological Medicine* **11**, 581–599 (1981).
36. R. Caetana, R. M. Suzman, D. H. Rosen and D. J. Voorhees Rosen, "The Shetland Islands: longitudinal changes in alcohol consumption in a changing environment" *British Journal Addiction* **78**, 21–36 (1983).
37. W. B. Miller, "Psychological and psychiatric aspects of population problems" **In**: (Arieti *et al.*, eds.) *American Handbook of Psychiatry*, Second Edition (Basic Books, New York, 1974).
38. D. Trichopoulos, K. Katsayanni, X. Zavitsanos, A. Tzonai, P. Dalla-Vorgia, "Psychological stress and fatal heart attack: the Athens (1981) earthquake natural experiment" *Lancet* **I**, 441–444 (1983).
39. P. A. Keller and J. D. Murray, *Handbook of Rural Community Mental Health* (Human Sciences Press, New York, 1982).
40. G. A. De Vos, "Cross-cultural studies of mental disorder" **In**: G. Caplan (ed.) *American Handbook of Psychiatry, Vol. 1*, Second Edition (Basic Books, New York, 1974).
41. A. Glorig, "Non-auditory effects of noise exposure" *Sound and Vibration* **5**, 28–29 (1971).
42. E. K. McLean and A. Tarnopolsky, "Noise, discomfort and mental health" *Psychological Medicine* **7**, 19–62 (1977).
43. F. Godlee, "Noise: breaking the silence" *British Medical J.* **304**, 110–113 (1992).
44. S. A. Stansfeld, "Noise sensitivity, depressive illness and personality" **In**: (B. Bergland, ed.) *Proceedings of 5th International Congress on Noise as a Public Health Problem, Vol III*, 339–344 (1988).
45. V. D. Sanua, "Immigration, migration and mental illness: a review of the literature with special emphasis on schizophrenia" **In**: (E. G. Brody, ed.) *Behaviour in New Environments* (Sage Publications, Beverley Hills, 1969).
46. J. Krupinski *et al.*, "Factors influencing the incidence of mental disorders among migrants" *Medical J. Australia* **2**, 269–277 (1965).
47. J. Krupinski *et al.*, "Psychiatric disorders in Eastern Europe refugees now in Australia" *Social Science and Medicine* **7**, 31–42 (1973).
48. B. Malzberg and E. S. Lee, *Migration and Mental Disease* (Social Science Research Council, New York, 1956).
49. J. Krupinski, "Psychological maladaptation in ethnic concentrations in Victoria, Australia" **In**: (I. Pilowsky, ed.) *Cultures in Collision* (Australian National Association for Mental Health, Adelaide, 1975).
50. E. W. Butler, *Urban Sociology: a systematic approach* (Harper and Row, New York, 1976).
51. L. Levy and L. Rowitz, *The Ecology of Mental Disorder* (Behavioural Publications, New York, 1973).
52. P. M. Higgins, "Stress at Thamesmead" **In**: (H. L. Freeman, ed.) *Mental Health and the Environment* (Churchill Livingstone, London, 1985).
53. J. Burke, "Supportive environments for health" *W.H.O. Workshop on Briefing Books* (Copenhagen, June 1990).
54. S. D. Webb, "Urban-rural differences" **In**: (H. L. Freeman, ed.) *Mental Health and the Environment* (Churchill Livingstone, London, 1985).
55. K. Wedmore and H. L. Freeman, "Social pathology and urban overgrowth" *Mental Health and the Environment* (1985).

56. D. Cappon, "Designs for improvements in the quality of life in downtown cores" *Int. J. Mental Health* **4**, 31–47 (1975).
57. R. Fox, "Of inhuman native and unnatural rights" *Encounter* **58**, 47–53 (1982).
58. H. M. Government, *Policy for the Inner Cities* (H.M.S.O., London, 1977).
59. S. Taylor, "Suburban neurosis" *Lancet* **i**, 759–761 (1938).
60. F. M. Martin, J. H. F. Brotherston and S. P. W. Chave, "The incidence of neurosis in a new housing estate" *British J. Preventive and Social Medicine*, **11**, 196–202 (1957).
61. M. Goodman, *Paper to Conference on Housing and Health* (University of Warwick, December 1986).
62. J. Hansen and W. Hillier, "The architecture of community" *Architecture and Behaviour* **2**, 20–29 (1984).
63. P. Willmott, "Population and community in London" *New Society* **27**, 357–358 (August, 1974).
64. H. L. Freeman, "Mental health and new communities in Britain" **In**: T. Nossiter, A. H. Hanson and S. Rokken, eds.) *Imagination and Precision in the Social Sciences* (Faber, London, 1972).
65. P. Dunleavy, *The Politics of Mass Housing in Britain 1945–75* (Clarendon Press, Oxford, 1981).
66. H. L. Freeman, "Housing" in *Mental Health and the Environment* (Churchill Livingstone, London, 1985).
67. H. L. Freeman, "Mental health aspects of high-rise housing" **In**: (R. Burridge and D. Ormandy, eds.) *Aspects of Unhealthy Housing* (Spon, London, 1992).
68. H. Sherlock, *Cities Are Good For Us* (Transport 2000, London, 1990).
69. H. M. Choldin, "Urban density and pathology" *Annual Review of Sociology* **4**, 91–113 (1978).
70. O. Newman, *Community of Interest* (Anchor, New York, 1980).
71. J. Magaziner, "Living density and psychopathology" *Psychological Medicine* **18**, 419–431 (1988).
72. R. E. Mitchell, "Cultural and health influences on building, housing and community standards" *Human Ecology* **4**, 297–330 (1976).
73. M. Fried, "Grieving for a lost home" **In**: (L. J. Duhl, ed.) *The Urban Condition* (Basic Books, New York, 1963).
74. P. Hall, "Some clinical aspects of moving house as an apparent precipitant of psychiatric symptoms" *J. Psychosomatic Research* **10**, 59–70 (1966).
75. H. L. Freeman, "Mental health in the inner city" *Environment and Planning* A **16**, 115–121 (1984).
76. J. S. Faria, "Notes on the promotion of mental health within the health promotion concept" *W.H.O. Working Group* (Dresden, June 1989).
77. A. Coleman, "Defensible space on the map" *Urban Design Quarterly* **3**, (Feb., 1983).
78. O. Newman, *Defensible Space* (Macmillan, New York, 1972).
79. S. Cohen, G. W. Evans, D. Stokols, and D. Krantz, *Behaviour, Health and Environmental Stress* (Plenum, New York, 1988).
80. M. L. Rutter, "Primary prevention of psychopathology" **In**: (M. M. Kent and J. E. Rolf, eds.) *Primary Prevention of Psychopathology* (University Press of New England, Hanover, N.H., 1979).
81. D. P. Mueller, "The current status of urban-rural differences in psychiatric disorder" *J. Nervous and Mental Disease* **169**, 18–27 (1981).
82. B. J. Gurland, "The comparative frequency of depression in various adult age groups" *J. Gerontology* **31**, 283–292 (1976).
83. M. L. Rutter "The city and the child" *American J. Orthopsychiatry* **51**, 610–625 (1981).
84. M. L. Rutter and D. Quinton, "Psychiatric Disorder" **In**: (H. McGurk, ed.) *Ecological Factors in Human Development* (North Holland, Amsterdam, 1977).
85. D. J. West and D. P. Farrington, *The Delinquent Way of Life* (Heinemann Educational, London, 1977).
86. J. Jacobs, *The Death and Life of Great American Cities* (Random House, New York, 1961).

6. STRESS IN THE WORK ENVIRONMENT

VALERIE J. SUTHERLAND* and CARY L. COOPER*

In today's world, stress and strain are inescapable facets of modern living and working life. This much-voiced statement seems strange and somewhat paradoxical when we consider all the benefits and comforts that new technology has bestowed on contemporary society. Through technological development man has succeeded in conquering all corners of our earth and now ventures into space to explore other planets. We have successfully created environments which permit us to live and work almost anywhere, surviving the harsh extremes of climatic conditions and crossing the boundaries of time and space. Nevertheless, there is increasing evidence to suggest that we have created an environment which is potentially very stressful and ultimately deleterious to society, the economy, and our own health and well-being. In this chapter, we will describe the stressors that exist in our working environment, and explain why it is vital that we understand the concept of occupational and psychosocial stress. We begin by defining what we mean by stress, how to recognise it and why it is such an important issue for consideration. Next, sources of environmental stress in the workplace are described. Examples are provided to show how exposure to stress has affected the population and/or those in certain occupations, and the steps that have been taken to try and manage the situation (i.e. to avoid or reduce the negative consequences of stress).

1. DEFINING STRESS

For most people stress has negative connotations, and an individual typically describes stress as an unpleasant and unwanted experience. Although it must be acknowledged that "any demand, either of a physical or psychological nature encountered in the course of living is known as a stressor",[1] it is also important to accept that stress can be both positive and negative in its consequences. Stress is not inherently deleterious; it provides stimulation, challenge, and is the spice of life essential to development, growth and change. As Hans Selye has suggested,[2] the only person without stress is a dead person. By establishing an acceptable definition of stress, it helps us to understand this difference between stress and distress (i.e. positive or negative stress), and why response to stress often varies greatly between one individual and the next, in seemingly identical circumstances.

Even though stress research and the concept of stress management are relatively new areas of interest, definitions of stress have evolved and developed

*Manchester School of Management, University of Manchester Institute of Science and Technology, PO Box 88, Manchester M60 IQD, UK

from the simplistic idea that strain, pressure or force exerts itself on the individual causing adverse consequences. Human beings are much more complex, and so an interactive definition of stress is needed. That is, stress is not simply "out there in the environment";[3] it has to be perceived and recognised by man as a stressor. It is a product of the interactions between the person, the source of stress and the total environment in which they exist. Thus, many factors will influence the outcome and manifestation of stress. It is true to say therefore that stress is in the eye of the beholder. This has significant implications for both stress researchers and stress management practitioners. It probably explains why we are more likely to resort to helping individuals cope with a source of environmental stress, rather than try to eliminate or minimise the problem. The weakness of this limited approach is further discussed later. A definition of stress provided by Levi[4] is consistent with the interactionist approach to understanding stress. He views stress as "poor person-environment fit": that is, "the interaction between, or misfit of, environmental opportunities and demands, and individual needs and abilities, and expectations, elicit reactions. When the fit is bad, when needs are not being met, or when abilities are over- or undertaxed, the organism reacts with various pathogenic mechanisms . . . and, in the presence or absence of certain interacting variables, they may lead to precursors of disease."

2. THE STRESS RESPONSE

Response to stress is well documented;[5] the consequences may manifest themselves in the form of behavioural changes (e.g. alcohol or palliative abuse, poor dietary habits, increased cigarette smoking, and accident proneness), physiological symptoms such as headaches, abdominal pains and constant tiredness, and/or psychological outcomes in the form of depression, anxiety and certain phobias (e.g. obsessionality, agoraphobia). The individual who experiences these consequences of exposure to undue stress and strain may also be less effective and productive at work, and so the individual and industry ultimately suffers the adverse impact of mismanaged stress. Figure 1 shows the potential interactive nature between physical, cognitive, behaviourial and environmental factors that might influence health and/or behavioural outcomes. A stressor may therefore impact directly and/or indirectly, thus making it difficult to understand the real cause and effect relationships between a stressor and the stress response.

A case history indicates the difficulties involved in understanding a stress problem. A 35 year-old married woman, with two young children, is reprimanded for repeated absenteeism from her job as data-input clerk at a large mail-order company. Close analysis reveals that she has complained about the constant pressures of work overload, and the headaches caused by the long hours of working at a visual display screen (VDU). The supervisor ignores these complaints and only seems interested in increasing the volume of work through his/her department. Mrs. X is constantly tired, finds the work very boring and tedious, but the hours suit her parenting arrangements as a single-parent family. When the children are in bed she often has a few glasses of sherry to drink in the evening (it helps her get to sleep!). Sometimes she has a bad migraine when she wakes up and is unable to face the prospect of going into work. Since her father was an alcoholic and treated her mother badly she feels guilty about this "little weakness", but it seems to help in getting her through this bad time in her life. Sometimes the children are ill and cannot go to school. She cannot leave them,

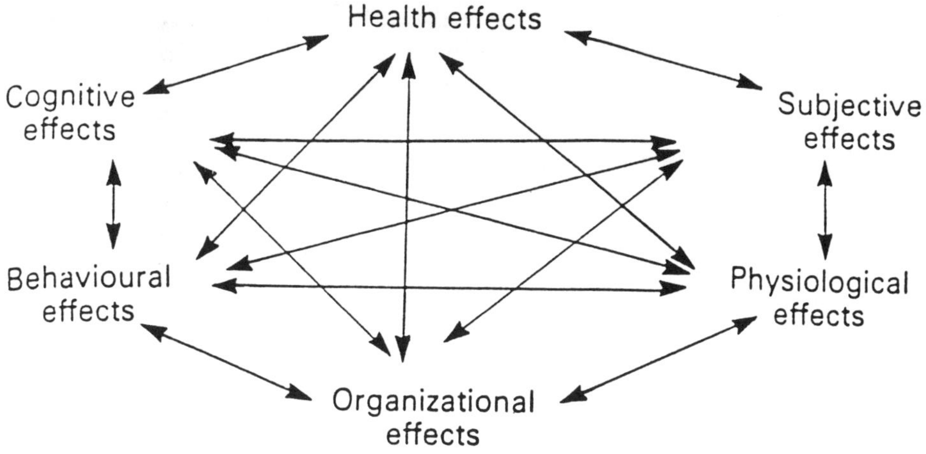

– an example.

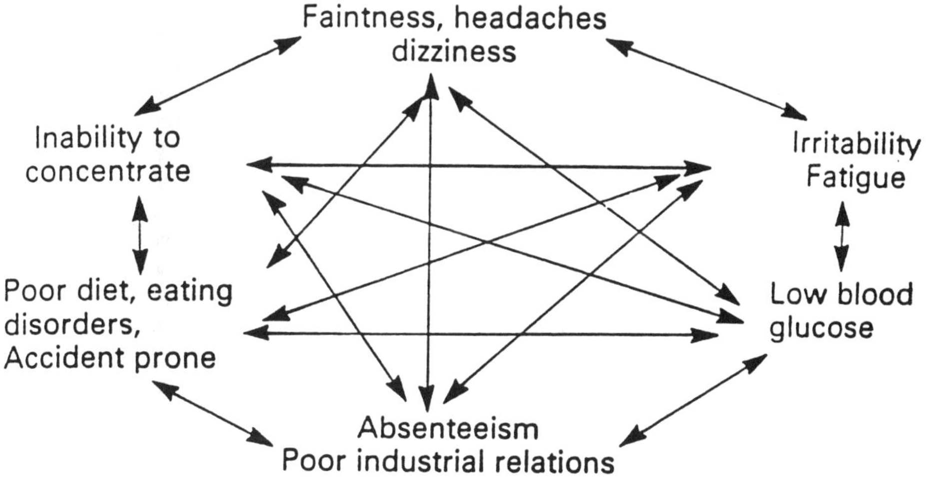

Figure 1 The interactive nature of the symptoms of stress.

and so is absent from work yet again . . . . What is/are her source(s) of stress? What can or should be done to help?

This example shows the circular nature of relationships between stress and the stress response, how easy it is to resort to maladaptive ways of coping with stress (i.e. alcohol) and to enter that vicious downward spiral to destruction. It emphasises the difficulties involved in the management of stress for both the individual and the company. It is also quite apparent therefore that the long-term effects of mismanaged stress are very costly to society; invariably, stress problems at work spill over into the home and social life (and vice-versa), and relationships are likely to suffer. Impaired and impoverished interpersonal relationships lead to disharmony at work and at home, often causing family breakdown and the many social problems associated with drug and alcohol abuse.

3. STRESS—THIRD WAVE PLAGUE OR WHIPPING BOY?

The topic of stress constantly appears in the media and lately has assumed a "flavour of the month" appeal. We have finally acknowledged that *stress* is a suitable subject for discussion and debate. Naturally there is still some unease and defensive attitudes towards the concept of stress and stress management. For some people, it is still regarded as a "four-letter word". The individual prefers to view it as "something which happens to other people", and although a few companies sensibly include stress management as part of their "employee assistance package", the majority adopt negative attitudes such as, "of course they (the workforce) are stressed, it is what we pay them for!" or, "if we provide stress management training they will all be convinced that they have a problem!". This negative attitude is not surprising (although it is a serious issue that should be addressed), because the high media image of the stress concept implies that it could or should be blamed for all our ills and problems. The layman readily and easily identifies with the concept and is eager to know more, but unfortunately tends to dwell on the more negative aspects of stress. Also, stress practitioners have in the past typically identified stress as "a problem", and so are seen to deal with problem people or problem organisations. Adopting a preventative approach in stress management education, (that is, one which raises self-awareness about positive and negative stress, and personal strengths and weaknesses in the stress-coping process) and conducting a *stress audit* in the organisation to identify stressors that could be reduced or eliminated by changes to the company structure (e.g. changes to the complex layers or divisions that inhibit communications), the work climate (e.g. highly competitive, very macho, or autocratic), systems and/or practices, would greatly help to overcome the negative attitude to stress management. In fact, to be really successful, stress management techniques need to consider both these options, rather than to simply adopt an "inoculation" approach to stress coping (i.e. where the workforce are only taught how to handle the stress that exists, rather than trying to remove the source of stress).

Perhaps we should play the devil's advocate at this point and ask if we really do have a stress problem in modern-day society, or are we, in fact, making a "psychological mountain out of a 'common or garden' molehill"? Increasingly the evidence suggests that the problems of stress at work, and for society itself, are most serious. It has become the focus of attention for governments and legislators at the highest levels, and is acknowledged by health professionals as a significant threat to the health and well-being of the workforce. Stress is implicated in the onset and development of certain forms of disease and illness (organic and psychological) either directly or indirectly.[5] For example, mental illness (depression, phobias), gastrointestinal disorders, diabetes mellitus, various forms of skin disease, certain allergies (such as asthma); even autoimmune malfunctions (e.g. ME, rheumatoid arthritis and multiple sclerosis) may be stress-related disorders. Indeed, the killer-diseases in contemporary society are the chronic disorders such as heart disease and cancer; both are associated with the interactive effects of lifestyle, physiological predisposition and environment variables, and so stress is an acknowledged risk factor for these diseases. As Knowles[6] states, "over 99% of us are born healthy and made sick as a result of personal misbehaviour and environmental conditions". It is clear that stress-related illness exacts a heavy payment on the individual and society. The real costs associated with social problems such as alcoholism, drug abuse, and the suffering of family breakdown poses a significant burden on our economy and threatens the stability of society itself. Industry also suffers because of high absenteeism levels, high labour turnover, accidents, forced early retirement, premature death, and poor performance and productivity (poor industrial relations, strikes, poor working relationships). Numerous studies have shown that the costs of mismanaged stress includes a loss of vitality which adversely affects job satisfaction, morale and motivation at work. Poor decision-making and communication breakdown are potentially costly outcomes for the organisation.[1]

4. CHANGE AND STRESS

In the next part of this chapter we will consider specific sources of stress in the work environment, but first we should consider one very important stressor which plays a significant and all-encompassing part in our lives today . . . "change". With a united voice, men of science, technology and philosophy, tell us that we are experiencing a period of change, which compared to events in the past, is nothing less than a revolution. Alvin Toffler[7] has described the startling acceleration of change in our society which has occurred in just one lifetime. There is not one sphere of human activity that has not been dramatically altered. We attempt to cope with this rapid pace of change in new technology by adapting to the new situation or circumstance. As Selye[8] suggested, all change necessitates adaptation, whether it is a positive or negative experience, which is welcomed, feared or resisted: and this adaptation or readjustment requires energy. Since our resources are not infinite, breakdown of the system, in part or total, ultimately occurs. In Selye's view, impairment of function and structural change are wholly, or in part, linked to this adaptation to a source of stress. This results in wear and tear on the body, and ultimately leads to exhaustion or the collapse (death) of the system. Indeed, we often try to cope with change by resorting to "maladaptive" ways of coping (e.g. we have a drink of alcohol because we believe it gives us confidence; we consume lots of strong coffee in order to provide the "buzz" which keeps us working long hours; we light a cigarette to calm our nerves; we use various pills and potions to calm ourselves down, or "pep" ourselves up), and in the long term this adds to our stress, either by exacerbating the problem, or becoming *the* source of stress itself.

In evolutionary terms the stress-response was adaptive and vital for survival. Various bodily functions/mechanisms are activated in order to prepare the body to respond to a stressor; that is, to either fight or flee from a threat situation. So when we are in the stress situation, the body is physiologically primed to take some form of action. For example, heart rate and blood pressure increase as blood is pumped to vital organs, carrying oxygen, and glucose for energy; blood supplies are redirected to the brain and skeletal muscle from the skin and the viscera; blood coagulation processes are enhanced to avoid blood loss and the supply of blood lymphocytes is increased to combat possible infection from our injuries. Thus we may experience some racing of the heart and increased respiration as adrenalin surges through our system; the mouth becomes dry; our hands and feet may feel cold; we may experience "butterflies" in the stomach. However, in contemporary society and especially in sedentary occupations, the stress response may be disadvantageous. Man is denied both the aggression release and the physical activity needed to quickly dissipate the hormone and chemical build-up, because he/she is not permitted to punch someone on the nose when the strains and pressures of work or life become intolerable; neither can he/she run away from the situation. In our industrialised society, we are expected to cope without showing signs of weakness. Therefore, we often add to the burden of stress by denying that a stress problem exists (this may be "self-denial", and/or denial towards the people around us).

Although we develop some resistance to a source of stress, the energy needed for adaptation eventually becomes depleted, especially if we remain in an intensely stressful situation for a long period of time. We cannot maintain this state of resistance indefinitely because biological activity causes wear and tear on the body, which leads to various forms of illness and disease, and/or weakens our resistance to disease. Since contemporary man lives in a world of rapid change, he lives and exists in a state of transience and impermanence which may be damaging, because energy is needed and expended by this constant need for adaptation to new situations and stimulation from the environment. In addition to this powerful stressor agent of "change", there are also a wide variety of potential sources of stress in the work environment which cause strain and pressure on the workforce. These are discussed in the next section.

5. STRESSORS IN THE WORK ENVIRONMENT

This section of the chapter will be in two main parts. Figure 2 illustrates the sources of stress included in this discussion. First we will consider aspects of the physical work environment that may be a source of strain and pressure in the workplace. Our physical surroundings can affect our mood and mental state. Sometimes we may not even be aware that the physical conditions at work are having an adverse impact on our well-being. Each occupation has its own environmental sources of stress and the burdens vary from job to job. These might include:

- noise
- temperature extremes of heat and cold
- ventilation problems
- exposure to outdoor climatic conditions
- vibration
- lighting
- hygiene
- crowding

Following this discussion we will review the "task" requirements of the job as potential sources of stress; these task demands include:

- hours of working: shift work; long hours
- workload
- new technology
- risk and hazard
- travel

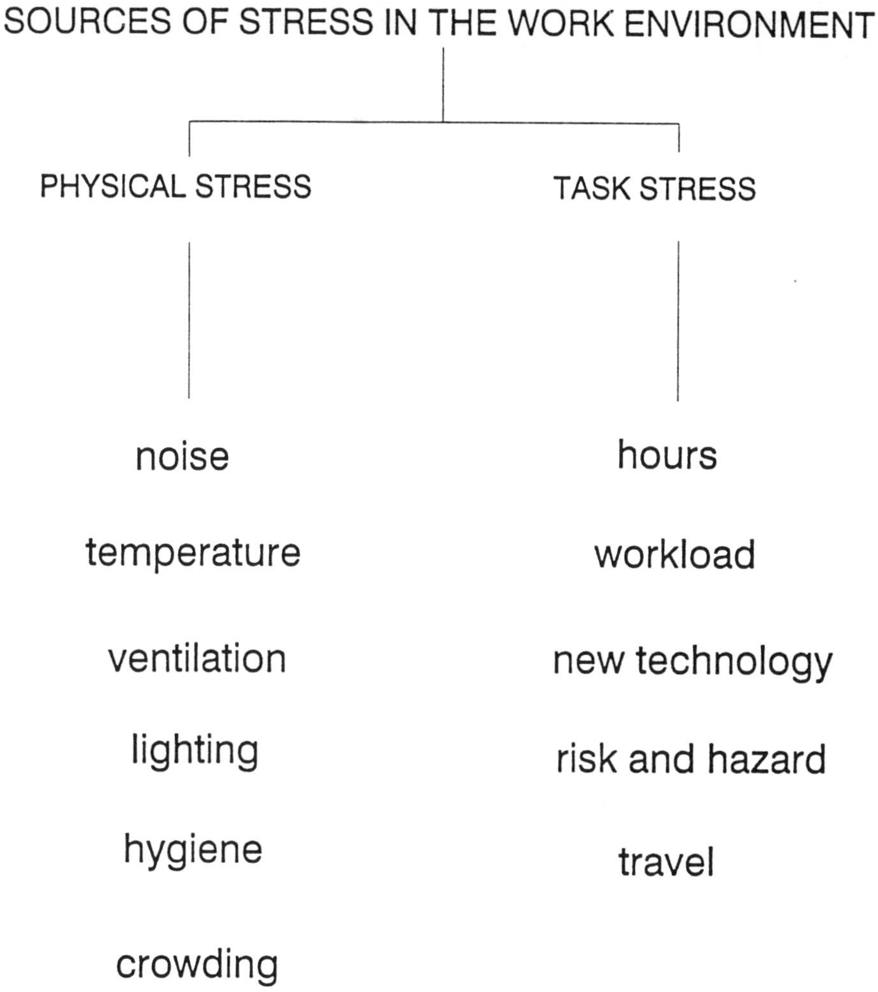

Figure 2 Sources of stress.

5.1. *Stress in the Physical Environment*

Early psychological research attempted to investigate the links between physical factors in the work place and productivity levels.[9,10] From this pioneering work, it was realised that "subjective experience" was a vital factor for consideration in the understanding of the relationships between emotional/social factors and performance and well-being. Objective measures of, for example, lighting levels on performance rates were not adequate to explain the differences observed in productivity; that is, the expected outcomes were not realised.[10] Although this experience taught researchers the power of the "Hawthorne Effect" in experimental situations, stress researchers also gained from this knowledge because it helped them to appreciate the need to take adequate account of individual subject experience/reactivity in response to strain and pressure. Thus, in order to understand stress and the stress response it is necessary to take account of:

a. The actual source of strain or pressure;

b. The individual(s) exposed to the stress; i.e. individual differences such as age, gender, past experience, values, beliefs, needs, and personality characteristics. Some or all of these will determine response to the stressor, the coping mechanism used, and thus the outcome of exposure to stress;

c. Background and situational factors; a stressor is not perceived in a social vacuum. The presence or absence of other people influences our response and this may be a negative or a positive effect. For example, other people can be supportive, helpful, and provide a boost to confidence in a stress situation. It sometimes helps to know that other people are in the same situation as oneself. Nevertheless, the presence of others can also be an irritation or a source of distraction, when we would rather quietly draw on our inner reserves of strength during a time of extreme pressure and personal crisis.

A study of response to noise at work is a good example of this need to understand stress in both objective and subjective terms (i.e. the interactive approach).

5.1.1. *Noise.* What is noise? Teenagers hardly seem to notice the screaming throb of sound in the fashion boutique or record shop as they congregate and choose their purchase; whereas it is likely to send the authors quickly, and empty handed to the nearest exit! Thus, Jones'[11] definition of noise as "unwanted sound" complies with our notion of "noise" as a source of stress. Exposure to noise at work can impede hearing ability and may be problematic in that the detection of a wanted sound is masked. For example, an accident may occur if warning sounds are not observed.[12] However, research evidence on "noise" as a stressor is complex. It is suggested that it causes an increased level of arousal and psychological imbalance, but the results are unequivocal. Ivancevich and Matteson[13] suggested that excessive noise (approximately 80 decibels) on a recurring, prolonged basis, can cause stress. In fact, they also say that noise seems to operate less as a stressor in situations where it is expected, than in those where it is unexpected, or at least unpredictable. The change in noise levels, more than absolute levels themselves, seems to be the irritant. This, or course, is simply another way of saying that noise, like other stressors, causes strain when it forces us to change. It seems that the main psychosocial impact of excessive noise and other physical demands, is to reduce worker tolerance to other stressors, and adversely affect motivation;[14] thus the impact may be additive, not primary.

Nevertheless, "noise" is reported by many groups of blue collar workers as a harmfully perceived stressor. For example, in the UK steel manufacturing industry,[15] for three-quarters of the seventy-minute process of casting liquid steel, the casters were unable to move away from high noise levels (reaching up to 110 decibels), in addition to the heat and fumes of the process. During this time the workers wore ear defenders, but this effectively isolated them from communication with the other members of the work crew. Unpleasant working conditions due to noise (and other factors intrinsic to the job) were significant predictors of job dissatisfaction among workers on drilling rigs and platforms in the North Sea.[16] In fact, the follow-up study[17] found that more than 40 per cent of offshore personnel (n = 245) rated noise in working conditions as a source of pressure, and 30 per cent considered that noise from other people was a problem for them in the living accommodation offshore, even though this industry has spent quite considerable resources on addressing this problem of noise.

Various studies have observed the relationships between noise in the workplace and productivity, and noise and rate of error.[18] Of particular importance is the factor of noise in increasing vulnerability to accidents. Kerr[19] found, for example, that mean noise levels correlated significantly with accident frequencies, but not accident severity. Cohen also found accidents to be more frequent in noisy areas (95 decibels or above), particularly among the younger and less experienced workers. Following his analysis of the problem, and the subsequent introduction of ear defenders, a significant reduction in accident rates was noted.[20,21]

Three hazards seem to be linked with exposure to noise in the workplace:

a. The risk of occupational deafness;
b. Noise increases arousal levels and might cause some psychological imbalance, i.e. the constant bombardment of stimuli overloads the system;
c. There is added risk if noise masks wanted sound (i.e. instructions or warnings).

However, it also appears that personality characteristics mediate response to to noise as a source of stress. Lader[22] suggests that those at risk from noise are those who habituate more slowly than normal; these individuals tend to be "anxious" personality types.

Although we can "get used to" noisy conditions (which is acceptable within certain limits, i.e. no subsequent hearing loss), it should be acknowledged that dealing with noise in the workplace is ultimately an issue of design. In the offshore environment, on drilling rigs and platforms, constant exposure to noise remains a major problem which is exacerbated because the workforce live and work in the same environment for extended periods of time without a break (7, 14 or 21-day tour of duty). Much has been achieved in reducing the problem of noise at the design stage,[23] but the focus tends to dwell solely on objective measurement[24,25] rather than also include the subjective impressions of the workforce. We must remember that noise can be expressed in many ways, for example, intensity, variability, frequency, predictability and the degree of worker control over the situation. Nevertheless, reaction to noise is still ultimately a subjective experience. Exposure to noise is associated with reported fatigue, headaches, irritability and inability to concentrate; the behavioral consequences are in terms of reduced performance, productivity and accident occurrence. Social behaviour is also influenced in noisy conditions; for example, a reduction in "helping behaviour", a more extreme or negative attitude to others, and more open hostility and overt aggression.[11] Thus, there are implications for impoverished relationships in the environment caused by the exposure to noise, the isolation due to excessive noise, the need to wear ear protectors, and/or the poor interpersonal interactions resulting from an accumulation of physical frustration and tensions.[26] However, research shows that the investigation of any single source of work stress is virtually impossible and may be essentially futile. For example, the extensive investigation of personnel working on the Statfjord Platform in Norwegian waters[27] indicated that between 20–25 per cent of respondents were dissatisfied with noise, heat, ventilation and humidity aspects of the environment, and the physical environment also influenced the perceived health condition of employees. But these authors also provided a reminder that state of health may, in fact, influence an individual's opinion of the work environment; that is, someone who suffers poor health may be less tolerant of the noisy conditions. This may also apply to the perception of thermal conditions in the workplace.

5.1.2. *Temperature.* Responses to thermal conditions varies greatly between workers and within the same individual from one occasion to the next.[28] Although temperature can be carefully controlled in most work places today, it seems that the work environment is frequently characterised as too hot, too cold, too stuffy, too draughty, etc., and thus it creates both physical and attitudinal problems.[14] It is suggested that work demanding critical decisions, fine discrimination and performance of fast or skilled action is impaired by thermal stress; manual dexterity and motivation levels are reduced in a cold environment and could be a significant factor in accident occurrence, due most likely to reduced sensitivity, slowed movement and interference from protective clothing, rather than the loss of impaired cognitive ability.[29] Research shows that subjective report of thermal comfort seems to be all important. It is likely that an inability to personally control one's physical environment could also be a significant factor in the perception of thermal comfort. As Ramsey suggests,[28] there seems to be a correlation between "comfort" and an individual's performance on perceptual motor tasks. In fact, a strong inverse correlation was observed between the perception of comfort and the level of accidents among shopfloor workers in a tractor factory.[26] It was suggested that heat and noise conditions led to "distractive" behaviour which resulted in reduced quality of work and increased vulnerability to accidents.

Noise, fumes and heat were the most commonly reported problems among casters in the steel industry.[15] Most frequently mentioned by Norwegian workers on offshore platforms were the draughts, uncomfortable temperatures and the dry humid air.[27] More dissatisfaction was expressed about overheated working conditions (25 per cent) than about the cold (12 per cent of the respondents). In overheated working conditions, we seem to adopt a negative reaction to our surroundings. For example, unpleasant working conditions, due to heat and/or cold, were significant predictors of job dissatisfaction among rig workers in the UK and Dutch sectors of the North Sea.[16] It is possible that workers might complain about these types of strain and pressure because they are "safe-subjects" for disgruntlement at work. Therefore, the real and more sensitive issues of pressure (for example, interpersonal relationships) may be masked if the stress audit (diagnosis) is not carried out in a confidential and discreet manner (i.e. there is a need to protect the self-esteem, confidence and dignity of the workforce). Another potential stress problem at work, and more recently topical, is related to that of poor ventilation in the work environment.

5.1.3. *Ventilation.* For the past few years we have heard quite a lot about "sick building syndrome" (or "tight-building syndrome"). It relates to indoor air-quality problems that occur in closed systems. That is, where the building is designed to be energy efficient, air is recirculated through air-conditioning systems (cooling and heating), and so it is not possible to open any of the windows. The problem seems to exist mainly in newly constructed or renovated property. As the building dries out, gas (air pollutants) released from the construction materials can not escape with the ease of the older and less airtight work environments. The situation is very difficult to diagnose, and the real cause of the problem often cannot be identified.[30] A wide variety of health complaints seem to be associated with "sick building syndrome"; for example, upper respiratory tract irritation (sore throat, runny nose), itchy eyes, headaches, fatigue, drowsiness and nausea.

The non-specificity of the illness observed, and the attention of the media, has led some researchers to suggest that the cause may be, in part, psychological rather than chemical; and incidents may increase in frequency due to suggestion (i.e. mass hysteria).[31] A case study reported by Alexander and Fedoruk illustrates this point:

> "An odour was reported in a building housing 153 telephone operators, following which 81 were treated in 11 emergency medical facilities for symptoms compatible with anxiety and hyperventilation on either of two occasions one day apart. The distribution of symptoms reported showed no consistent pattern. The precipating factor was shown clearly to be a specific employee who had a military history of involvement in psychological operations and who for reasons unknown precipitated by suggestion a mass psychogenic illness among employees in the building. Extensive environmental analysis was conducted . . . and conclusively ruled out toxic exposure. No changes were made in the building's ventilation, operations, or in the use of chemicals for maintenance. A rechallenge of the same population under the same conditions did not result in another outbreak."[32]

As Guidotti *et al.*[33] suggested, it was not an infectious agent being transmitted, but anxiety. Therefore, it is necessary to exercise care in the diagnosis of these ventilation problems in order to identify and resolve the real issue: it could be an incident of psychogenic origin, one of chemical toxic exposure, and/or a combination of the two. The projects reported by Whorton *et al.* indicate that the problem can be successfully resolved if the matter is investigated thoroughly and the correct diagnosis is obtained.[30]

For some workers the problem of noxious stimuli is caused by inadequate ventilation systems. This is a common complaint for health workers; nurses have identified bad smells as potential stress agents, although they often acquire an immunity to them.[34] On a Norwegian offshore oil production platform, the workers most frequently complained about draughts and the dry humid air caused by the artificial ventilation in their living and working accommodation.[27] They believed that contagious diseases/viruses were spread very quickly by these conditions, and their perceived lack of control over this artificially created environment probably exacerbated the problem. Again, this indicates some of difficulties in isolating any single particular stressor. Workers forced to work in an environment with a constant need for artificial ventilation, lighting and temperature control, tend to report various forms of discomfort caused by the dry air conditions and the noise and constant whine of air-conditioning units and generators. As Smith *et al.*[14] state, this will result in lowered tolerance to other stressors and affect worker motivation. The presence of these sources of stress "uses up" some attention capacity[35] and, therefore, limits our ability to attend to task-relevant information. Although certain occupations do still seem to suffer from the problems associated with temperature variation and poor ventilation on some worksites, the outdoor-worker naturally faces uncontrolled climatic conditions which can add to the strains and pressures of the job itself.

5.1.4. *Climate.* A nation-wide sample of workers polled in the United States in 1971 rated "unpleasant working conditions" the third source of discontent (compensation rated first; health and safety hazards rated second).[36] So, the worst thing was "the weather" and the need to work outside in all conditions. This affects a significant number of blue collar workers, including those on construction sites, in dockyards, on highways, seafarers and agricultural workers.

Members of the police force, traffic wardens, postal delivery workers, and the thousands of people engaged in other delivery service work also spend a lot of time working outdoors. Unpleasant climatic conditions can affect our physical well-being and reduce morale and levels of motivation. It may also affect our vulnerability to accident involvement. So if work routines and environments can not be altered to minimise the adverse impact of poor climatic conditions, then adequate protective clothing is vital, and attempts to address morale and motivational problems should be seriously addressed.

5.1.5. *Vibration.* This is acknowledged as a major source of stress resulting in elevated catecholamine levels and alterations to psychological and neurological functioning.[37] Health hazards include nausea, loss of balance and fatigue. Vibrations from rotary or impacting machines could be problematic, for example, in steel casting,[15] and where pneumatic drills or riveting hammers are used. Vibrations that transfer from physical objects to the body may adversely affect performance; hands and feet are particularly vulnerable, but the "annoyance" factor is also a major psychological consideration. Frankenhaeuser and Gardell[38] observed significantly elevated catecholamine levels among assembly-line workers in a sawmill and compared these workers to maintenance and repair men in the same plant. The differences were attributed to the perceived degree of control reported by the two groups. However, it is possible that the machinery in use, with a vibratory action, affected the hands and could have contributed to the physiological response observed.

For personnel working in the offshore environment on drilling rigs and production platform installations, vibration is a potential stressor. High levels of vibration can disturb sleep and recreational activity.[25] Certain guidelines are set to account for the 12 hour working shift which is currently the norm. These categories specify a "just detectable" level recommended for sleep and recreation areas, up to the highest permitted level where exposure duration should be restricted to four minutes; but these are only guidelines, not legislation. The activity of helicopters landing on helidecks positioned over living accommodation, and the drilling operations cause vibrations which may be significantly more damaging in the long term, because the individuals must live and work in the same environment for 7 to 14 days (or more) without respite. Although the workers claimed that "you get used to it", "unpleasant working conditions due to vibration" and disturbance in the living accommodation were rated as significant stressors by 37 and 27 per cent of workers, respectively. The long-term effects of this exposure are not known.[16]

5.1.6. *Lighting and Illumination Levels.* Poor lighting and glare leads to eye strain, damaged vision, headaches, visual fatigue, tension and frustration because the task may be more difficult and time consuming.[12] Many work environments require constant artificial lighting. It is therefore important that a system creates a pleasant environment which facilitates performance and promotes safety. Adequate illumination is also an obvious factor associated with safe working. One study has shown that nurses found hospital lighting "monotonous, artificial, too bright and/or garish". In fact, nurses working in intensive care units reported that the oppressive visual environment became particularly stressful over a period of time.[39] Again, workplace designers should be sensitive to this problem, but rarely do they seek the opinions or attitudes of the end-users when planning colour schemes and decor. The equipment used in many workplaces, factories and offices has changed quite dramatically in recent years: companies have probably invested considerable capital in new equipment, but they are less likely to have acknowledged that the lighting and illumination levels are no longer compatible with the new mode of working. For example, glare from a visual display keyboard caused by poor overhead lighting causes strain. Indeed, inadequate protection from glare produced by the sun shining through the windows of our "all glass" modern offices increases the burdens for those engaged on a boring or monotonous job, and where they are confined to a workstation for the major part of the day.

5.1.7. *Hygiene.* Too many of the world's workforce are employed where hygiene conditions are poor and exposure to accidents and/or disease causes a constant threat to health.[40] Shostak[36] reports blue-collar grievances of generally neglected working conditions and the "double standard" that exists. The "all glass" lavish front offices are compared to the noise and lack of windows or air conditioning in dirty factories and workshops. For example, workers in the steel industry described dirty, dusty conditions, the poor accommodation provided for the rest periods, and the lack of toilet facilities nearby.[15] These features of the physical environment were rated as high stressor variables among casting crews. A clean and orderly place of work is important for both safety and hygiene reasons and for the morale of the workforce, especially where a situation is already acknowledged as hazardous, and where the worker might be exposed to substances which are potentially dangerous, i.e. chemicals that are skin irritants, or where the dust and vapours from certain materials or gases could be harmful.

5.1.8. *Crowding.* Evidence suggests that there is an association between "crowding" and psychological stress which leads to an increase in both contagious and non-contagious illness.[41] Where individuals do not have adequate workspace, work performance suffers, increased blood pressure is observed, and job dissatisfaction is reported.[42] Today, legislation is intended to control potential problems of overcrowding in terms of actual physical space required for each employee. However, the situation is more complex because each individual has a varying need for interpersonal space and distance. In fact, it may be perceived as too much, or too little; and a source of social stimulation to one individual may be perceived as aggravation and annoyance to another. Assessment of personality characteristics and individual differences would therefore play an important part in the selection and recruitment of personnel to work in known conditions where either crowding and/or isolation are significant features of the job. For example, crane drivers reported the isolation of work in the cab;[43] and casters in the steel industry, forced to wear ear defenders, complained that they felt isolated from their peers (that is, it restricted conversation and communication between workers).[15] As Quick and Quick suggested, a balance in social density is desirable;[1] this suggests that job rotation might also be appropriate in these situations in the workplace.

Observation of life and work in the offshore oil and gas industry in Norway found none of the usual stress reactions to crowding, even though 300–400 personnel would be working on these installations. These researchers did not observe formation of cliques, isolated individuals, struggle for territory and/or unfriendly, aggressive behaviour as expected.[27] However, it may be difficult to assess the problem of crowding in this environment because of other contributing factors; for example, the enforced 24-hour, multi-cultural society, and the lack of privacy, etc. These workers realise that their employment situation is unique and they must work together for long periods of time without a break, and so they may make a special effort to avoid problems. Nevertheless little is known about individual differences and the reaction to violation of personal space and territory; this is very important where individuals work and live together in close proximity for long periods of time. Although the concept of burnout (stress associated with a high degree of contact with others) is normally applied to professional occupations, there might be a need to understand burnout in terms of the blue collar worker. The consequences are equally important to all occupational groups, because Hartman and Perlman[44] defined the three components of being burned out as:

a. Emotional and/or physical exhaustion;

b. Lowered job productivity;

c. Over depersonalisation.

Behavioural response to job burnout includes increased absenteeism from work, high labour turnover and drug abuse. The consequences are therefore extremely deleterious for the individual and costly for industry.

While it may be impossible, impractical or unnecessary to modify or eliminate the impact of some aspects of stress associated with the physical demands of a job, it is vital to understand the stress situation from a total perspective before any changes can or should be made. It is also necessary to realize that task demands may also be potential sources of stress: and the actual work pattern or hours of work may be a significant source of strain for some employees.

5.2 *Task Related Stress*

5.2.1. *Shiftwork.* In order to maximise the usage of expensive capital equipment and plant, many organisations operate a twenty-four hour business. This means that a considerable proportion of the workforce have jobs requiring them to work in shifts, some of which involve working staggered hours. We are increasingly becoming a "24-hour society" and so the need to work shifts is now a common occupational stress factor. Despite considerable research effort, it is not possible to make overall generalisations regarding the "best" shift system, but it is clear that shiftwork represents a major source of stress for many workers.[45] Shiftworkers complain more frequently of fatigue and gastrointestinal troubles than day workers.[46] In fact, the impact of shiftworking on eating habits may be responsible for the gastrointestinal disorders observed.[46] Influences are both biological and emotional, due to circadian rhythm disruption of the sleep/wake cycle, temperature fluctuation and adrenalin excretion rhythms. Shiftwork affects blood temperature, metabolic rate, blood sugar levels, mental efficiency and work motivation.[47] Monk and Folkard[48] believe that there are three factors which have got to be right for successful coping with shiftwork; sleep, social and family life, and circadian rhythms. These factors are interrelated, so that a problem with one can negate the positive effects of success achieved in the others. Ultimately, roster designs should aim to maximise the positive effects and minimise the negative impact of shiftwork on health and social life caused by the disruption to circadian rhythms. Research evidence suggests that certain individuals, with steeper adrenalin curves than normal are those who tolerate shiftwork best; that is, those with a pattern of very high adrenalin levels during the day, and very low levels at night. The steeper rhythm is less vulnerable to distortion, and the rationale is that non-adaptation is better than partial adaptation. Akerstedt's study of railroad workers supports this observation.[49] Three weeks of daywork were followed by three weeks of nightworking. Very little adjustment in the pattern of adrenalin excretion was observed after three shifts of nightwork, with adrenalin remaining high during the day when the workers should have been sleeping. Even after three weeks of nightworking, no significant adjustment was observed. In fact, Akerstedt[49] suggested that high adrenalin levels may be the source of disrupted sleep patterns. In spite of the observation that some people may be physiologically better suited to shiftwork, many blue collar workers have no choice. Therefore, Knauth and Rutenfranz[50] suggest that rapidly rotating shift systems with few nightshifts in succession may be best because this causes less circadian rhythm disruption.

Although Selye suggested that individuals do habituate to shiftwork and it becomes physically less stressful with time,[37] some work patterns might prevent habituation occurring; for example, the offshore rig worker who may work a mix of 12-hour day and night shifts for 7 or 14 consecutive days without a break, and then has 7 or 14 days at home on leave.[16,17] The long term effects of this continual change from one environment to another are not known, although disrupted sleep over a period of time can cause a deterioration in the health and well-being of the individual. Hellesøy *et al.*[27] found that sleep problems were indicative of the strains of the offshore work environment. Personnel claim that they wake up tired, or wake up and have trouble going back to sleep. Although this industry is making a significant attempt to provide single cabin accommodation for the workforce, it is still common practice for an individual to share sleeping quarters. Disturbed sleep is thus a common complaint among these workers. A study of air-traffic controllers also indicated the strains associated with shift-working. Although other major job stressors were present, this group had four times the prevalence of hypertension, and more mild diabetes and peptic ulcers than did a control group of US Air Force personnel.[51]

Disturbance of nocturnal sleep leads to daytime fatigue, and sleepiness which impairs motivation and vigilance, thus affecting safety at work. Lavie *et al.*[52] found that industrial workers (n = 1502) complaining about their sleep had significantly more work accidents, repeated work accidents, and significantly more sick days per work accident. Furthermore, complaint of daytime sleepiness was related to multiple work accidents independent of age and physical effort required at work. Although it must be acknowledged that observed daytime sleepiness can be symptomatic of narcolepsy or breathing disorders, this behaviour may also be a response to stress associated with reduced mental well-being (including symptoms of depression and somatic anxiety). Individuals forced to work nightshifts or early morning shifts suffer from sleep disturbance in that the quantity and quality of their sleep differs as a consequence of shift working.[53] Daytime sleep is both more fragile and unstable than nocturnal sleep. However, the nightshift worker may be able to take cat-naps to catch up on sleep loss, thus reducing the sleep debt which accumulates (it is suggested that over a 7-day period the employee could effectively lose the equivalent of at least one night's sleep). This source of stress for shift workers may exacerbate other stress factors in the work environment; for example, irritability with other employees and intolerance of the physical conditions at work.

Although there are obvious problems associated with the need to work shifts on a regular basis, employees who habitually work long hours may also experience this as a considerable source of strain and/or pressure.

5.2.2. *Long Hours of Working.* A link between working long hours, stress and ill-health has been established. For example, Breslow and Buell[54] observed a relationship between the number of hours worked and death from coronary heart disease (CHD). Their study of "light industry" workers in the USA found that subjects of less than 45 years of age, but working more than 48 hours per week, had twice the risk of death from CHD, compared to similar individuals working 40 hours or less a week. Also, a study of 100 young coronary patients showed that 25 per cent of them had been working at two jobs, and an additional 40 per cent had worked for more than 60 hours a week.[55]

Sutherland and Davidson[56] found that long hours of working were part of the culture for on-site managers in the construction industry. It was clear that certain peer-group pressures operated to encourage the manager to arrive early and stay late. It was commonly believed that promotion prospects and assignment to the more lucrative contracts were contingent on being highly visible at work and demonstrating a strong commitment to the company (by working very long hours). However, this pattern of behaviour was observed to have an adverse impact on the workforce. Managers who had worked in excess of 20 hours overtime in the month prior to taking part in the health and stress audit screening were significantly worse off in terms of poorer psychological health (depression), job dissatisfaction and the perception of more stress and strain at work and home, when compared to managers who had worked less than 21 hours of overtime that month. Overall, this occupational group tended to work long hours (10.3 hours on average per day, sd 1.3), and only 20 per cent reported that they never worked at weekends. This need to work long hours also has other negative consequences: the individual consequently spends less time in social relationships, and so the benefits of social support as a buffer in a stressful job are reduced. It may also cause strain in his/her relationship with a spouse or partner, thereby creating home/work overspill stress. Indeed, this was observed to be a problem among top executives. Cooper and Sutherland[57] found that senior and top executives (CEOs) across Europe also experienced "work-spillover" problems which adversely affected their home and social lives. This survey of 118 chief executives found that two thirds worked nearly 12 hours each day; some up to 16 hours a day. Almost all of them worked at the weekends, thus restricting time spent with the family. Many of these senior executives seemed to realise the potentially damaging consequences of their lifestyle: 25 per cent believed that they were at substantial risk of job burnout (physical and emotional exhaustion) and more than a quarter of the executives responding to the survey perceived themselves to be at high risk from heart disease (especially the younger CEOs). From the results of this survey it was clear that these individuals were questioning their lifestyle, and many were actively thinking about quitting "life at the top". Nevertheless, it should be acknowledged that some individuals regard work, and working long hours, as a psychological haven and a way of escape from the pressures of home and family and/or unsatisfactory personal relationships. Distress only exists when demand exceeds or fails to satisfy abilities or expectations (real or perceived): but we may also decide to accept a stress situation for a given period of time because the "pay-off" is highly desired. Unfortunately, we often fail to balance the true costs and the gains, and thus become casualties of stress at work.

Although the need to work long hours appears to be a potent stressor in a job, the actual workload conditions may cause this problem. Simply having too much to do in the normal working day or shift results in the need to work overtime. However, the worker who struggles to do a job that is too difficult will take more time to finish the task, and may need to work extra hours in order to complete the job to a satisfactory standard. In the next section the differences between "quantitative" and "qualitative" work overload as sources of stress or pressure are explored further.

5.2.3. *Work Overload.* Having too much to do (quantitative overload) or work that is too difficult (qualitative overload) are forms of stress in the work environment. Various symptoms and behavioral malfunctions are associated with the experience of job overload,[58] although the stress associated with workload is complex. French and Caplan[64] suggested that quantitative and qualitative load produces many different symptoms in addition to psychological and physical strain; these include job dissatisfaction, job tension, lowered self-esteem, threat, embarrassment, high cholesterol levels, increased heart rate, skin resistance and more cigarette smoking. Nevertheless, Cooper and Marshall[58] have emphasised that objective workload should not be viewed in isolation, because it is necessary to consider perceived demand and ability in relation to the individual's actual capacity, personality, expectations, needs and the levels of social support in the work environment.

a. *Quantitative Overload.* Both physical and mental job overload are potent stressors at work. Having to work under time pressures in order to meet deadlines is an independent source of stress, and studies show that stress levels increase as difficult deadlines draw near.[59] There is also a relationship between objective, quantitative overload and cigarette smoking, a risk factor in coronary heart disease and cancer.[60] Linked to the concept of overload is the issue of "workpace". Rate of working has been shown to be a significant factor in blue collar ill-health, especially when the worker is not able to control the pace.[61] In a national survey in the USA, Margolis *et al.*[62] found that quantitative overload was significantly related to a number of symptoms or indicators of stress, including poor motivation, low self-esteem, absenteeism, escapist drinking and an absence of suggestion to employers.

b. *Qualitative Overload.* This occurs when the individual does not feel able, or capable, of performing the given task or duties. There is evidence that qualitative overload as a source of stress is strongly related to perceived low levels of self-esteem. In fact, this form of overload can occur when a worker is promoted to a supervisory or managerial job on the grounds of superior work performance: but his/her lack of experience or training in the supervision of others, and/or work delegation, can result in considerable strain and pressure for the individual. This stress situation may be compounded if the individual has to take disciplinary action against a previous co-worker.

5.2.4. *Work Underload.* Although most of us readily agree that having too much to do at work causes tension and strain which manifests in a wide variety of symptoms, it is also acknowledged that work underload is also a potentially damaging stress factor in the workplace. This relationship between the amount of work and health and performance is explained in terms of the Yerkes-Dodson Law,[63] "the inverted-U hypothesis". Figure 3 illustrates the associations between performance and underload/overload demands.

If the employee does not have enough work to do he/she may become bored; this also occurs when the task itself does not utilise the skills or potential of the worker (a real or perceived situation).[64] Therefore, in a "job underload" state, either quantitative or qualitative underload, the workforce is likely to be apathetic and bored: levels of motivation are probably low, and high levels of absenteeism are the costly consequences suffered by industry. Gardell[65] suggested that the effects of underload are aggravated by a lack of control over the work situation.

Paced assembly lines are often characterised as unstimulating work environments: the pace may be high (quantitative overload), but the nature of the task itself could be classified as a "qualitative underload" situation. Benyon and Blackburn[66] reported the feelings of tension resulting from boredom among workers on a packing-line system, because their work provided no sense of achievement or satisfaction. The job stress associated with passive, low skill demands, lack of variety, repetitiveness, and low decision latitude factory work also spills over into leisure time, and so negatively affects life outside of work.[65] New technology and the increasing automation of industry often reduces the quantitative overload condition, but produces jobs that are both highly repetitive, dull and monotonous, and this is detrimental to physical and mental well-being.

a. *Quantitative Underload.* Boredom in the daily work routine as a result of too little to do may result in inattentiveness. This is potentially hazardous if the employee fails to respond appropriately in an emergency.[67] Cheliout *et al.*[40] reported a high incidence of deactivation episodes among electronic assemblers (monitored by continuous electro-encephalography over the entire day). The theta rhythm observed, referred to as "micro-sleep", is indicative of the boredom

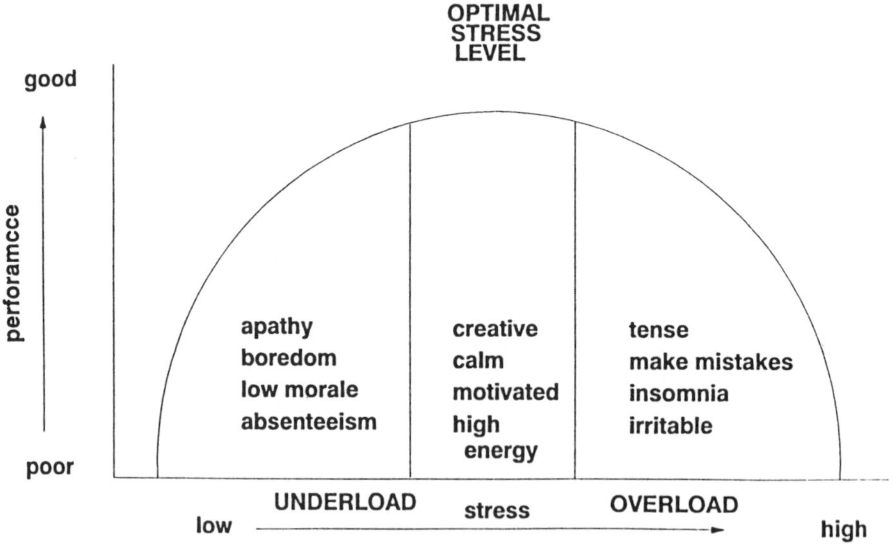

Figure 3 The relationships between workoverload/underload and performance.

and tedium experienced by the workers, and may be responsible for accident occurrence. Therefore, lack of stimulation may lead to dangerous conditions for employees. Boredom was identified as a significant source of stress among crane operators.[43] However, the long periods of inactivity are in the nature of their job and so job redesign would be necessary to alleviate this problem, because "boredom" and "lack of challenge" were found to be significant predictors of raised levels of anxiety, depression and reported job dissatisfaction among crane operators. Indeed, lack of stimulation may be potentially more damaging for shift workers at night when employees could have difficulty adjusting to the changes in sleep patterns, but do not have enough work in order to keep alert.[12]

Job overload and/or underload may result from the irregular flow of work which is not under the control of the worker. This is not restricted to paced assembly lines; many outdoor occupations are paced by climatic conditions. Pace may therefore vary from complete shutdown to "very hectic" as workers try to keep to contract deadlines in unpredictable conditions. In addition to financial gain, the promise of the next contract may be dependent on successful completion of the current job. Thus the workforce is exposed to strain and pressure which they cannot control or change.

b. *Qualitative Underload.* This may be as damaging as overload, in that an individual is not given the opportunity to use acquired skills or to develop full potential ability. As with quantitative underload, boredom and "shifts" or lapses in attention may have serious consequences. Underload due to lack of stimulation leads to low morale and poor motivation to work. The individual feels that he/she is "not getting anywhere", and is powerless to show perceived and/or actual skills and talent. Udris[40] suggested that qualitative overload was associated with dissatisfaction, tension, low self-esteem, whereas qualitative underload was linked to dissatisfaction, depression, irritation and psychosomatic complaints.

However, some studies which have used unobtrusive monitoring of heart rate and blood pressure have provided greater insights on the issue of work overload and underload conditions. For example, O'Brien *et al.*[68] demonstrated that the heart rate of medical nurses was "sensitive" to emotional stress, whereas the response among truck assembly-line workers was a function of the physical work and activity. Also, Salvendy and Knight[69] used on-line monitoring of blood pressure to demonstrate that industrial workers in machine-paced tasks did not display any greater stress response risk than in self-paced tasks. They stated that CHD risk is a complex equation of individual characteristics (age, expectations, attitudes and personality), perceived stress level in the job environment and the interaction between these factors. Use of objective measures in the form of physiological monitoring may help to discover the extent to which stress is implicated as a cause of the prominent killers in modern society. However, the perception of a stress agent and the subsequent psychological response are also critical: for example, Broadbent and Gath[70] found an association between reported levels of job dissatisfaction and somatic symptoms among assembly-line workers. Those engaged in paced work reported high levels of anxiety and depression.

Various programmes of job enlargement, job enrichment, job rotation, increased participation in decision-making and shared ownership may be introduced to alleviate some of these problems. But they also create separate problems and different sources of stress in the work environment; for example, disruption to social relationships, role ambiguity and status incongruence in the workplace.

5.2.5. *New Technology.* Lack of stimulation, under-utilization of skills and boredom characterises many blue collar occupations; but, in rapidly changing work environments, skills may also quickly become obsolete. The need to constantly become familiar with new equipment and systems may pose a threat to the individual. Unless adequate training is provided, potentially stressful situations may develop when new technology is introduced into the workplace, and the individual feels unable to do the given task. This is known as "technostress". Change introduced by computerisation demands flexibility; it requires an ability to alter former practices, reorientate and adapt. Older, established staff may find it a strain to make such dramatic changes and these feelings of pressure may be exacerbated when they see the younger generation of workers, educated with computers, quickly grasp the new ways of working. It seems that we also try to protect ourselves against "the unknown", and so tend to resist computerisation of the work environment. New and unfamiliar work patterns cause worry and anxiety, and pose a threat and challenge to our perception of autonomy and self-control. The individual may also experience a threat to job security if the introduction of new technology brings (or threatens!) a reduction in manpower requirements in the organisation. The employee often faces an extended period of time at work in the knowledge that he/she may ultimately become redundant when the computerisation process has become established. This time of uncertainty is very stressful. Often the stress caused is quite unnecessary because the fears of job loss are not realised. However, this constant threat hangs over the workforce like the Sword of Damocles during the time of transition from one system to another.

Threat of job loss is a potent source of stress associated with several serious health problems, including ulcers, colitis and alopecia,[71] and increased muscular and emotional complaints.[72] The morale and motivation of the workforce is affected, with subsequent negative impact on productivity and efficiency. Indirectly, fear of job loss and insecurity in times of high unemployment adversely affects both the individual and the organisation. A keen, competitive job market may threaten the quality of peer relationships at a time when social support is of particular importance. The stress associated with feelings of insecurity may otherwise be reduced by the buffering effect of good supportive relationships at work; and this may be broken down if the workforce perceives that competition is necessary to retain the job. Other individuals may stay in a job that is disliked or unsuitable because no alternative for change exists. A study of personnel working for one offshore contractor showed that 41 per cent of workers reported the stress of "feeling trapped into offshore work because no suitable onshore work was available."[16] Indeed, the stress associated with the need to change and/or retrain is also likely to manifest at a time of life when the individual is most vulnerable. An individual under the strain of impending job loss realises that, in middle age, learning seems to take longer, energy is more scarce, opportunities are less, and the threat of a keen, younger workforce competing for jobs are formidable obstacles. Fear of failure, rejection, and rejection itself, are damaging to morale, self-esteem and confidence. Overall, the perception of stability in the organisation and employment affects the well-being of employees, and this has obvious implications for productivity and performance levels.

An added source of strain in the rapidly changing work environment, and related to "job overload", is having a boss schooled in the "old way". The new employee, trained in the latest methods, and perhaps educated to a higher standard in order to compete in a keen job market, may experience overload stress if adequacy of supervision is questionable. Confidence and respect in the ability of those responsible for the efficient and safe operation of a system or plant is vital to good interpersonal relationships at work. Introduction of new technology may expose a supervisor to conflict and the experience of qualitative overload, and threaten the subordinate with overload stress because supervision is perceived as inadequate. This ultimately has an adverse impact on the quality of relationships in the workplace.

In addition to the problems associated with "technostress" there is some evidence to suggest that the "hardware" of new technology may also cause strain and tension. One particular area of interest has been the usage of computerised work methods especially in the use of video display terminals/units (VDT/VDU) and the possible risk of developing adverse conditions related to vision, muscle and joint aches and pains, colds and sore throats, and headaches.[72,73] More recently, concerns have been focused on the psychosocial factors in health, for example, workload, stress and health; motivation, job satisfaction and development; cognitive processes in human/computer interactions; and social and organisational structures and functions.[74] The results from these studies on the psychosocial impact of computerisation are complex. For example, Johansson and Aronsson[75] reported that VDU workers in an insurance company felt that their work was less varied after the introduction of VDUs, although they did not agree that the work had become monotonous or too simplistic. When the work of data coding (the biggest source of decrease in job variety) was redistributed, in order to eliminate the pure data entry jobs, the perceived unsatisfactory changes resulting from the computerisation were eliminated.[76]

In many types of clerical work, the introduction of new technology has actually alleviated the problem of qualitative mental underload, rather than de-skilling the job: Kalimo and Leppönen[74] reported that VDU typesetters assessed their work as more challenging than it was under the old system which used perforated typesetters. The new system allowed them to enlarge the job (they had feedback on quality, so they could make corrections to the work, thus the perceived degree of personal control was greater within the new system). Similarly, the computerisation of salary calculation was viewed as a positive experience by the workforce, and the expected monotony and simplification of job demands were not realised. Many of the laborious manual tasks were eliminated after computerisation and the payroll clerks reported that they subsequently had a better understanding of the whole process of salary calculation.[74] Nevertheless, studies have shown that equipment breakdown, leading to disturbances and delays, cause considerable stress to the users, resulting in irritation and frustration at the situation, which tends to be outside the control of the workers. Loss of data (resulting in added workload burdens), extra workload when the breakdown is over, and customer complaints are some of the stress problems of the "hi-tech" work environment. There are also various health related worries associated with work computerisation, especially radiation fears.

The health effects of working with VDUs have been studied extensively, but it is not possible to draw firm conclusions because of methodological differences and difficulties in the published studies. According to Rossignol *et al.*, this includes the problems associated with the lack of a control group; use of a non-comparable control group; disruptions or disputes in the workplace at the time of the study, and sample bias.[77] Their large scale study attempted to address these issues. In a controlled cross-sectional study of 1545 clerical workers in the USA, they recorded an increased prevalence of adverse conditions relating to vision, musculoskeletal discomfort, and headaches among VDT workers. The magnitude of this was dose dependent, that is, dependent directly on the duration of daily VDT use. They suggested that workers who used VDTs four or more hours per day may decrease the frequency of discomfort and prevent any associated chronic effects if they reduced the daily numbers of hours at a VDT to three or fewer. Kalimo and Leppönen agree; they state:

"The combination of routine, non-autonomous work, time constraints and long periods in front of VDUs is a stress provoking combination that should be avoided. If long delays, breakdowns and ergonomic problems are added, one can be sure that the workers both experience stress and have objective short-term stress reactions. The same problems can be observed in many workplaces without VDUs. Thus, the organisation rather than the equipment has the decisive role."[74]

In certain industries the workforce faces a "hi-tech" environment and the stress burdens associated with possible exposure to dangerous or hazardous conditions, either intrinsic to the task itself, or by virtue of the unescapable environmental conditions.

5.2.6. *Exposure to Danger and Hazard.* The risk and hazard associated with danger in certain occupations may be a source of stress. Blue collar workers in the USA rated "health and safety hazards" as a highly significant source of job discontent.[78] Various occupational groups are identified as "high-risk" in terms of physical danger, for example, police officers, mine-workers, soldiers, prison personnel, firefighters and workers on oil and gas exploration and production installations.[16,17,67,79,80,81] However, it is not known if the special risks associated with these occupations are perceived as sources of stress by the competent and trained employee. Nevertheless, it is possible that a continued emphasis on the need for safety in a hazardous environment might be a greater source of stress. Bohemier[82] suggests that it is human nature to avoid thinking about danger or death in a hazardous or risky environment, and that it is necessary to block out some of the realities which the worker must otherwise continually face. As Cherry[36] declares in a study of high-building iron workers, "many workers learn to combine a hard-boiled veneer about job hazards with abiding private anxieties."

Various studies indicate that some workers do perceive risks and hazards associated with the job as a source of stress. Casters in the steel industry and crane operators acknowledged the dangers of the job:[15,43] awareness of danger and the consequences of making a mistake were significant predictors of depression and anxiety among crane operators. In a Norwegian study, 36 per cent of offshore platform personnel felt unsafe about helicopter transport; 34 per cent felt unsafe about evacuation facilities, and 24 per cent were concerned about the risk of fire and explosion.[27] Risk of exposure to certain chemicals is frequently reported as one of the most harmfully perceived stressors among blue collar workers.[40] This includes the inhalation of vapours and dust, and exposure to chemicals that are irritants to the skin. Adequate protective clothing is vital, but training and education are equally important in reducing the stress associated with working in a potentially hazardous environment. The perceived adequacy of medical facilities and the ability to cope with an emergency situation will mitigate the potentially stressful risky situation.

Indeed, for the offshore oil and gas worker, part of the hazard and risk associated with the job is the *travel* to and from work. In addition to the stress associated with long-distance travel, and the time lost from shore leave if the individual is required to stay overnight to meet early "check-in" requirements, the mode of travel, by helicopter, is itself a potential source of stress.[16,17] Previous helicopter accidents, high impact media attention, and the need for special training to deal with emergency routines, heightens the level of risk awareness and can be additional sources of stress. In the Statfjord study in Norway, employees stated that members of their family worried about helicopter travel, but *they* got used to it and did not experience much fear or worry.[27] However, they may deny their real feelings and emotions in order to cope with the situation that they cannot control: and repression of true feelings might ultimately manifest in the form of psychosomatic illness and/or nervous debility, etc. In fact, the need to travel by helicopter is believed to be the most common reason for resignation from offshore employment in the Norwegian sector of the North Sea.[83] A reported helicopter incident from one of the hundreds of daily flights to and from offshore installations, always prompts a high number of resignations from this way of life and work.

The risks and dangers associated with many occupations cannot be changed, but the perception of risk can be reduced by training and education (to ensure rational and logical reasoning about the risks involved). This is very important, because anxious, obsessional and phobic workers are less motivated to work, have low morale and may be more vulnerable to accidents.[17] In the long term, they may also suffer the consequences of stress-related illness, including heart disease and ulcers.

5.2.7. *Travel.* Although relatively few people travel to work each day by helicopter, statistics show that our daily commuting by road and rail is in reality more dangerous than helicopter travel. Each year there are more deaths in road traffic accidents than in the offshore environment. Nevertheless, the high media image and publicity surrounding incidents on drilling rigs and oil and gas platforms heightens the sense of danger for those working in the industry. The rest of us tend not to dwell on such risks each day, but do appear to suffer from the strains of commuting to and from work. Traffic jams on the roads cause delays and upset our busy time schedules; delayed trains and other public transport causes high levels of stress because the situation is unpredictable. Our lack of control in such circumstances tends to magnify the problem.

For many individuals travel to and from the job also absorbs a considerable part of the day and so is a potential source of stress. For example, on-site construction managers identified "travel" as a significant stressor.[56] It was rated as fifth in their top ten sources of pressure, but was also related to the other major stress factors reported by this group. For example, "time pressures", "working long hours" and "an insufficient time to pursue leisure interests", the top three stressors, were all exacerbated by the need to travel for two to three hours each day. Many of the managers used public transport so they could spend this time productively dealing with the high volume of paperwork in the job: but the crowded trains, delays in their journey time and cancelled services, caused considerable pressures before they had even arrived at work to face the hassles of the day.

6. STRESS MANAGEMENT AND PREVENTION

In the previous section we have considered sources of stress in our work environment. From the discussion it is quite clear that a wide variety of potential stressors exist, but many factors mediate in the stress process. Thus it is acknowledged that the first step in the stress management process must be:

6.1. *Awareness*

It is impossible to deal with stress unless we can effectively recognise the problem. Ideally, we should seek to identify a stress agent in order to eliminate it, but we live in a real world not an ideal one. So we must also try to manage stress in the work environment by adopting adaptive coping strategies, rather than use maladaptive coping. Inappropriate coping strategies may actually add to the stress experienced rather than help us to deal with difficult pressures or threats. Table I describes "adaptive" and "maladaptive" coping strategies. Adaptive behaviour deals directly with the stressful situation by seeking and implementing solutions; maladaptive behaviour is defensive in nature and usually removes a stressful situation only on a temporary basis. Sometimes the actual behaviour in response to stress renders the individual less able to cope in the future; for example, dependence on drugs, or procrastination, could result in a bigger, or more complex problem to face later; and it is more likely that the

Table I Coping with Stress: adaptive and maladaptive behaviour

Adaptive	Maladaptive
—accept that a potentially stressful situation exists	—deny the problem/situation
—deal directly with the situation in an objective and rational way	—procrastinate; "it" might just go away
—use active coping behaviour strategies (i.e. be in control)	—passive acceptance of the situation (relinquish control)
—set priorities and deal with them accordingly	—rush around; try to tackle several things at once; if you are busy it must be OK!
—use constructive self talk—"be positive"	—dwell on negative thoughts; "destructive of self"
—develop a social support network; at work and with family and friends (seek help—emotional and instrumental)	—adopt the "John Wayne, go-it-alone" technique of coping
—express feelings and emotions openly	—"bottle" things up inside —anger —frustration —guilt
—accept that you have learned from a situation and will cope even better next time	—live in dread and fear of "next time"
—reward yourself for coping with a difficult situation	—continually hold post mortems and punish yourself for your less than perfect performance
—have well balanced work and home-life outside interests	—be a workaholic
—take regular exercise (an outlet for stress build-up)	—be a "couch potato"
—learn to relax/take yoga or meditation etc. Use techniques on a daily basis	—depend on tranquilisers and sleeping pills in order to relax; smoke too much or drink too much alcohol to help you to relax, or give yourself confidence to face the day
—remember, you need adequate energy to cope with difficult situations; so eat regularly (a balanced diet) and get plenty of rest/sleep	—exist in a state of anxiety and tension so that you cannot eat or sleep properly
—acknowledge that you work best in an optimal stress state. Mismanaged stress is destructive	—become a "stress seeker"—i.e. hooked on the stress "buzz"

health and well-being of the individual is impaired by reliance on maladaptive ways of coping, and so he/she will be less able, or equipped to deal with any subsequent threat or pressure. It is therefore very important that an individual has a well-developed sense of self-awareness in order to recognise their own strengths and vulnerabilities in the stress coping process. This information is crucial to the next step in the management of stress, that is:

6.2. *Analysis*

Correct identification of a stressor is often difficult because of the interactive nature of stress. As we have seen, certain forms of stress may combine to form a stress-chain. This occurs when small, minor daily hassles combine to exert considerable pressure on the individual. Often the total is greater than the sum of the parts. As we say, "it is the straw that breaks the camel's back", and this applies in the stress scenario. A colleague may explode in a rage or trantrum at a seemingly minor request or during some interaction with another person. We may look on in bewilderment because we have failed to realise that this was "the last straw" for the other person. It is difficult to be logical and to objectively analyse the situation when conditions are overheated and distraught. Therefore, we should adopt a preventive approach to stress management to avoid the "dis-tress" situation. Responsible organisations have adopted this "stress audit" approach to stress management as a preventive strategy. The result of a stress audit provides each individual with a personal stress profile of strengths and weakness, which can be used as part of a personal development programme, and the organisation is able to consider the most suitable and practical forms of intervention and/or organisational development. This leads to the next step in the stress management process:

6.3. *Action*

We have already discussed certain "action" options for the individual in Table I: Table II shows some of the action strategies that may be introduced as part of a health promotion and stress management intervention by the organisation. Provision of these programmes are common in America, whereas the UK and other European companies have adopted a more sceptical and cautious approach. However, the introduction of these activities continues to grow as evaluation studies show that stress management programmes are cost effective, aside from the moral and humanistic arguments. Indeed, if employers are forced to accept more of the share in employee health care costs in the future, and the fears of stress litigation continues, we can expect to see a significant surge of interest in stress management in the workplace. Stress in the work environment is ultimately the responsibility of management and the workforce. It is not cost effective to adopt the view that employees should be able to cope on their own: they (the workforce) are the company's most expensive asset and should be valued accordingly. By adopting an *Awareness–Analysis–Action* approach to stress management, both the individual and the organisation can realistically begin to tackle occupational stress in the work environment. As Levi[84] suggests, "much can be accomplished through two universally available measurements and intervention tools, namely listening and speaking. People know their own problems best and they should be encouraged to speak for themselves."

Table II Managing stress and health in the organisation

—Stress education
—Stress management programmes
 —individual stress profiling (with action plan and formulation progress follow-up sessions)
—Skills training
 —managing type A behaviour
 —time management
 —assertion training
 —social skills development
—Keep fit/healthy lifestyle programmes
—Relaxation/meditation classes
—Counselling
—Job re-design
 —job rotation/job share
 —career sabbatical
 —flexi-hours
 —parental leave
 —home-working
 —creche
—Increased participation
 —quality circles
 —team briefing

References

1. J. C. Quick and J. D. Quick, *Organisational Stress & Preventive Management* (New York, McGraw-Hill, 1984).
2. H. Seyle, *The Stress of Life* (New York, McGraw-Hill, 1956).
3. R. S. Lazarus, *Psychological Stress and the Coping Process* (New York, McGraw-Hill, 1966).
4. L. Levi, "Definitions and the conceptual aspects of health in relation to work" **In**: (R. Kalimo, M. A. El-Batawi and C. L. Cooper, Eds.), *Psychosocial Factors at Work and Their Relation to Health* (Geneva, WHO, 1987) pp 9–14.
5. V. J. Sutherland and C. L. Cooper, *Understanding Stress: A Psychological Perspective for Health Professionals* (London, Chapman and Hall, 1990).
6. J. H. Knowles, *Doing Better and Feeling Worse: Health in the United States* (New York, Norton 1977).
7. A. Toffler, *Future Shock* (London, Pan, 1970).
8. H. Selye, "Correlating Stress and Cancer" *Amer. J. Proctology, Gastroenterology, Colon and Rectal Surgery* **30**(4), 18–28 (1979).
9. H. Munsterberg, *Psychology and Industrial Efficiency* (New York, Houghton-Mifflin, 1913).
10. F. Roethlisberger and J. J. Dickson, *Management and the Worker* (Cambridge, Mass., Harvard University Press, 1939).
11. D. M. Jones, "Noise" **In**: (R. Hockey, ed.), *Stress and Fatigue in Human Performance* (London, John Wiley, 1983).
12. E. C. Poulton, "Blue Collar Stressors" **In**: (C. L. Cooper and R. Payne, Eds.), *Stress at Work* (London, John Wiley, 1978) pp. 51–81.
13. J. M. Ivancevich and M. T. Matteson, *Stress at Work* (USA, Scott, Foresman, 1980).
14. M. J. Smith *et al.*, "Characteristics of successful safety programs" *J. Safety Research* **10**, 5–15 (1978).
15. M. Kelly and C. L. Cooper, "Stress among blue collar workers. A case study of the steel industry" *Employee Relations* **3**(2), 6–9 (1981).
16. V. J. Sutherland and C. L. Cooper, *Man and Accidents Offshore: the costs of stress among workers on oil and gas rigs* (London, Lloyd's List/Dietsmann International, NV, 1986).
17. V. J. Sutherland and C. L. Cooper, *Stress and Accidents in the Offshore Oil and Gas Industry* (Houston, Gulf Publishing Company, 1991).
18. D. E. Broadbent and F. A. Little, "Effects of noise reduction in a work situation" *Occupational Psychology* **34**, 133–140 (1960).
19. W. A. Kerr, "Accident proneness of factory departments" *J. Applied Psychology* **34**, 167–170 (1950).

20. A. Cohen, "Industrial noise, medical absence and accident record data on exposed workers" **In**: (W. D. Ward, ed.) *Proceedings of the International Congress on Noise as a Public Health Problem* (Washington, US Environmental Protection Agency, 1974).
21. A. Cohen, "The influence of a company hearing conservation program on extra-auditory problems in workers" *J. Safety Research* **8**, 146–162 (1976).
22. M. H. Lader, "Response to repetitive stimulations" **In**: (L. Levi, ed.), *Society, Stress and Disease, Vol. 1* (London, Oxford University Press, 1971) pp. 417–432.
23. H. E. Bustin, "Environmental health requirements on offshore oil and gas exploration and production installations" *International Symposium Proceedings, Safety Health in the Oil and Gas Extractive Industries* Luxembourg, April 1983 (London, Graham and Trotman, 1983).
24. B. C. Ross, "Noise on offshore platforms" *Occupational Health* **30**(11), 525–527 (1978).
25. N. A. Grundy, "Offshore Noise and Vibration. Problems and Solutions" **In**: (Catherine O'Keefe, ed.), *Certification and Safety of Offshore Installations* (London, Scientific and Technical Studies. Conference Transcript, May 1981).
26. V. Keenan and W. Kerr, "Psychological climate and accidents in an automotive plant" *Journal of Applied Psychology* **35**(2), 108–111(1951).
27. O. H. Hellesøy, (ed.), "Work Environment Statfjord Field" (Universitetforlaget, Bergen, 1985).
28. J. D. Ramsey, "Heat and Cold" **In**: (Robert Hockey, ed.), *Stress and Fatigue in Human Performance* (London, John Wiley, 1983) pp. 33–57.
29. J. Surry, *Industrial Accident Research: A Human Engineering Appraisal* (Canada, Ontario Department of Labour, 1968).
30. M. D. Whorton *et al.*, "Investigation and Work-up of Tight Building Syndrome" *J. Occupational Medicine* **29**(2), 142–147 (1987).
31. A. S. Robertson, "Building Sickness" *Practitioner* **229**, 531–534 (1985).
32. R. W. Alexander and M. J. Fedoruk, "Epidemic Psychogenic Illness in a Telephone Operators' Building" *J. Occupational Medicine* **28**, 42–45 (1986).
33. T. L. Guidotti *et al.*, "Epidemiologic Features that may Distinguish between Building-associated Illness Outbreaks due to Chemical Exposure or Psychogenic Origin" *J. Occupational Medicine* **29**(2), 148–150 (1987).
34. P. Hingley and C. L. Cooper, *Stress and the Nurse Manager* (London, John Wiley, 1986).
35. G. R. Hockey, "Effect of loud noise on attentional selectivity" *Quarterly J. Experimental Psychology* **22**, 28–36 (1970).
36. A. B. Shostak, *Blue-Collar Stress* (Reading, Mass., Addison-Wesley, 1980).
37. H. Seyle, *Stress in Health and Disease* (London, Butterworths, 1976).
38. M. Frankenhaeusser and B. Gardell, "Underload and overload in working life: Outline of a multidisciplinary approach" *J. Human Stress* **2**, 35–46 (1976).
39. D. Hay and D. Oken, "The Psychological Stress of Intensive Care Nursing" *Psychosomatic Medicine* **34**, 109–118 (1972).
40. ILO, "Psychosocial factors at work: Recognition and control" *Report of the Joint ILO/WHO Committee on Occupational Health. Ninth Session. 1984* (Geneva, International Labour Office, 1986).
41. V. C. Cox *et al.*, "The relationship between crowding and health" **In**: (A. Baum and J. Singer, eds.), *Advances in Experimental Psychology, Vol. 4* (Hillsdale, NJ, Lawrence Erlbaum, 1982) pp. 271–294.
42. G. W. Evans, "Behavioural and physiological consequences of crowding in humans" *J. Applied Social Psychology* **9**(1), 27–46 (1979).
43. C. L. Cooper and M. Kelly, "Stress among crane operators" *J. Occupational Medicine* **26**(8), 575–578 (1984).
44. E. A. Hartman and B. Pearlman, "Burnout: summary and future research" *Human Relations* **35**(4), 283–305 (1982).
45. T. M. Monk and D. I. Tepas, "Shift Work" **In**: (C. L. Cooper and M. J. Smith, eds.), *Job Stress and Blue Collar Work* (London, John Wiley, 1985).
46. J. Rutenfranz *et al.*, "Biomedical and psychosocial aspects of shift work" *Scandinavian J. Work Environment and Health* **3**, 165–182 (1977).
47. M. J. Smith *et al.*, "Health and Safety Consequences of Shift Work in the Food Processing Industry" *Ergonomics* **25**, 133–144 (1982).
48. T. H. Monk and S. Folkard, "Circadian rhythms and shiftwork" **In**: (R. Hockey, ed.), *Stress and Fatigue in Human Performance* (London, John Wiley, 1983).
49. T. Akerstedt, "Invasion of the sleep-wakefulness pattern: Effects on circadian variation in psychophysiological activation" *Ergonomics* **20**, 459–474 (1977).
50. P. Knauth and J. Rutenfranz, "Development of criteria for the design of shiftwork systems" *J. Human Ergol.* **11**, Suppl. 337–367 (1982).

51. S. Cobb and R. H. Rose, "Hypertension, Peptic Ulcer and Diabetes in Air Traffic Controllers" *J. Australian Medical Association* **224**, 489–492 (1973).
52. P. Lavie *et al.*, "Sleep disorders and safety at work in industry workers" *Personality and Individual Differences* **6**(3), 305–311 (1982).
53. A. J. Tilley *et al.*, "The sleep and performance of shift workers" *Human Factors* **24**(6), 629–641 (1982).
54. L. Breslow and P. Buell, "Mortality from coronary heart disease and physical activity of work in California" *J. Chronic Diseases* **11**, 615–626 (1960).
55. H. I. Russek and B. L. Zohman, "Relative significance of heredity, diet and occupational stress in CHD of young adults" *Amer. J. Medical Sciences* **235**, 226–275 (1958).
56. V. J. Sutherland and M. J. Davidson, *Using a Stress Audit as a Precursor to Stress Management Interventions: the Construction Site Manager Experience* (Unpublished Report, UMIST, 1990).
57. C. L. Cooper and V. J. Sutherland, "The Stress of the Executive Lifestyle: Trends in the 1990s" *Employee Relations* **13**(4), 3–7 (1991).
58. C. L. Cooper and J. Marshall, *Understanding Executive Stress* (London, Macmillan, 1978).
59. M. Friedman *et al.*, "Changes in the serum cholesterol and blood clotting time in men subjected to cyclic variation of occupational stress" *Circulation* **17**, 852–861 (1958).
60. J. R. P. French and R. D. Caplan, "Psychosocial Factors in Coronary Heart Disease" *Industrial Medicine* **39**, 383–397 (1970).
61. M. Frankenhaeuser and G. Johansson, "Stress at work: psychobiological and psychosocial aspects" *Int. Rev. Applied Psychology* **35**, 287–299 (1986).
62. B. Margolis *et al.*, "Job Stress: an unlisted occupational hazard" *J. Occupational Medicine* **1**(16), 659–661 (1974).
63. R. M. Yerkes and J. D. Dodson, "The relation to the strength of the stimulus to the rapidity of habit formation" *J. Comparative Neurology and Psychology* **18**, 459–482 (1908).
64. J. R. P. French and R. D. Caplan, "Organisational Stress and Individual Strain" **In**: (A. J. Marrow, ed.), *The Failure of Success* (New York, Amacon, 1973) pp. 30–66.
65. B. Gardell, "Technology, Alienation and Mental Health" *Acta Sociologica* **19**, 83–94 (1976).
66. H. Benyon and R. M. Blackburn, *Perception of Work: Variations within a Factory* (Cambridge University Press, 1972).
67. M. J. Davidson and A. Veno, "Stress and the Policeman" **In**: (C. L. Cooper and J. Marshall, eds.), *White Collar and Professional Stress* (London, John Wiley, 1980) pp. 131–167.
68. C. O'Brien *et al.*, "A study of the strains associated with medical nursing and vehicle assembly" **In**: (C. Mackay and T. Cox, eds.), *Response to Stress: Occupational Aspect* (London, IPC Science and Technology Press, 1979) pp. 52–60.
69. G. Salvendy and J. Knight, "Circulatory responses to machine-paced and self-paced work. An industrial study" *Ergonomics* **26**, 713–717 (1983).
70. D. E. Broadbent and D. Gath, "Symptom levels in assembly line workers" **In**: (G. Salvendy and M. J. Smith, eds.), *Machine Pacing and Occupational Stress* (London, Taylor and Francis, 1981) pp. 243–253.
71. S. Cobb and S. V. Kasl, "Termination—The Consequences of Job Loss" (USA, HEW Publication 77-224, NIOSH 1977.
72. M. J. Smith *et al.*, "An investigation of health complaints and job stress in video display operations" *Human Factors* **23**, 389–400 (1981).
73. U. O. V. Bergqvist, "Video Display Terminals and Health: A Technical and Medical Appraisal of the State of the Art" *Scandinavian J. Work Environment and Health* **10**, 1–87 (1984).
74. R. Kalimo and A. Leppönen, "Visual Display Units—Psychosocial Factors in Health" **In**: (M. J. Davidson and C. L. Cooper, eds.), *Women and Information Technology* (London, John Wiley, 1987) pp. 193–227.
75. G. Johansson and G. Aronsson, "Stress Reactions in Computerised Administrative Work" *Reports from the Department of Psychology, Supplement 50* (Stockholm, University of Stockholm, 1980).
76. G. Aronsson, "Work Content, Stress and Health in Computer-Mediated Work: A Seven Year Follow-up Study" *Proceedings of the International Scientific Conference on Work with Display Units* (Stockholm, May 12–15, 1980) pp. 401–404.
77. A. M. Rossignol *et al.*, "Video Display Terminal Use and Reported Health Symptoms among Massachusetts Clerical Workers" *J. Occupational Medicine* **29**(2), 112–118 (1987).
78. F. Wallick, *The American Worker: An Endangered Species* (New York, Ballantine, 1972).
79. R. Kalimo, "Stress in work: Conceptual analysis and a study on prison personnel" *Scandinavian J. Work Environment Health* **6**(3), 148 (1980).
80. S. V. Kasl, "Mental health and work environment. An examination of the evidence" *J. Occupational Medicine* **15**, 506–515 (1973).

81. D. H. Elliott, "The offshore worker" *The Practitioner* **229**, 576–571 (June, 1985).
82. A. Bohemier, "Mar-Tech, 1985, Montreal". Reported *Lloyds List* (May 23, 1985).
83. A. Sunde, (1983), "Psychosocial aspects of offshore work" *Int. Symposium Proceedings* Luxembourg, April 1983 (London, Graham and Trotman, 1983).
84. L. Levi (ed.), *Emotions: Their Parameters and Measurement* (New York, Raven Press, 1975).

7. STRESS, HEALTH, AND THE OFFICE ENVIRONMENT

LISA DORN*

The relationship between working environments and worker's health has been the subject of much research for some years. However, the office landscape has radically altered in recent years with the introduction of new technology and the design of more energy efficient office buildings. There has been comparatively little research which examines the effect of these changes on worker's health. The office can be regarded as a source of stress for the worker which incorporates many potential stressors, i.e. the nature of the job, badly designed office furniture, office layout etc. This chapter reviews the literature on the relationship between health and two potential sources of stress and ill health among office workers: the visual display terminal (VDT) and 'Sick Building Syndrome' (SBS). Health concerns for the VDT user range from visual, musculoskeletal, and even reproductive problems. For those inhabiting a so-called 'sick building', workers report suffering sinus problems, eye irritation, headaches, etc. There is an ever-increasing labour force working in offices and a widespread use of new technology. The office has had an impact not only on the office workers' health but also on the subjective feeling of stress at work. The office as a source of stress must be acknowledged in the design of both technology and the office building itself.

1. INTRODUCTION

The centrality of work in Western industrial societies has lead many researchers to examine whether working environments may adversely affect worker health and well-being, resulting in increases in job-related stress and in such withdrawal behaviours as absenteeism and turnover. Job stress and health are significantly related[1] and 'daily hassles' experienced at work are significantly associated with absenteeism.[1] Currently the most typical working environment is the office. In the United States, for example, almost half the labour force work primarily in offices,[2] and the number of office workers is increasing rapidly, both in developed and developing countries. This chapter is concerned with two areas which have become prominent in recent research on the health and well-being of office workers: first, the use of visual display terminals (VDTs) and second, what has become known as 'Sick Building Syndrome'. Both these factors can be regarded as 'stressors' within the working environment.

*Applied Psychology and Organization Studies, Aston Business School, Aston University, Aston Triangle, Birmingham, B4 7ET, UK

2. THE INTRODUCTION OF VISUAL DISPLAY TERMINALS TO THE WORKPLACE

There has been a rapid spread of computer usage within various fields of working life during the last decade. It is estimated that by the year 2000 up to half of US employees will be making daily use of VDTs.[3] This technology has the effect of reducing working space, emitting constant noise, radiating heat and reflecting lighting glare from its screens in an office which is often ill-equipped in terms of ventilation. Research shows that VDT users report more negative features of the work environment than several other office job categories.[4] The introduction of VDTs has radically changed the office landscape and appears to have paid little regard to human factors and industrial and organizational psychology.

Computers have assumed pole position in the office layout and become the focus of the working day. However, in many cases VDTs, printers, keyboards, external disk drives, etc., have been 'slotted in' to existing offices not designed for their inclusion. These units often sit uneasily in such environments, giving an untidy appearance caused by a profusion of electric cables and sockets. Since the advent of technology the office and furniture design has become a political issue, as unions are increasingly helping their members to be recompensed for adverse environmental conditions at work.[5] Unions are concerned with the failure of employers to take the health of the office worker into account in the design, positioning and operation of VDTs. Ergonomic design of the workstation and adequate instruction on how to operate new technology help prevent physical and mental strain.[6] Yet there is often inadequate training on the new technology, leading to errors, confusion and stress. The Manpower Services Commission[7] recommends that a period of three months is necessary before an experienced typist becomes a competent word processor. However, three days' training is more typical of most organizations.[7]

2.1. *The Nature of VDT Work*

VDTs have created the electronic equivalent of the repetitive, boring and highly paced work of the factory 'hand'. The new computerised work no longer requires creation, storage and distribution of documents; rather, employees carry out their work at the touch of a button. It is therefore important to acknowledge that the nature of VDT work may be contributing to worker ill-health. It is the lowest stratum of VDT users who process data that report heightened intensity demands.[8]

Inadequate working conditions and features of the task such as workload and/or work pressure are likely contributors to the VDT/health debate.[9] It appears that VDT operators contend with a more highly paced job, less work control and more deadline pressures compared with other types of office work.[10] Lack of job autonomy, supervisory control and opportunities for career advancement, along with increased work load are characteristics of the nature of VDT work. A factor analysis of a 40-item questionnaire administered to 34 VDT users found that the most significant factor predicting ill-health at work was 'Lack of Role Clarity' (44.2% of variance).[11]

Research finds that VDT users see work overload and time pressure as major stressors at work.[12] Overload can result in:[13]

- job dissatisfaction
- job tension

- lower self-esteem
- psychological threat
- embarrassment
- increased heart rate
- high cholesterol levels
- increased smoking

Both stimulus underload and overload can occur in highly mechanised automated work.[14] Combined with a lack of personal control, such stimulus conditions can trigger physiological changes such as adrenal secretions. Frequent or long-lasting elevations of catecholamine levels may result in structural changes of the blood vessel. Such changes may lead to cardiovascular disease.[15] Even small environmental changes can have an effect.[16,17] VDT users who spent more than 50% of their time at a terminal entering data, had higher levels of catecholamine than controls.[15] This difference increases in the evening, suggesting that VDT users take longer to unwind. Slower unwinding implies a greater wear on the biological system.[15]

Not surprisingly then, several studies report that VDT operators may be more stressed generally than non-VDT users.[12,18] There are many possible explanations for this increased level of stress. For example, reports on the impact of new technology on workers who are familiar with older non-technological systems show significant differences in health and well-being, particularly on mental health measures, after being 'redeployed' to advanced technology.[19]

2.2. *VDTs and Psychological Well-Being*

There is growing concern over the change VDT technology has brought to the office, which may cause psychological disturbances. Anxiety, depression and irritability are common complaints among VDT operators compared with other job categories.[4] Several studies have shown high rates of self-reported anxiety and depression,[20] irritability and sleep disorders,[21] and mental fatigue, anger and confusion[22,23] among VDT users. A link has been established between incidence of stress symptoms with total hours worked at the terminal.[24] It has been reported that hours using a VDT has a moderate though inconsistent impact on well-being[25] and VDT users report more boredom and lower self-esteem compared with controls.[22,23] Amount of time spent working at the VDT screen correlates not only with job satisfaction but also with physical stress.[26] For some office workers, the automated office may represent a stress-inducing environment. Continued high levels of physical stress can be detrimental not only to workers' productivity but also to their physical and mental health and may be a factor in the development of psychological stress.[18]

'Technophobia' cannot be ruled out as contributing to the subjective feeling of stress. Operators often report a loss of control and a fear of the unknown.[27] Up to one third of office workers have a basic mistrust of computers and as many as 5% are cyberphobic (a fear of computers) to such an extent that they suffer adverse physiological reactions such as dizziness, nausea and increased heart rate when asked to use a computer.[28] Such reactions may have developed due to an inability to meet the demands of the task.

Another possible contributor to the high rates of self-reported feelings of stress may be the unusual ease with which VDT workers can be monitored. According to the National Institute for Occupational Safety and Health (NIOSH), about two thirds of VDT operators in the USA are electronically monitored in order to measure job performance. Measurements of speed, accuracy and volume for work completed can be automatically logged. Other less obvious causes of heightened stress levels may be due to media interest in VDTs and their effects on health. Increased anxiety, or simply the awareness of a possible link between health and exposure, may themselves lead to stress-related health problems.

The installation of VDTs has also caused a disruption in the role of the office worker.[29] Role ambiguity, responsibility for people, and organisational boundary conflicts are related to the subjective feeling of occupational stress.[30] Office relationships have responded to such changes with a shift from person contact to increasing computer-mediated communication.[8] Technological changes may disrupt not only the nature and meaning of work, but also social networks. Such disruptions may themselves be implicated in health complaints.

There is a decrease in opportunities for contacts and co-operation as a function of the intensive cognitive requirements of the job.[8] It is likely that contacts are reduced so that communications do not disrupt performance. The role of the open-plan office for the VDT user cannot be disregarded as a feature of the technological office. There are the obvious disadvantages of lack of privacy and overheard conversations. Lack of peace and seclusion in an open-plan office may be a more important factor in the cause of mental strain than repetitive data-entry, typical of VDT work.[31] Such a deterioration of relationships at work can remove the cushion of social support between the VDT user and their environment.

Little is known about the mechanisms by which stress can lead to disease and negative psychosocial conditions which may otherwise have helped protect workers from stress-induced physical or mental impairment. Attributing *cause* of symptom to VDTs *per se* is problematic. When the VDT users report problems and attribute cause, cognitive processes such as memory and perception come into play—processes which are subject to bias. It is also important to note that the ability to contend with the office environment depends on individuals' ways of coping. Some individuals may adopt a less efficient way of coping, perhaps by escaping the threat to health that the office may pose (i.e. by being 'off sick'), rather than concentrating on solving the problem effectively (i.e. by alerting union leaders and employers). The ability to exert control over work and working conditions not only increases satisfaction and well-being and hence reduces neuroendocrine stress responses, it can also help develop more effective ways of coping with stress.[14]

3. THE VDT AND HEALTH

The impact of VDT technology on health has far-reaching implications both in terms of worker satisfaction and performance, and economic costs of absenteeism and human error. This has stimulated research about the possibility of a link between the new technology and ill-health, investigating the relationship between various physical and biological factors and exposure to VDTs. In a large-scale survey, after controlling for demographic factors, tenure and role strain, VDT exposure had a significant effect on VDT users' well-being.[25] Workers who operated a VDT for 10 hours or more a week experienced greater psychosomatic distress, job tension, job dissatisfaction and strain than non-operators or employees who operate VDTs infrequently.

Recent research has focused on visual problems, musculoskeletal problems and psychological stress.[10,32] Early concerns centred on radiation effects which may in turn affect the reproductive system. However, the National Academy of Science[33] has criticised many studies on methodological grounds. Despite hundreds of studies and several international conferences on the subject, the debate is far from settled.

3.1. *VDT-related Visual Complaints*

There is no scientific evidence that VDTs cause cataracts or other serious eye pathologies.[34,35] However, a large percentage of VDT users regularly and consistently complain of relatively minor visual problems such as visual fatigue, visual discomfort and dysfunction.[10,24,36] Several studies show small short-term effects of VDT use such as changes in accommodation and vergence to dark focus.[37,38] Yet, other studies[39,40] have failed to show any changes in visual functioning including accommodation, vergence, pupillary responses, eye movements and saccade pattern under prolonged experimental VDT use conditions.

Visual complaints have been identified as caused by overuse of the eyes, poor screen images affecting focussing, screen glare, fluctuating brightness, improper illumination and high contrast sources.[10,26,41] Vision problems may be partly caused by long intensive periods of work and partly by shifting focus from black characters on white paper to the light letters on a dark background displayed on the screen.[5] The majority of VDT users may suffer visual problems because of inadequately designed workstations and VDT technology.[35] Users who worked for more than two hours at the terminal complained of strained eye muscles.[12] Although, after rest, most visual problems disappear.[12]

At worst then, VDTs seem to cause visual discomfort. But what of the long-term effects? A longitudinal study by the Association of Opthalmologists[42] found that over a five-year period there were no changes in measures of visual acuity, phorias, muscle balance and vergence. However, VDT users did show significantly greater incidence of eye irritation than non-users. Visual complaints of this nature do not represent a serious threat to health, but they are nevertheless a nuisance and may lead to worker dissatisfaction and under-performance.

3.2. *VDT-related Musculoskeletal Complaints*

Musculoskeletal problems have been identified as a source of dissatisfaction among VDT users ever since their introduction to the workplace.[43] Unergonomically designed workstations cause workers to adopt poor postures in static seated positions whilst carrying out repeated motions. Industry uses the term 'repetition injury' to describe some of the problems resulting from repeated tiny traumas caused by overstressing a joint or muscle. The symptoms often develop over a 5 to 10 year period, but aches and pains can appear within a very short time. Grandjean and his associates[44,45] show that ergonomic features such as keyboard thickness, height of keyboard from floor, inability to rest hands on desk or keyboard and turning of the head whilst keying are associated with various musculoskeletal problems. Such postures give rise to complaints of aches and pains in workers' backs, necks, shoulders and wrists. There may be an epidemic of repetitive strain injuries among electronic keyboard operators—affecting forearm, wrists and hands.[46] In 1984, 10,000 cases for industrial compensation for musculoskeletal injuries were filed in Australia.[46] Moreover, 20% of keyboard operators were absent from work at any one time with such injuries.[46]

Current research on musculoskeletal complaints has focussed on repetitive strain injuries, chronic trauma and diseases and syndromes such as carpal tunnel syndrome. Such conditions are becoming increasingly prevalent amongst VDT users.[47] Chronic traumas represents immediate health risks among VDT users—especially back and wrist disorders.[10] Whilst back disorders are caused by bad postures and/or inadequate workstations, wrist disorders have been identified by several researchers[48,49] as being caused by improper wrist action during repetitive keying. This may produce tenosynovitis and/or carpal tunnel syndrome. Neck problems have been reported by between one and two thirds of the VDT users.[50,51] As with visual complaints, aches and pains may rapidly improve with rest.[52]

VDT work demands constraining postures. Undue loading on the spine and/or back muscles caused by inadequate posture has resulted in up to a half of VDT users reporting back problems.[23,50,53] Such complaints have been confirmed in the measurement of electromyographic (EMG) activity during VDT use.[54] Static load on the trapezius muscle was twice that of acceptable load levels. In a study of preferred seating arrangement, VDT users did not prefer an upright posture—but a backward inclination (97–121 degrees).[45] Such preferences do not agree with standards and recommendations normally adopted, but are nonetheless likely to give greater musculoskeletal comfort to the VDT user.

3.3. *VDT Radiation and Effects on the Reproductive System*

Ionizing radiation contains sufficient energy to cause a neutral atom to acquire a positive charge, or to 'ionize'. The X-ray is the most well-known kind of ionizing radiation. In the VDT, the cathode ray tube which projects an electron beam on to the phosphorescent screen produces X-rays. Other ionizing radiations are also emitted by the VDT (mostly the screen) including optical, infra-red and ultraviolet light. Such VDT radiation has been shown in a variety of studies to be well below acceptable levels of safety.

VDTs also emit non-ionizing radiation such as radiofrequency and microwave radiation that do not cause changes to the atom. Microwaves are minimal but radiofrequency wavelengths with frequencies between 0 Kilohertz (kHz) and 3030 kHz and particularly very low frequency (0 to 30 kHz) have been implicated in the VDT health debate. Difficulties in measurement have led to inconclusive reports. However, the Canadian Centre for Occupational Health and Safety reports that the VDTs pulse-modulated very low frequency electromagnetic fields sometimes exceed exposure limits.

In the USA, trade unions have been dissatisfied with explanations of clusters of birth defects and miscarriages amongst female VDT users. There have been 15 alleged clusters around the USA and Canada where high percentages of female workers have had reproductive problems such as foetal abnormalities and miscarriages.[55] The possibility of a link between VDT radiation and birth defects and miscarriages is obviously a cause for concern. Studies in the USA have not conclusively implicated VDTs as a causal factor, but several research bodies have not been able to eliminate them either.[56,57]

When controlling for age, smoking, etc., VDT users had an observed to expected ratio of 1.21 of birth defects.[58] This compares with a non-user ratio of .89. This study also shows that length of exposure is not related to reproductive problems, although a methodological limitation of this study was that exposure to VDTs was estimated from job title. In a review of the literature it was concluded that 'most experts agree that the biological effects of VDT radiation are inconsequential to health.'[59] However, research still questions the role of pulsed magnetic fields as a possible factor for reproductive problems, especially since animal studies show abnormal pregnancies amongst mice exposed to pulsed electromagnetic fields.[36,60] However, researchers have failed to confirm these findings in humans.[59]

Currently, most researchers agree that biological effects of radiation are not significant. Non-ionizing radiation levels measured from VDTs are very low levels indeed.[61] The emission has been recorded as one one-thousandth of the least allowable exposure of ultraviolet and infra-red, one one-thousandth visible, one one-billionth microwave and two one-thousandths radio frequency. These are very low levels indeed. Observations of reproductive problems may be a statistical quirk or a serious health risk.[3]

4. RECOMMENDATIONS AND COUNTERMEASURES

The scientific evidence is incomplete or even contradictory in the early stage of the research into the link between VDT technology and health. It would therefore be prudent to consider several measures in order to alleviate some of the likely adverse physical effects of VDT work.

4.1. *Psychological Well-Being*

Many workplaces have now introduced health promotion, employee assistance stress management and stress reduction programmes.[1] Health promotion focuses on life-style changes such as diet, smoking, alcohol intake and exercise. This approach assumes that it is the non-work factors which are mainly responsible for the costs associated with bad health. However, evidence presented here shows that the physical working environment plays a major role in sickness and injury. As work stress is associated with factors at work rather than at home[1] perhaps the focus of stress management programmes should concentrate on the reduction of work-related stress. For example by improving working conditions both at an occupational and environmental level.

4.2. *Visual Problems*

In order to counteract some of the effects of VDTs on vision, many changes have been suggested to the environmental aspects of the VDT and workstation.[3] Adjustable VDT screens and keyboards should be installed which can help reduce glare from screen and help the worker to position the equipment for maximum comfort.[3] Antiglow screens and hoods can similarly reduce the effects of glare and reduce eyestrain.[3] Ergonomically designed workstations should be introduced so that factors such as distance from screen and position of documents during data entry can be taken into account.[3] Document holders which can be attached to the VDT are useful in reducing eye strain. VDT users should also be permitted to take breaks from continuous operation.

4.3. *Musculoskeletal Problems*

A comparison of a well-designed and poorly-designed VDT workstation indicated an almost 25% improvement in performance among workers using the ergonomically designed workstation.[24] More importantly, musculoskeletal complaints decreased. Large reductions have been reported in the incidence of musculoskeletal symptoms among VDT users by implementing a control strategy based on the following:[62]

- education programs for managers and operators
- workstation assessment and refurbishing
- health screening
- work-load restrictions
- treatment and rehabilitation

Over a period of 3 years based on year two figures, a reduction of 50% was predicted.[52] Black and Decker also reported a 50% reduction in musculoskeletal problems after instituting a programme to evaluate and modify the workers' posture and the workplace itself. It is necessary to take account of psychological as well as physical factors in an attempt to reduce rates of musculoskeletal problems.[52]

4.4. *Radiation*

The introduction of metal shields to reduce radiation emissions has been recommended,[3] and equipment should be regularly monitored for all different types of radiation. Some experts, in the light of studies reviewed here, suggest a cautious approach if pregnancy occurs amongst VDT operators and recommend that pregnant workers should reduce their exposure to VDTs.

Many of the environmental problems under investigation with regard to VDT use have always existed in offices.[63] Sceptics argue that a secretary will suffer the effects of a bad office environment as much as a VDT operator will. It is important therefore, that research focuses on the comparisons between similar work environments both with and without the VDT.

5. SICK BUILDING SYNDROME: EFFECTS ON HEALTH

Traditionally, occupational medicine has focussed on the effect of high level exposure to various substances on workers' health in the chemical or coal industry for example. More recently, the effects of relatively low-level exposure to substances in the working environment have increasingly been the focus of attention, particularly in the last decade. Given that the volumes and levels of measures taken in the workplace are usually well within standards and guidelines, such investigations have led to much controversy and debate.

The office building and office equipment cost comparatively little when compared with the enormous costs of salaries and other staff benefits.[64] Yet it appears that due to many cost-saving exercises in the construction, layout and operation of a new building, employers suffer far greater costs of employee ill-health. However, the extent and overall effects of this so-called 'Sick Building Syndrome' (SBS) are difficult to assess as most employees are likely at some time or other to suffer from the relatively minor complaints typical of SBS.

Reports of SBS first appeared in the mid-1970s in Scandinavia. SBS refers to the presence of chronic symptoms associated with occupying a specific building. Such symptoms often dissipate when the individual ceases occupancy for a period of time. The World Health Organization (WHO) differentiates between 'temporarily' and 'permanently' sick buildings. 'Temporarily sick buildings' are newly built or remodelled and occupancy is followed by an outbreak of health problems. Symptoms usually dissipate within 6 months. This outbreak may be caused by the evaporation of volatile compounds from various materials. 'Permanently sick buildings' show symptoms over a much longer period. The cause is usually assumed to be poor indoor air quality and bad thermal conditions.

A typical 'sick building' has been identified by the WHO as having a number of common features, in particular, forced ventilation, a light construction and indoor surfaces may be covered with textiles. Such buildings are often energy-efficient and have windows which cannot be opened. NIOSH[65] finds that symptoms range from eye irritation, dry throat, headaches, sinus problems, fatigue, skin problems, coughs, dizziness, nausea and shortness of breath. Some physical symptoms are clear-cut, such as eye irritation and sinus problems, whilst others are less clear-cut, such as lethargy and headaches. Work-related lethargy is one of the commonest symptoms of 'sick building'. Headaches are often described as bilateral and frontal, and either supra-orbital or infra-orbital. They are rarely described as migrainous.

6. WHAT CAUSES SICK BUILDING SYNDROME?

SBS is a complex phenomenon with a number of possible causes, related to a variety of factors and exposures. The aetiology of this syndrome is not fully understood, leading some researchers to doubt its existence. Almost every aspect of the work environment could be associated with symptoms of SBS, as investigations have uncovered several possible environmental parameters which may be responsible for the syndrome.

Investigations have extended the concept of SBS to include not just poor air quality but other factors which may adversely affect workers health. These can be summarised as:

- Climatic effects (i.e. air quality)
- Chemical pollution (i.e. ozone and formaldehyde)
- Microbiological factors (i.e. bacteria, mould)
- Electrical environment (i.e. electromagnetic fields caused by VDTs)
- Psychological factors (i.e. work relationships)
- Organisational (i.e. job type) and demographic factors (i.e. sex)

Some environmental stressors have a direct impact on physiology, i.e. temperature, humidity, noise, light, etc., resulting in the elicitation of the physiological coping mechanism which in turn may lead to ill-health and/or disease. However, the same environmental stressors at lower levels may act indirectly in a more subtle way, resulting in psychological and/or behavioural adjustment. Such reactions may be exacerbated by a lack of control over the environment. Persistent exposure to subtle environmental annoyances may also give rise to physiological responses and eventual ill-health and/or disease.

Researchers have argued that the phenomenon may be physical, psychological or psychosomatic. However, with regard to a psychosomatic explanation, the symptom pattern for SBS is not typical of psychogenic illness as SBS is endemic rather than epidemic which contrasts with the normal course of mass hysteria.[66] Most studies conclude that building sickness is related to physical characteristics of the building, but often highlight the importance of psychological factors.

6.1. *Climatic Factors*

One of the main physical causes of SBS is thought to be poor air quality due to inadequate mechanical ventilation systems. Ventilation systems are known to transmit airborne diseases such as Legionnaires' Disease, Humidifier Fever and other infections. It is quite likely that ventilation systems can distribute contaminants from different areas of the building. However, such diseases are distinct from the kind of symptoms typically reported with SBS. If airborne contaminants are responsible for SBS, then it may be due either to toxicity levels, irritation, infection or allergy, or to a combination of these factors. Infection has been ruled out as the cause of SBS as symptoms are mainly irritant in nature rather than allergic.[67]

Many offices are ventilated by some type of mechanical ventilation system. Such systems treat the air by filtration and control its temperature and humidity by cooling or heating, and by humidifying and de-humidifying it.[68] Since the energy crisis of the 1970s there has been a reduction in the volume of air delivered to the occupants of a building. With the increase in the use of synthetic material and finishes, it is likely that the quality of indoor air is far from satisfactory in many buildings.

Mechanical ventilation in buildings differs from natural ventilation in the following ways:

- Allows greater environmental control but little personal control
- Air supply into the system may be restricted. This increases the recirculation of air and reduces the volume of fresh air
- System components may be susceptible to failure, poor design and/or installation
- System may harbour organic growth

Inadequate ventilation may account for the health problems referred to as SBS.[69] However, some studies[70,71] have failed to find evidence for a difference in symptom prevalence between mechanically ventilated and naturally ventilated offices, with the exception of mucosal irritation which correlated with several measures of air quality.

An epidemiological study was conducted in the UK of nine office buildings with various types of ventilation systems.[72] Eye, nose and mucous membrane symptoms, lethargy, headaches, and dry skin were significantly more likely to occur in air conditioned buildings. A survey of six office buildings with different ventilation systems found that incidence of eye irritation and respiratory complaints were higher in air conditioned offices.[73,74]

A higher prevalence of work-related reports of dry eyes and throat, congested nose, itchy/watery eyes, runny nose, lethargy and headaches have been found in air conditioned buildings than in naturally ventilated buildings.[75] Type of humidifier (none, evaporation/spray or steam) has also been related to itchy eyes, congested nose, chest tightness, lethargy and breathing problems.[75]

It is possible that the stability of the temperature in sealed buildings may cause some of the problems associated with SBS. Human beings are adaptable organisms and it could be that the uniform temperature is contrary to their expectation of a fluctuating thermal environment. Moreover, office planners often fail to take account of the wide range of individual differences in thermal and lighting needs. A high room temperature may enhance the prevalence of SBS.[71,76]

6.2. *Chemical Factors*

The potential range of pollutants in the office is enormous. Many sources of airborne pollutants have been identified.[73] Building occupants release CO_2, water vapour and microbial organisms and matter. Smoking is also a considerable source of contamination releasing many toxic substances into the atmosphere. Passive smoking may cause irritation to the mucosal membrane. Incidence of *new* SBS has been related to smoking.[78]

Building fabrics and furnishing can release toxic substances such as the irritant formaldehyde. Several studies have found a relationship between formaldehyde concentration and SBS.[79,80] 'Temporarily sick buildings' may house organic and solvent vapours and dust and fibres from such materials. Studies have found that wall-to-wall carpeting is related to chronic SBS.[78,81] Removal of carpets led to significant decreases in symptoms. In another study, the removal of a man-made, acoustic ceiling and the installation of mechanical ventilation led to a significant decrease in symptoms after one year, even though there was no physical evidence to suggest that the ceiling or ventilation system was responsible.[82] Perhaps the chemical stimulation of the trigeminal or olfactory nerves caused by some of the chemical releases from fibres and furnishing may explain some of the symptoms of SBS,[83] even though they remain at levels not thought to cause health problems.

Chemical substances are rarely found in isolation in work settings, but are usually present as part of a chemical mixture.[84] There may be a pattern of substances, rather than a group of substances, which may be responsible for the outbreaks of SBS.[84] Patterns can be found in sick buildings, although the exact definition has yet to be identified. However, 40 chemical constituents have been identified so far and research is continuing to try and uncover the pattern of chemicals which may be forming a lethal mixture. Future studies on temperature, ventilation and air humidity should consider how these factors may affect the chemical composition of the air.

6.3. *Microbiological Factors*

Microbiological contaminants can be found in humidifiers, chillers or duct works, primarily because these systems are often located in inaccessible places making servicing difficult.[85] Low relative humidity can also lead to irritation and an increase in the incidence of respiratory infection possibly due to the presence of bacteria and viruses. High relative humidity can encourage growth of larger micro-organisms. However, as most infections are caused by smaller micro-organisms, if humidity is responsible for some of the symptoms of SBS, it is likely to be the presence of low rates of humidity which give cause for the greatest concern. As SBS is not caused by infection,[86] humidity levels may be largely irrelevant, but low levels may cause some symptoms such as Erythema (facial dermatitis) and eye irritation (caused by 'dry air'). Problems with rhinitis and eye irritation are more common if a humidifier is present. Perhaps the droplets cause an irritant or allergic reaction.

6.4. *Electrical Factors*

Office machinery, particularly photocopiers can release ozone in poorly ventilated photocopying rooms. There is also some evidence to show that charged ions can affect health and physiology. An increase in negative ions has been reported to reduce the incidence of many symptoms similar to those reported from a 'sick building'. Ionisers are essentially air cleaners and may well be beneficial to workers. Exposure to static electricity and SBS symptoms has been demonstrated in a study by Norback *et al.*[87] However, an earlier study failed to find such a relationship.[85]

6.5. *Psychological, Organizational and Demographic Factors*

There are some sick naturally ventilated buildings and some healthy air-conditioned buildings, so that physical features of the building alone cannot be responsible for SBS.[85] SBS is likely to be exacerbated by employees' reactions and attitude towards the workplace environment. Psychological factors, such as attitude towards air-conditioning, are likely to play an important role in the occurrence of such health complaints.[75] Symptom reporting depends on workers' perception of health, satisfaction with job and satisfaction with the working environment, and peer group reaction to work-related illness.[88] Work-related symptoms were more commonly found among clerical workers than technical and professional workers, and both these groups report more symptoms than managers.[85] This may be related to their better accommodation within the workplace or their greater control over their job and environment.

Sex, job category, work functions (such as VDT operation) and psychosocial factors (such as dissatisfaction with superiors) are associated with work-related mucosal irritation and other symptoms.[66] Other studies have shown that the climate of co-operation at work[89] and work stress[90] have been related to symptoms of SBS. Analysis has revealed that SBS is associated with individual characteristics (age, sex), occupational factors (job stress, job type) and psychological processes (perceived environmental control, perceived ambient conditions, etc.).[75] A path analysis showed that psychological processes mediate the association between these factors and SBS. The findings from such studies has led researchers to conclude that non-atmospheric factors such as working conditions, person variables and organizational factors influence the workers' good health.[85,91,92]

Many studies suggest then that occupational factors need to be considered in the SBS/health debate. However, other researchers would disagree. A study compared two adjacent buildings housing employees who were working for the same employer and who were employed in the same types of jobs. One building was mechanically ventilated with sealed windows and cold water humidifier, and the other building was naturally ventilated. Symptoms of itchy eyes, stuffy nose, dry throat, lethargy and headaches were significantly more likely in the mechanically ventilated building. In a comparison of clerically-based public and professionally-based private sector employees working in the same building, using the same ventilation units results showed that there was no significant difference in the prevalence of symptoms between the two groups of employees.[72] This suggests that the occupational basis of the work undertaken is not an important factor in the occurrence of SBS. Finally, a survey of nine office buildings revealed a direct association between the type of ventilation and the reporting of SBS-related symptoms. It is the physical properties of the building which are the important determinants of whether SBS is likely to occur rather than organizational factors, according to Finnegan *et al.*[72]

However, more recently, several researchers have disputed this claim. In a review of the literature, it was found that clerical staff are more likely than managerial staff to suffer from SBS.[77,93] Complaints are also more frequent in the public than the private sector and when buildings house many staff. Symptoms are also more prevalent in the afternoons than the mornings. Some of the reported sickness may be due to a perceived lack of control over the environment.[77,93]

Studies have found a relationship between personal factors such as hyperreactivity and SBS.[78,87] Sufferers of 'building sickness' may belong to a group of people who tend to fall ill with airway diseases due to outside air factors which make them hypersensitive to small environmental pollutants at work. Sensitivity to psychological, organizational and environmental factors varies from individual to individual, causing difficulties in establishing a cause and effect relationship. Perhaps the way forward is to study the type of individuals who suffer from its effects rather than to concentrate on the health complaint and aspects of the physical environment.

7. RECOMMENDATIONS AND COUNTERMEASURES

The evidence available thus suggests that at least some of the causes of SBS may be due to an inadequate fresh air supply. Many researchers suggest that to increase the ventilation rate remedies most symptoms of both temporary and permanent SBS. The standards that have been set pertaining to the supply of fresh air to buildings has changed considerably over the past few decades. Recently, new higher standards have been set in the US increasing, for example, the supply of fresh air from 2.5 litres per person per second to 7.5 litres per person per second.

Several recommendations have been made which may help to minimise SBS.[77] For example, by implementing an increase in fresh air supply to mechanically ventilated offices and ensuring adequate air movement and controllable ambient temperature. Humidity levels should be maintained between 40% and 70% and natural lighting should be provided.[77] There should be regular checks for airborne pollutants and the ventilation system for organic matter. Finally, employers should attempt to improve worker morale and environmental satisfaction.

In 1988 the American Society of Heating, Refrigerating and Air-conditioning Engineers (ASHRAE) produced a document which provides a series of strict guidelines. The following guidelines will be implemented in the construction of new buildings:

- A full documentation of all ventilation systems must be provided
- All ventilation systems must be designed for ease of maintenance
- Demonstrations should be given prior to installation
- Outside air used for air supply will have to satisfy Environmental Protection Agency pollution standards
- Sources of indoor air pollutants will have to be removed rather than relying on ventilation to dilute them

Systematic monitoring of employee symptoms before and after corrective action should also be undertaken.[82] Follow-up studies may uncover more facts underlying the cause of SBS.

Thermal comfort and air flow can be simply remedied in naturally ventilated buildings, i.e. by opening or closing windows and doors, or adjusting radiators, window blinds or seating arrangement. However, such flexibility is often impossible in mechanically ventilated open-plan offices. Office planners fail to consider individual differences among workers when providing uniform environments for all. Perhaps a focus on the solutions to the problem of SBS may be more productive than trying to find the cause of the problem. It is likely that the individual effects of the 'sick building' may be operating in an additive way, so that searching for a single cause may be fruitless.

8. A MULTIPLICATION OF EFFECTS

Perhaps a synergist analysis may be appropriate to the understanding of the effects of VDTs and 'sick building' on workers' health. Synergy postulates that a combination of various effects may have a greater and more different effect than any of the individual elements taken in isolation. A synergist analysis, for example, would combine the effects of visual problems, physical pain and discomfort, radiation and/or work-related stress. This combination may be more detrimental to health than its individual components. It could be that work-stress, anxiety and other psychological symptoms are exacerbated by physical symptoms or vice versa. For example, any observed reproductive problems may not be due to radiation *per se*, but to a combination of several different physical and stress-related problems characterised by the nature of the work and/or the environment of the VDT user. It is important to note that during stressful episodes hormonal changes occur. As we have seen, VDT users suffer greater overall stress than other occupations. Therefore the combined stress load may well be a major contributor in observed pregnancy abnormalities.

An interesting aspect of research in the area of SBS is that studies often show that health complaints are significantly higher in an air-conditioned office than a naturally ventilated office, yet no single environmental parameter has consistently been pinpointed as the cause of ill-health. Furthermore, alteration of one or more isolated parameters has not so far been shown to consistently improve the incidence of symptoms of SBS. Studies appear to uncover different sources of SBS.

Despite a prevalence of typical symptoms being reported from air-conditioned offices, measures of air temperature, mean radiant temperature, relative humidity, air velocity, positive and negative ion levels, carbon monoxide, ozone and formaldehyde have been shown in one study to be not significantly different in naturally and mechanically ventilated offices.[86] In a later study,[87] no significant difference between buildings was found for some of the above measures and other measures of illumination levels, sound levels, carbon dioxide, total oxidants, volatile organic compounds and respirable particulates. Furthermore, with the exception of concentrations of total oxidants correlating with nasal irritation, no single environmental parameter was found to be associated with health problems.

It could be that there is a buildup of pollutants in certain built environments which as yet has remained undiscovered. Recently, a novel approach has been adopted by taking a measure of the sum of hydrocarbons present in the working atmosphere.[87] The complex mixture of hydrocarbons may be responsible for the syndrome, as total indoor hydrocarbon concentration was significantly related to symptoms of SBS.

In summary, it is important to adopt a holistic approach to the office envi-

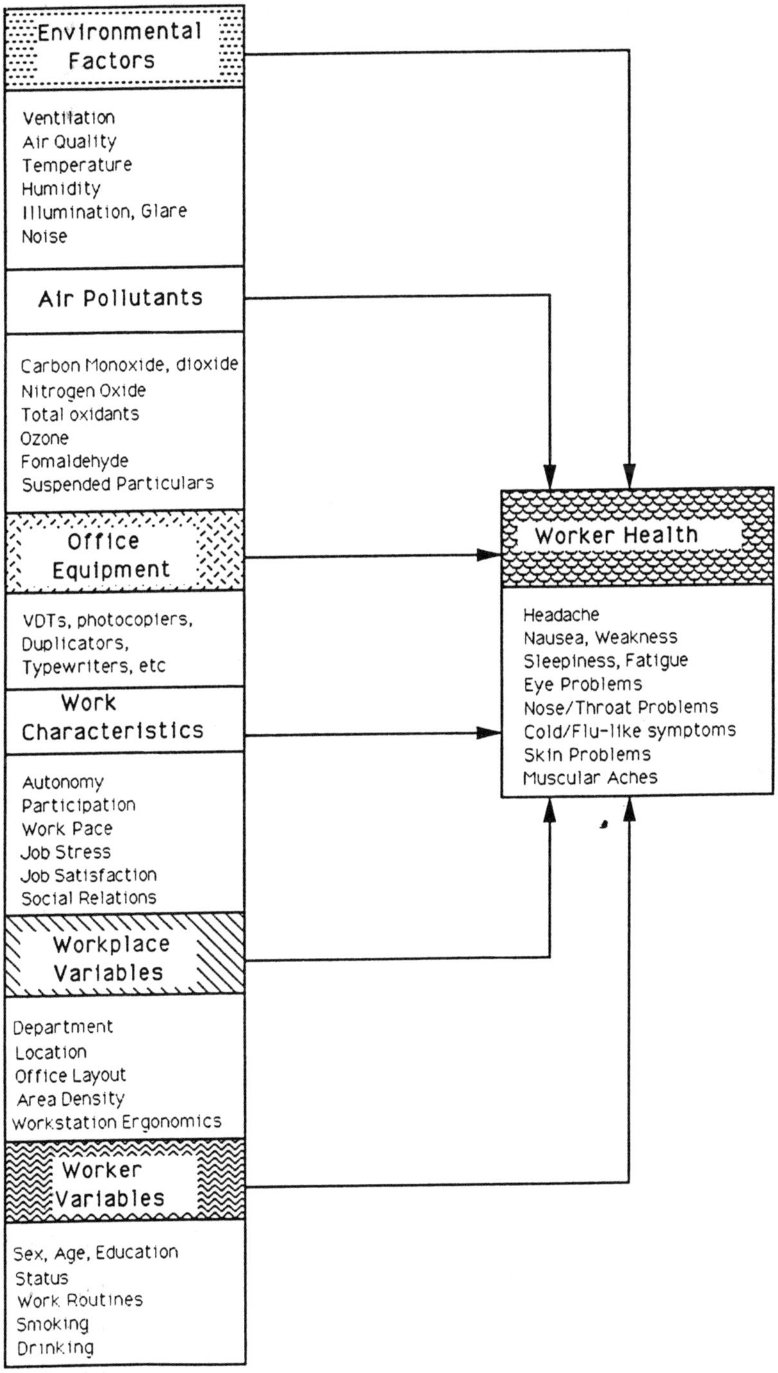

Figure 1 Model of factors contributing to office worker ill health (A. Hedge *et al*, "Evaluating Office Environments" In: *Human Factors in Organizational Design* (Eds. O. Brown and H. W. Hendrick Jr.) North Holland, Amsterdam, 1986, pp. 419–424).

ronment in order to develop an integrated model which incorporates both interactions between people, organizations and the built environment,[94] although relatively few studies have attempted to do this. The model shown in Figures 1 and 2 shows how the combined effects of the office environment may have a bearing on the office worker's health.

Any model of office environmental effects must include direct (i.e. lighting) and indirect (i.e. office layout) influences which, added to the employees' appraisal of effects, equates to a total stress load. It is the culmination of such effects which may lead to various health-related complaints. The simultaneous operation of a number of complex systems such as ventilation and lighting are now a common feature of modern offices.[94] Some researchers advocate a macro-ergonomic systems approach in which outcomes such as worker comfort, performance and health are viewed as the product of the dynamic relationship between a variety of sub-systems such as environment, work, technology and organization.[94]

Technology and the needs of commerce to equip an office building to house more and more employees has ignored the plight of the office worker. With an ever-increasing number of VDTs at work, housed within buildings which are ill-equipped to cope with the human inhabitant, the future may well see a new group of high-risk jobs. Research needs to focus not just on the effects of heavy industry on the workers' health, but also on the hitherto 'safe' office job as a potential source of stress and ill health, given the increasing role of this type of

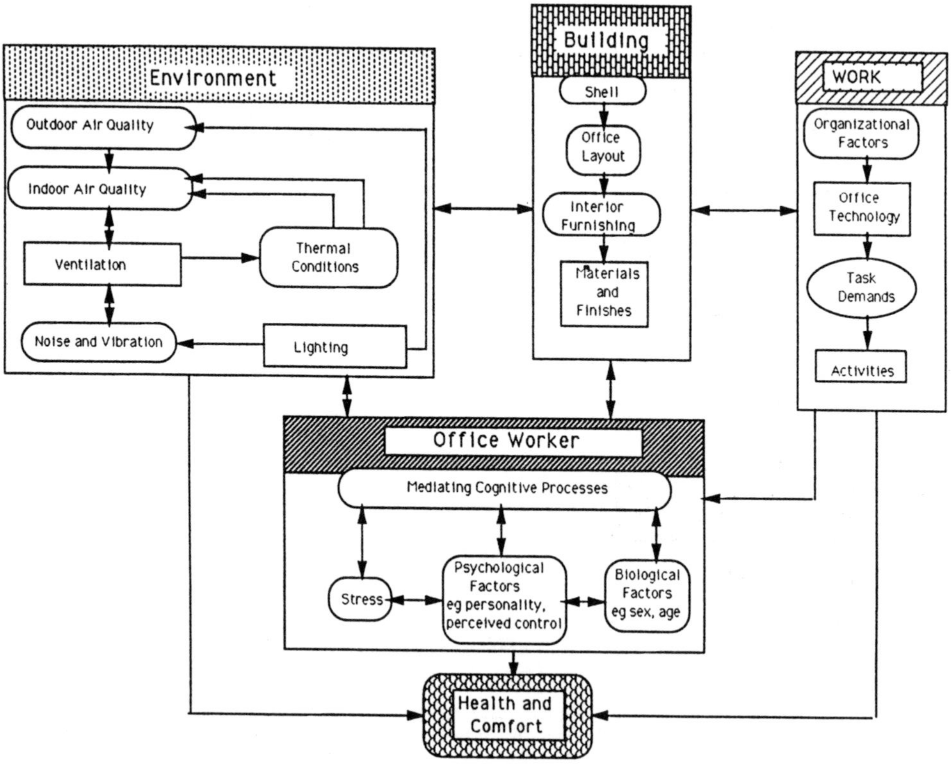

Figure 2 Systems model of office worker's health and comfort.[75]

worker in our society. Environmental quality and health are inextricably linked. Health needs must be incorporated in the design of both technology and the office building.

References

1. S. Klitzman *et al.*, "Work stress, nonwork stress, and health" *J. Behavioural Medicine* **13**(3) 221–243 (1989).
2. B. Christie (ed.), *Human Factors of Information Technology in the Office* (J. Wiley & Sons, Chichester, 1985).
3. J. Parry *et al.*, "Are VDTs hazards to your legal health?" *Mental and Physical Disability Law Reporter* **8**(4), 342–360 (1984).
4. P. C. Sainfort, "Perceptions of work environment and psychological strain across categories of office jobs" *Proceedings of the Human Factors Society 34th Annual Meeting* 849–853 (1990).
5. W. B. Kleeman Jr., "The politics of office design" *Environment and Behaviour* **20**(5) 537–549 (1988).
6. A. Boner, "Inadequate instruction in the electronic workplaces as a cause of mental strain" *Behaviour and Information Technology* **6**(3) 337–341 (1987).
7. Manpower Services Commission, *A study of the training provided by manufacturers/suppliers of word processors* (Sheffield, MSC Training Services, 1981).
8. G. Aronsson, "Changed qualification demands in computer-mediated work" *Applied Psychology: An International Review* **38**(1) 57–71 (1989).
9. S. Shindell *et al.*, "The healthy worker: Effect-fact or artifact?" *J. Occupational Medicine* **20**, 807–811 (1978).
10. M. J. Smith, "Health issues in VDT work" **In**: (J. Bennett, ed.) *Visual Display Terminals, usability issues and health concerns* (Englewood Cliffs, NJ Prentice-Hall, 1984) pp. 193–228.
11. C. L. Cooper and A. Cox, "Occupational stress among word process operators" *Stress Medicine* **1**, 87–92 (1985).
12. G. Bradley, "Effects of computerization on work environment and health: from a perspective of equality between the sexes" *Occupational Health Nursing* **31**, 35–39 (1983).
13. J. R. P French and R. D. Caplan, "Organizational stress and individual strain" **In**: (A. Marrow, ed.) *The Failure of Success* (New York, Amacom, 1973).
14. M. Frankenhauser and G. Johansson, "Stress at work: psychobiological and psychosocial aspects" *Int. Rev. Applied Psychology* **35**(3), 287–300 (1986).
15. J. P. Henry and P. M. Stephens, *Stress, Health, and the Social Environment. A sociobiological approach to medicine* (New York, Heidelberg, Berlin, Springer, 1977).
16. M. Frankenhauser, "Psycho-neuroendocrine approaches to the study of emotion as related to stress and coping" **In**: (H. E. Howe and R. A. Deistbeir, eds.) *Nebraska Symposium on Motivation 1978* (University of Nebraska Press, Lincoln, 1979).
17. M. Frankenhauser, "Psychobiological aspects of life stress" **In**: (S. Levine and H. Ursin, eds.) *Coping and Health* (New York, Plenum Press, 1980).
18. H. Kahn and C. Cooper, "Computing Stress" *Current Psychological Research & Reviews* 148–162 (Summer, 1986).
19. J. Cullen *et al.*, *Impact of New Technology on Experienced Workers* (physical and psychological stress) (Luxembourg: European Foundation for the improvement of living and working conditions: EEC, 1986).
20. L. Ghiringhelli, "Collecting subjective opinions on the use of VDUs" **In**: (E. Grandjean and E. Vigliani, eds.) *Ergonomic Aspects of Cathode Ray Tube Displays* (London, Taylor & Francis, 1980) pp. 227–232.
21. R. Elias *et al.*, "Investigations in operators working with CRT display terminals: relationship between task content and psycho-physiological alterations" **In**: (E. Grandjean and E. Vigliani, eds.) *Ergonomic Aspects of Visual Display Terminals* (London, Taylor & Francis, 1980) pp. 211–218.
22. M. J. Smith *et al.*, "Video display operator stress" **In**: (E. Grandjean and E. Vigliani, eds.) *Ergonomic Aspects of Visual Display Terminals* (London, Taylor & Francis, 1980) pp. 201–210.
23. M. J. Smith *et al.*, "An investigation of health complaints and job stress in video display operations" *Human Factors* **23**(4), 389–400 (1981).
24. J. Evans, "VDU operators display health problems" *Health and Safety at Work* **7**, 33–37 (1985).
25. B. D. Steffy and J. W. Jones, "The psychological impact of video display terminals on employees' well-being" *Stress Management* **4**(2), 101–107 (1989).

26. M. J. Dainoff, "Occupational stress Factors in Visual Display Terminal operation: a review of empirical research" *Behaviour and Information Technology* **1**, 141–176 (1982).
27. L. S. Sproull *et al.*, "Encountering an alien culture" *J. Social Issues* **40**, 31–48 (1984).
28. S. B. Weinberg, 1983, "Conquering computer fear" *Impact OA* (4–5 May, 1983).
29. J. Bennett *et al.*, *Visual Display Terminals, usability issues and health concerns* (Englewood Cliffs, New Jersey, Prentice-Hall, 1984).
30. C. L. Cooper and J. Marshall, "Occupational sources of stress" *J. Occupational Psychology* **49**, 11–28 (1976).
31. G. Johansson and G. Aronsson, "Stress reactions in computerised administrative work" *J. Occupational Behaviour* **5**, 159–181 (1984).
32. B. G. Knave, "Work with display terminals among office employees" *Scandinavian J. Work, Environment and Health* **11**(6), 447–494 (1985).
33. NAS, *Video Displays, Work and Vision*, National Research Council, National Academy of Sciences (Washington, DC, National Academy Press, 1983).
34. A. B. Smith *et al.*, *Cross-sectional Survey of VDT Users at the Baltimore Sun* (Cincinnati, OH, National Institute for Occupational Safety and Health, 1982).
35. F. M. Grignolo *et al.*, "Ocular tonometry in VDT operators" **In**: *Proceedings of International Scientific Conference: Work with Display Units* (Stockholm, Sweden, National Board of Occupational Safety and Health, 1986) pp. 578–581.
36. U. O. Bergqvist, "Video Display Terminals and Health: a technical and medical appraisal of the state of the art" *Scandinavian J. Work and Environmental Health* **10**, 1–87 (1984).
37. O. Ostberg, "CRTs pose health problems for operators" *Int. J. Occupational Health and Safety* 24–26, 46, 50, 52 (Nov./Dec., 1975).
38. S. Yamamoto *et al.*, "VDT operators' variation of the accommodation of the eyes during VDT work" **In**: *Proceedings of International Scientific Conference: Work with Display Units* (Stockholm, Sweden, National Board of Occupational Safety and Health, 1986) pp. 878–881.
39. L. Stark, "Visual fatigue and the VDT workplace" **In**: (J. Bennett, ed.), *Visual Display Terminals, usability issues and health concerns* (Englewood Cliffs, NJ Prentice-Hall, 1984).
40. S. E. Taylor and B. A. Rupp, "Display image characteristics and visual response" **In**: (J. Bennett, ed.), *Visual Display Terminals, usability issues and health concerns* (Englewood Cliffs, New Jersey, Prentice-Hall, 1986).
41. E. Grandjean, *Ergonomical and Medical Aspects of Cathode Ray Tube Displays* (Zurich, Switzerland, Federal Institute of Technology, 1979).
42. Association of opthalmologists of Quebec, *Cathode Ray Tube Display Terminals and their Effects on Ocular Health* (Quebec, University of Laval, 1982).
43. W. Hunting *et al.*, "Constrained postures of VDU operators" **In**: (E. Grandjean and E. Vigliani, eds.) *Ergonomic Aspects of Cathode Ray Tube Displays* (London, Taylor & Francis, 1980) pp. 175–184.
44. E. Grandjean *et al.*, "VDT workstation design: preferred settings and their effects" *Human Factors* **25**, 161–175 (1983).
45. E. Grandjean (ed.), "Ergonomics of Computerised Offices" (Taylor & Francis, London, 1984).
46. P. Ellis, "Keyboard-induced injuries" *Facilities*, **2**(12), 2 (1984).
47. G. Bammer, "VDUs and musculo-skeletal problems at Australian National University: a case study" **In**: *Proceedings of International Scientific Conference: Work with Display Units* (Stockholm, Sweden, National Board of Occupational Safety and Health, 1986) pp. 678–682.
48. R. Arndt, "Working posture and musculoskeletal problems of video display terminal operators: review and appraisal" *Amer. Industrial Hygiene Association J.* **44**, 437–446 (1983).
49. S. L. Sauter *et al.*, "Wrist trauma in VDT keyboard use: evidence, mechanisms and implications for keyboard and wrist rest design" **In**: *Proceedings of International Scientific Conference: Work with Display Units* (Stockholm, Sweden, National Board of Occupational Safety and Health) (1986) pp. 230–234.
50. A. Cakir *et al.*, *The VDT Manual* (Darmstadt, Germany, Ince-Fry Research Association, 1979).
51. K. Kemmlert *et al.*, "Musculoskeletal trouble in the neck and shoulder: relationship with clinical findings and workplace design in offices" **In**: *Proceedings of International Scientific Conference: Work with Display Units* (Stockholm, Sweden, National Board of Occupational Safety and Health, 1986) pp. 174–177.
52. S. L. Sauter *et al.*, "Chronic neck-shoulder discomfort in VDT use: prevalence and medical observations" **In**: *Proceedings of International Scientific Conference: Work with Display Units* (Stockholm, Sweden, National Board of Occupational Safety and Health, 1986) pp. 154–157.
53. S. L. Sauter *et al.*, *The Well-being of Video Display Terminal Users* (Cincinnati, OH, National Institute for Occupational Safety and Health, 1983).
54. M. Hagberg *et al.*, "Muscular load on the descending part of the trapezius muscle when working

at a VDT work processor" **In**: *Proceedings of International Scientific Conference: Work with Display Units* (Stockholm, Sweden, National Board of Occupational Safety and Health, 1986) pp. 336–337.
55. M. Sandstrom *et al.*, "Effects of weak pulsed magnetic fields on chick embryogenesis," **In**: *Proceedings of International Scientific Conference: Work with Display Units* (Stockholm, Sweden, National Board of Occupational Safety and Health, 1986) pp. 60–63.
56. NIOSH, *Health Hazard Evaluation of Southern Bell, Atlanta, Georgia, HETA-83-329-1498* (Cincinnati, OH, National Institute for Occupational Safety and Health, 1983).
57. CDC, *Cluster of Spontaneous Abortions, Dallas, Texas EPI-80-113-1* (Atlanta, GA, Centers for Disease Control, 1981).
58. A. D. McDonald, "Birth defect, spontaneous abortion and work with VDUs" **In**: *Proceedings of International Scientific Conference: Work with Display Units* (Stockholm, Sweden, National Board of Occupational Safety and Health, 1986) pp. 669–680.
59. M. J. Smith, "Mental and Physical Strain at VDT workstations" *Behaviour and Information Technology* **6**(3), 243–255 (1987).
60. J. M. R. Delgano *et al.*, "Embryological changes induced by weak, extremely low frequency electromagnetic fields" *J. Anatomy* **134** (3), 533–551 (1982).
61. R. H. Hirsch, "VDTs and the human factors community; tipping the iceberg" *Human Factors Society Bull.* **27**(6), 1–3 (1984).
62. S. A. Rowe, "Management involvement—a key element in preventing musculoskeletal problems in visual display unit users in Australia" *Ergonomics* **30**(2), 367–372 (1987).
63. P. A. Howarth and H. O. Istance, "The association between visual discomfort and the use of visual display units" *Behaviour and Information Technology* **4**, 131–149 (1985).
64. J. D. Wineman (Ed.), "Behavioural Issues in Office Design" (Van Nostrand Reinhold, New York, 1986).
65. NIOSH, *Guidance for Indoor Air Quality Investigations* (National Institute for occupational safety and health, Cincinnati, OH, 1987).
66. P. Skov *et al.*, "Influence of personal characteristics, job-related factors and psychological factors on the sick building syndrome" *Scandinavian J. Work Environment and Health* **15**, 286–295 (1989).
67. K. Thestrup-Pedersen *et al.*, "Allergic investigations in patients with the sick building syndrome" *Contact Dermatitis* **23**, 53–55 (1990).
68. D. J. Croome and D. M. Roberts, *Airconditioning and Ventilation of Buildings*, 2nd Edition, Vol. 1 (Pergamon, Oxford, 1981).
69. T. A. Seitz, *NIOSH Indoor Air Quality Investigations: 1971 through 1988* (National Institute for Occupational Safety and Health, Cincinnati, OH, 1989).
70. P. Skov *et al.*, "The 'sick' building syndrome in the office environment" *Environment International* **13**, 339–349 (1987).
71. O. Valbjorn *et al.*, "Influence of indoor climate on the sick building syndrome prevalence" **In**: (B. Siefert *et al.*, eds.) (West Berlin, Institute for Water, Soil and Air Hygiene, Vol 2, 1987) pp. 593–597.
72. M. J. Finnegan *et al.*, "The sick building syndrome: prevalence studies" *Brit. Med. J.* **289**, 1573–1575 (1984).
73. A. Hedge, "Suggestive evidence for a relationship between office design and self-reports of ill-health among office workers in the United Kingdom" *J. Architectural and Planning Research* **1**, 163–174 (1984).
74. A. Hedge, "Ill health among office workers: an examination of the relationship between office design and employee well-being" **In**: (E Grandjean, ed.) *Ergonomics and Health in Modern Offices. Section A: The ambient environment in offices* (Taylor & Francis, London, 1984) pp. 46–51.
75. A. Hedge, "Environmental Conditions and Health in Offices" *Int. Rev. Ergonomics* **3**, 87–110 (1989).
76. J. J. K. Jaakkola *et al.*, "Mechanical ventilation in an office building and sick building syndrome; short term trail" **In**: *Indoor Air 87, Proceedings of the 4th International Conference on Indoor Air Quality and Climate* (B. Siefert, *et al.*, eds.) (West Berlin, Institute for Water, Soil and Air Hygiene, Vol 2, 1987) pp. 566–571.
77. J. M. Sykes, "A contemporary review of the sick building syndrome: Part 2" *Australian Safety News* **60**(7), 30–38 (1989b).
78. D. Norback *et al.*, "Volatile organic compounds, respirable dust, and personal factors related to prevalence and incidence of sick building syndrome in primary schools" *Brit. J. Industrial Medicine* **47**, 733–741 (1990).
79. D. M. Main and T. J. Hogan, "Health effects of low level exposure to formaldehyde" *J. Occupational Medicine* **25**, 896–900 (1983).

80. J. H. Olsen and M. Dossing, "Formaldehyde induced symptoms in day care centers" *Amer. Industrial Hygiene Association J.* **43**, 366–370 (1982).
81. D. Norback and M. Torgen, "A longitudinal study relating carpeting with sick building syndrome" *Environment International* **15**, 129–135 (1989).
82. L. Hansen, "Monitoring of symptoms in estimating the effect of intervention in the sick building syndrome: a field study" *Environment International* **15**, 159–162 (1989).
83. B. Berglund and T. Lindvall, "Sensory reactions to 'sick building' " *Environment International* **12**, 147–159 (1986).
84. J. C. Baird and B. Berglund, "Thesis for environmental psychophysics" *J. Environmental Psychology* **9**, 345–356 (1989).
85. P. S. Burge *et al.*, "Sick Building Syndrome: a study of 4373 office workers" *Annals of Occupational Hygiene* **31**, 493–504 (1987).
86. A. S. Robertson *et al.*, "Comparison of health problems related to work and environment measures in two office buildings with different ventilation systems" *Brit. Med. J.* **291**, 373–376 (1985).
87. D. Norback *et al.*, "Indoor air quality and personal factors related to the sick building syndrome" *Scandinavian J. Work Environment and Health* **16**, 121–128 (1990).
88. A. S. Robertson and P. S. Burge, "Building sickness—all in the mind?" *Occupational Health*, 78–81 (March, 1986).
89. I. Michel *et al.*, "An epidemiologic study of the relation between symptoms of fatigue, dental anagram and other factors" *Swedish Dental J.* **13**, 33–38 (1989).
90. L. Morris and L. Hawkins, "The role of stress in the sick building syndrome" **In**: *Indoor Air '87, Proceedings of the 4th International Conference on Indoor Air Quality and Climate* (B. Siefert, *et al.*, eds.) (West Berlin, Institute for Water, Soil and Air Hygiene, Vol 2, 1987) pp. 566–571.
91. A. Hedge, "Office Health Hazards: An annotated bibliography" *Ergonomics* **30**, 733–772 (1987).
92. A. Hedge *et al.*, "Environmental, psychological and organizational correlates of employee health in offices: a proposed model" **In**: *Proceedings of the 31st Annual Meeting of the Human Factors Society* (Human Factors Society, Santa Monica, Human Factors Society, CA, 1987) pp. 736–740.
93. J. M. Sykes, "A contemporary review of the sick building syndrome: Part 1" *Australian Safety News* **60**(6), 33–40 (1989).
94. A. Hedge and D. Ellis, "Computer-aided facilities diagnostics: A new software tool for the armory of the macro-ergonomist" **In**: *Proceedings of the Human Factors Society 34th Annual Meeting, 1990* pp. 859–863.

8. STRESS AND PUBLIC CONCERN OVER HAZARDOUS WASTE

JUDITH PETTS*

This chapter deals with stress and concern over hazardous wastes impact on public confidence in the management system, on siting decisions and on health. It identifies the causes of stress and public concerns, including health effects. Case studies of concern about specific UK hazardous waste facilities are presented. The potential for different strategies to ameliorate stress are discussed, i.e. risk communication, improved health effects understanding and public involvement in decision-making.

1. INTRODUCTION

Hazardous wastes are some of the many products of an industrialised society that impact on man and his surroundings. The potential for negative impact in terms of damage to health and the environment from hazardous wastes, both accumulated from the past, and the production of hazardous wastes in the future, is recognised. However, the full nature and the extent of the potential effects are not always easily assessed. Hazardous waste is often a complex mixture of chemicals, and specific health or environmental effects can be difficult to establish given the multiple sources of wastes and multiple pathways by which a chemical could reach a sensitive target.

Public concern about hazardous waste within the general question of environmental protection is growing. A recent survey by the Department of the Environment[1] showed that the disposal of certain hazardous substances caused greater concern than radioactive wastes, chemical pesticides, acid rain, or vehicle exhaust fumes. The subject of hazardous waste is a highly emotive one. Technical concerns, in terms of the effectiveness and appropriateness of certain treatment and disposal methods, are often subsumed within issues of social values, quality of life, economic impact, psychological impact, trust and confidence in industry and regulators, fairness in risk distribution, and acceptance of political decisions. The result of such concerns has been increasing demand by the public, and specifically local communities, for direct involvement in risk decision-making and management. The NIMBY (Not in My Back Yard) syndrome represents perhaps the strongest manifestation of concern, and in many countries it is having a negative impact on the provision of required treatment and disposal facilities.

Increasingly, industry and regulators have recognised that one of the most important factors in establishing a successful risk management strategy for hazardous waste is the understanding of the stresses that are experienced by indi-

*Centre for Extension Studies, University of Technology, Loughborough, Leics., LE11 3TU, UK viduals and communities in dealing with the issue. The effectiveness of any risk management system where public concerns and interests are represented will be highly dependent upon the ability of the risk manager to communicate in a meaningful way. However, it is also increasingly being recognised that improving risk communication will not in itself lead to public acceptance of waste management strategies. Attention to risk mitigation, risk compensation, and public involvement in risk control, may also be required in order to improve the legitimacy of management strategies.

This chapter describes the background to hazardous wastes and discusses the causes of stress and public concerns. Topics addressed include the perception of risk and real health effects as factors influencing stress, as well as other causes of stress and the ways in which the public respond. The chapter concludes with a summary of the issues of concern and a discussion of the potential for different risk management strategies in reducing or ameliorating stress.

2. WHAT IS HAZARDOUS WASTE?

2.1. *Definitions*

An immediate problem with the concept of hazardous waste is that of definition. The term can mean different things to different people and is often used emotively and without precision. Hazardous waste is frequently taken to be synonymous with all types of industrial waste. However, certain waste substances which may not in themselves be hazardous may in certain circumstances become so. For example, ordinary household refuse does not fall into most definitions of hazardous waste, however, it may pose a health hazard if it is handled inadequately. Certain waste materials will give rise to more significant effects than others and many countries have identified specific categories of hazardous industrial wastes, which have particular controls imposed upon their production, transport, and disposal. For example, in the UK 'Special waste' refers to controlled wastes which may be dangerous to human health. However, the Department of the Environment has also promoted the term 'difficult wastes'[2] to include wastes which could in certain circumstances be harmful in either the short or long term to the environment. This definition is considered to be a good description of the general concept of wastes which require special provision for handling or treatment.[3]

A further problem arises with the definition and use of the term hazardous itself. Almost any material can be hazardous in particular circumstances. Equally materials described as hazardous do not necessarily always cause damage to health or the environment. It is generally felt that categorisation of waste should move away from descriptions such as hazardous to enable more precise definition. The UK government has confirmed the need for a 'consistent vocabulary' in the management of hazardous waste.[4]

2.2. *Origins*

Hazardous waste can originate from a variety of industries and embrace a wide range of chemicals, often as complex mixtures. Waste arises from all stages of industrial processes, and can include unused or spoilt raw or other process material, emissions and by-products from unit processes and waste arising at the end of the production cycle. It can also include contaminated or unwanted product.

Current estimates, for England and Wales, of waste arisings per annum, are of 77 million tonnes of industrial wastes, compared with 21 million tonnes of household waste. Of these arisings 2.2 million tonnes are 'special wastes' and 4.8 million tonnes are 'hazardous wastes' as defined by the Department of the Environment.[5,6]

2.2.1. *Contaminated Land.* Most industrial countries have a legacy of uncontrolled wastes production and disposal which requires attention and which is increasingly an issue of public concern. The potential problems include pollution of both water and land and result from:

- Storage and transport
- Stack emissions
- Disposal of waste materials (on or adjacent to the site)
- Demolition of buildings or plant
- Application of sewage sludge or other materials to land

It has proved difficult to estimate the amount and extent of contaminated land or the extent to which water resources may be affected. Recent estimates suggest that 50 000–100 000 sites might be considered to be contaminated in Britain, affecting perhaps 50 000 hectares. Only a small proportion of these, however, are likely to pose an immediate threat to public health or the environment.[7] In Britain the major industrial conurbations of the North, the Midlands and South Wales have large areas of land formerly used by industry and now derelict or reclaimed for other uses. However, industries or trades with the potential for contamination are numerous, for example, the sites of former town gas works, scrap yards, and waste disposal facilities, are found throughout the country.

2.3. *Management and Control Strategies*

Where waste material can be collected, it requires safe handling, treatment and ultimate disposal. There are a number of treatment and disposal options to deal with hazardous waste arisings including chemical, biological or physical treatment, thermal treatment by incineration or vitrification, and stabilisation and solidification. Disposal of many materials is usually to landfill, and all of the above treatment options produce residues which are ultimately landfilled. In the UK currently approximately 85% of controlled wastes are landfilled, 4% incinerated, 7% treated, with the remainder going for sea disposal.[5] Strategies for effective management of wastes include encouragement of the use of clean technology, waste minimisation, recovery and recycling.[8]

There are general features of disposal techniques which make some more appropriate than others. The Royal Commission on Environmental Pollution in its Twelfth Report[9] developed the concept of Best Practicable Environmental Option (BPEO) defined as a combination of criteria involved in making a waste management decision:

'The consideration of what constitutes the best practicable means for a particular process involves a balance of factors, including the present state of development of the relevant control technology, the financial implications, current knowledge about the effects of pollutants and local considerations and circumstances.'

2.3.1. *Environmental Impacts of Waste Disposal and Contaminated Land.* Perceptions of potential environmental impact of waste disposal and treatment activities represent the major source of stress and public concern. Potentially significant environmental impacts of waste disposal and treatment activities arise from a number of sources (Table I) and relate both to the materials handled—water reactive substances, oxidising materials, biological wastes (e.g. clinical), flammable materials (e.g. organic chemicals and solvents), toxic materials (e.g. cyanides, mineral acids), odiferous wastes (e.g. mercaptans)—and to the methods of handling. The hazards may be immediately apparent, e.g. a fire involving waste tyres, or may take many years to become apparent, e.g. groundwater pollution from slow seepage of leachate from a landfill. A recent study[10] for the Department of the Environment which examined 100 UK landfills, found that nearly half had experienced some surface or groundwater contamination.

In relation to contaminated land the hazards often relate to the intended land use. Table II, taken from guidance issued by the Interdepartmental Committee on the Redevelopment of Contaminated Land (ICRCL)[11] summarises some commonly encountered contaminants, the sites where they are likely to occur, and the principal hazards they produce.

2.3.2. *Management Strategy.* The implementation of a wastes management strategy requires not only controls over industrial processes to ensure that production and disposal of hazardous wastes is managed effectively, but also the establishment and consolidation of the technology and infrastructure for wastes disposal. A strategy has to be developed in the light of factors such as:

a. Developing policy objectives, e.g. commitment to increased recycling of waste.
b. Tightening standards, e.g. in relation to the engineering and operation of landfills as reflected in the EC Draft Directive on Landfill,[12] or the control of stack emissions from incinerators.
c. Changing disposal and treatment price differentials, e.g. as a result of increasing landfill costs due to engineering upgrading of sites, or the changing economics of energy recovery from incinerations.
d. The remaining length of operational life of existing facilities.
e. Commercial development of innovative technologies.
f. Changes in the balance of 'in-house' treatment of wastes by industry compared to use of the merchant sector, for example as a result of increasing disposal prices, or increasing awareness of potential liabilities.
g. Changes in the geographical spread of waste arisings.
h. Development of an integrated approach to pollution control which stresses the need to prevent or solve problems rather than transferring them to other environmental sectors.
i. Integration of environmental considerations into other broader aspects of public policy.
j. NIMBY attitudes to the siting of new facilities.

In the public's mind the questions asked over siting of the industry will start with needing to know what the potential impacts of the industry on its environment will be. An examination of NIMBY attitudes to siting hazardous waste treatment facilities in the USA identifies fears of public health effects as the strongest correlate of opposition.[13] In the UK (and elsewhere in Europe) such

Table I Some potential environmental impacts of waste handling

Environmental impact	Source
Water pollution (ground and surface)	Accidental spillages; Landfill leachate; Run-off from surfaces; uncontrolled effluent disposal
Air pollution (odour, air quality impact, human health)	Incinerator stack emissions; Fugitive emissions from storage and process areas; Landfill gas; inadequate cover of landfilled wastes
Damage to flora and fauna	Stack emissions; fugitive emissions from storage and process; loss of habitat from land take; dust from landfill
Noise and vibration	Plant operation; traffic
Road traffic (Accidents, noise, loss of amenity)	Waste transport and residue removal from facilities
Fire and Explosion	Waste storage areas; landfilling of incompatible wastes; spontaneous combustion in landfills; handling of water reactive wastes
Visual impact	Plant operations; incinerator stack; landfilling activities

Table II

Type of contaminant	Likely to occur on	Principal hazards
'Toxic' metals, e.g. cadmium, lead, arsenic, mercury. Other metals, e.g. copper, nickel, zinc.	Metal mines, iron and steel works, foundries, smelters. Electroplating, anodising and galvanising works. Engineering works, e.g. shipbuilding. Scrap yards and shipbreaking sites.	Harmful to health of humans or animals if ingested directly or indirectly. May restrict or prevent the growth of plants.
Combustible substances, e.g. coal and coke dust.	Gasworks, power stations, railway land.	Underground fires.
Flammable gases, e.g. methane.	Landfill sites, filled dock basins.	Explosions within or beneath buildings.
'Aggressive' substances, e.g. sulphates, chlorides, acids.	'Made ground' including slags from blast furnaces.	Chemical attack on building materials, e.g. concrete foundations.
Oily and tarry substances, phenols.	Chemical works, refineries, by-products plants, tar distilleries.	Contamination of water supplies by deterioration of service mains.
Asbestos	Industrial buildings. Waste disposal sites.	Dangerous if inhaled.

questions are often addressed by means of an Environmental Assessment carried out before new industry is located (Town and Country Planning (Assessment of Environmental Effects) Regulations, 1988). For existing industries there are also controls on the identification of certain sites where there are particular risks (Control of Industrial Major Accident Hazards Regulations, 1984; The Planning and Hazardous Substances Regulations, 1990), and there are increasingly stringent controls on the emissions of industry to the environment as a whole (e.g. Environmental Protection Act, 1990, Part 1). Many of these procedures require public participation in the decision-making process and even where they do not, there is increasing pressure for consultation with, and provision of information to, interested parties.

In relation to contaminated land many countries, including the UK, have embarked on a strategic approach to identifying the problems. Some countries (e.g. Denmark, USA, Netherlands, Germany) have already begun to compile lists of sites or types of sites which are actually contaminated. In the UK, Section 143 of the Environmental Protection Act 1990 provides for a duty to be placed on local authorities to compile public registers of sites where potentially contaminative land uses are, or have been, present.

Determination of the significance of contamination, the development of clean-up technology, and the allocation of limited resources to those environmental problems that pose the most serious risk, requires the development of an integrated management framework. The framework must encompass consideration of environmental protection and use of resources; development aims and land-use planning needs; economic priorities; objectives and priorities for environmental and public health protection; and social concerns and perceptions.[14]

2.4. *Effects of Public Concerns*

Public concern over hazardous wastes is having a number of impacts on development of the management strategy:

- Tightening of regulatory control.
- Tighter controls over waste imports.
- Imposition of stricter environmental quality standards.
- Problems in siting of new facilities.
- Investigation of health effects from existing and closed sites.

As a starting point in the assessment of public reaction and stress to hazardous waste facilities or contaminated land, it is useful to review the information available on observed health effects. In scientific terms it is important when dealing with risk perception to have an understanding of the real (as opposed to perceived) risks to humans or other parts of the environment.

3. WHAT ARE THE PROBLEMS IN DETERMINING HEALTH EFFECTS?

3.1. *The Need to Determine Health Effects*

In general terms, the need to determine the health effects (actual and potential) of hazardous wastes arise in relation to:

- The Environmental Assessment of a proposed new facility.
- The auditing of an operational disposal or treatment site.
- The investigation of public and/or worker claims of effects from operational sites.
- The investigation of a closed or potentially contaminated site to ascertain required remedial action.
- The investigation of public claims of health effects from a closed or potentially contaminated site.

Evidence of actual health effects of hazardous wastes include only a small number of clinically proven cases. Most claims of effects are not supported by substantiated data.

3.2. *Approaches Adopted*

For the purposes of assessing the potential for health effects the assessment would commonly consist of the following activities:

a. Detailed characterisation of the site and/or activity.
b. Selection of the key, indicator, contaminants/ chemicals of concern.
c. Identification of potentially sensitive receptors.
d. Analysis of potential rates of release and identification of potential exposure pathways.
e. Review of the physio-chemical and toxicological properties of the chemicals of concern.
f. Estimation of probability of exposure and probability of receiving a toxic dose, including dose-response assessment.
g. Comparison of contaminant concentrations with any applicable or appropriate standards.
h. Evaluation of the significance of the exposure risk.

In the US, the Environmental Protection Agency has adopted a four-step public health risk assessment process which has been used to evaluate risks associated with abandoned waste disposal sites.[15] Assessment of the actual health effects of a site will incorporate similar assessment steps, particularly a–e, in the list above, although the evaluation of the seriousness of exposure will be based upon actual epidemiological surveys.

3.3. *Problems in Health Risk Studies*

Problems in establishing unequivocal proof of direct cause and effect relate to the robustness of the methods available, and the complexity of the sources of the problem. Evaluation of health effects involves the use of toxicological data, environmental modelling, chemical analyses, biological data, and epidemiological surveys. 'Many of the methods may produce conflicting or incomplete results and cannot give a categorical answer about the nature or extent of the risk.'[16]

3.3.1. *Epidemiology.* Epidemiology describes the distribution of disease in populations and analyses the factors that influence these distributions. The purpose of epidemiology is to assist interpretation of patterns of occurrence of disease and to develop quantitative assessments of conditions which enable personal risks to be related to environmental characteristics. This is achieved by direct measurement of the health status of people and of the variables associated with health and disease. Thus it is directly relevant to, and useful for, measurement of real effects of chemical contamination at real locations.

Many factors influence the value and validity of an epidemiological study including:

- Consistency of association—can the study findings be replicated in diverse circumstances?
- Strength of association—the incidence of the disease increases with exposure, i.e. confidence in the causal relationship is high because a large proportion of exposed people have the disease and a large proportion of the non-exposed do not.
- Temporal relationships—the exposure must pre-date the onset of symptoms.
- Specificity—chronic exposure to multiple substances rarely allows high specificity to be established.
- Plausibility—the cause-effect relationship should not conflict with established knowledge.

One of the problems of public health relates to clusters—areas where the prevalence of a disease is, in statistical terms, significantly higher than expected. Problems arise because clusters are difficult to detect if a disease is rare, and diseases are usually most common where most people live. Further it can be difficult to determine whether people in a cluster were exposed to the suspect source in that location or to a source in a different location where they might have lived or work previously.

Most epidemiological work relates to occupational exposures. There have been very few epidemiological studies of the effects of waste disposal sites and contaminated land.

3.3.2. *Complexity of the Source.* In addition to the general difficulties of carrying out a statistically valid study of health effects, contaminated land and waste treatment and disposal facilities present a number of specific problems for any investigation of health effects or risks. Such sites commonly contain a range of chemicals in complex forms with the potential for synergistic (interactive) effects which are difficult to determine. The potential for a contaminant to enter the environment (its availability) and the pathways by which it might reach a person may be highly variable and migration or transport rates very difficult to estimate. A substance may be present in air, soil, foodstuffs and in water, making the effects of particular exposures very difficult to isolate. At old sites even the nature of the contaminants or their possible concentrations may be unknown. Brett *et al.* in reporting an assessment of health risks associated with excavation of a hazardous waste site[17] summarize the major uncertainties:

a. Limitations in sampling data.

b. Uncertainties in toxicity data, e.g. use of information from relatively high doses of materials in small numbers of test animals to predict what will happen when large numbers of humans are exposed to small concentrations.

c. General limitations in determining degree of exposure, e.g. uncertainties in emission estimates; uncertainties in hypothetical exposure assumptions, which are often based on worst case scenarios.

d. Limitations in migration or transport analysis.

3.3.3. *Limitations of Assessment of Potential Health Risks.* Assessment of potential (as opposed to actual) health effects of either operational or closed waste disposal sites is in its comparative infancy in the UK, certainly compared with the US where risk assessment is established in many federal programmes as a necessary tool for demonstrating the validity of siting and design criteria. In relation to proposed new plant or sites there is a requirement for an Environmental Statement (ES) of any facility which will handle hazardous wastes (Town and Country Planning (Assessment of Environmental Effects) Regulations, 1988). An analysis of incinerator applications over the period 1989–91[18] reveals the highly variable quality of the Statements produced to date and problems in assessing potential health risks from incinerator emissions. Particular areas of uncertainty are in the characterisation of waste streams; in the models for dispersion and subsequent deposition of the constituents of the stack gas; in attempts to quantify indirect exposure, e.g. through the food chain and dermal exposure; the relationship between dose and response for carcinogens when assessing toxicity; the quantification of non-carcinogenic effects (currently there is no widely accepted methodology).[19] Generally there is a tendency to overestimate risk, e.g. through the use of linear multi-stage dose-response models to predict human cancer risk at low doses. Some risk assessors are so concerned about the political or statistical sensitivity of quantified results that they shy away from the process. This reluctance does little to assist public discussion of health effects. 'What is important is that those who produce and those who use risk numbers are aware of the inherent uncertainties, so that any judgements or decisions which result are taken in this knowledge.'[20]

3.4. *Results of Studies in the USA*

Grisham described health investigations at 28 US waste disposal sites and reported the comments of his expert group.[21] There were few published scientific reports of health effects clearly attributable to chemicals from uncontrolled waste disposal sites, and only one report offered convincing evidence of a serious effect resulting from exposure to a specific chemical—arsenic poisoning via contaminated groundwater at Perham, Minnesota. The studies were commonly confounded by a lack of knowledge of the materials present, the exposure, the dose, and of the contribution of exposures from air, water, food, in the home and in the workplace. The Grisham review concluded that:

a. Evidence for causal association between the sites and diseases was weak. However, the absence of demonstrable effects on human health should not be taken as proof that no such effects exist.

b. The epidemiological investigations to date had been retrospective and relied heavily on questionnaire surveys. Prospective studies involving biological indices of exposure would be more useful.

c. Studies had been complicated by the inherent scientific difficulties of carrying them out as well as by the climate of public concern in which the studies were performed.

d. The issue of health effects in populations near chemical disposal sites is so emotionally charged, and the economic and social consequences of the findings so great, that specific attention must be paid to the quality of investigations.
e. Although there was no direct causal evidence of health effects from either active or closed disposal sites this should not result in postponement of remedial action.
f. Monitoring of emissions from disposal sites must be maintained.
g. There is great disparity between the public perception of health effects and current scientific evidence.

The study identified the need for:

a. Standards for protocols to be used in surveys of health effects.
b. Results of investigations to be made public promptly.
c. Local communities to be kept informed and involved in health effects studies.
d. An independent mechanism for peer review of epidemiologic investigations.

Grisham noted the problems of investigations being undertaken under situations of stress, often at the request of the residents around a site. The effects of stress on health have to be considered. Many effects have been suggested, including headaches and migraine, skin rashes, chest and back pains, insomnia and dyspepsia.[22] A failure to cope with stress-provoking stimuli may have serious patho-physiological consequences.[23] Higher levels of stress have been linked to an increased susceptibility to infectious diseases,[24] cancer,[25] and excessive inflammatory reactions. The relatively new field of psycho-immunology is producing a large body of evidence which indicates that a link exists between stress and the functioning of the immune system.[26] Thus residents around or on a site who face an elevation in stress levels may be subject to an increased vulnerability to a range of diseases which are not specifically related to exposure to any chemical contaminants.

Grisham concluded that 'prediction of health effects in waste site environs is an extremely complex and inexact process. Uncertainties arise because of the need for a large number of assumptions and inferences to complete each of the steps.'

3.5. *Health Studies in the UK*

A study on hazardous wastes and the public carried out for the European Foundation for the Improvement of Living and Working Conditions, concluded that there has been relatively little coordinated statistical analysis of health effects on a national basis in the UK, either for contaminated land or waste disposal facilities or for any localised source of pollution.[27]

The reasons given for this lack of studies included the absence of an accepted procedure for the study of health effects during the life a contaminated site, and the fact that at some old sites it is impossible to substantiate claims of ill-health made after the source of contamination has been removed. Where studies have taken place, they have resulted not from any systematic approach but from local public and political pressure. However, as a result of the recommendations of the Independent Advisory Group which investigated the possible increased incidence of cancer in West Cumbria around the Sellafield nuclear power plant,[28] the Government announced the formation of the Small Area Health Statistics Unit (SAHSU) at the London School of Hygiene and Tropical Medicine. The Unit has been given a remit to study available statistics in order to identify incidence of unusual diseases and, in collaboration with relevant bodies, build up reliable information on the distribution of disease in small areas so that specific clusters can be placed in proper context. SAHSU is currently looking at specific industrial sectors including an investigation of the incidence of mortality and cancer using retrospective data in the neighbourhood of waste incinerators.

Waste incineration has for some years been a source of public concern about dioxin emissions and possible health effects. There has also been questioning of the effectiveness of plants, particularly municipal incinerators, in minimising these emissions.[29] Openshaw and colleagues, in a study of the incidence of childhood leukaemia in the Northern Health Region,[30] confirmed a cluster of occurrence in the Gateshead area. This was linked speculatively to an incinerator and work is in progress to monitor levels of environmental pollution in the area. SAHSU is currently investigating the incidence of cancer of the larynx in people near incinerators following some site specific work in Lancashire by a team from Lancaster University. The results of the latter study by Diggle *et al.*[31] point to the problems in proving a causative relationship between pollution from a specific source and an observed health effect. Specifically in this case no information was available on other causative factors which might have influenced the development of cancer, e.g. smoking. The BMA has stressed that studies which generate hypotheses in a general trawl of evidence need always to be followed by studies which actually test the hypotheses.[16]

Gatrell and Lovett, in a paper identifying the need for further work on the possible health effects of incineration,[32] identified problems in using available government statistics which are based on returns by district Medical Officers, and called for investigation of the value of geographical information systems which link morbidity data to models of emissions dispersion around facilities. Knox *et al.* in their study of the quality of notification of congenital malformations similarly pointed to the lack of accuracy and reliability in official statistics.[33]

The Department of the Environment Pollution Paper No. 27 'Dioxins in the Environment' examines the potential for harm from dioxins. The only data available were from studies of people known or assumed to have been exposed to unusually high levels of PCDDs and PCDFs as a result of occupational exposure or accidents. A crucial problem in reaching any firm conclusions was that the subjects were never exposed to pure PCDDS or PCDFs, but rather to mixtures including these compounds, and in none of the studies was it possible to determine the actual amount received.

In the UK few official, detailed, and specific, investigations to assess health effects have been undertaken. Two in relation to (i) the hazardous waste incinerators operated by Rechem, and (ii) to lead contamination in Shipham, Somerset are worth summarising.

3.5.1. *Rechem.* The waste incineration company Rechem Ltd. was founded in 1968 and operated three high temperature incinerators—at Bonnybridge, north-west of Glasgow, Pontypool in South Wales, and Fawley south of Southampton. Bonnybridge (now closed) and Pontypool have been licensed to incinerate PCBs. In 1977 a farmer reported high mortality in livestock on a farm near Bonnybridge and this was followed by a similar report by another farmer in 1983. Independent investigations of the claims of mortality by the Department of Agriculture and

Fisheries, Scotland, and Glasgow University Veterinary School found no linkage to the Rechem plant. The reports found that the livestock deaths were attributable to specific diseases or to causes related to husbandry. Anecdotal evidence suggested a high level of twinning in humans and cattle. A retrospective study of records of human births showed an increase in twinning over the period 1975–83. The BMA Report[16] suggests this would need further investigation.

In the early 1980s two babies were born with rare congenital eye defects (anophthalmos—absence of an eye; and microphthalmos—reduction in the size of the eye) in the Bonnybridge area. Welsh Office officials were instructed to examine health data for the area around Pontypool. An inquiry set up by the Scottish office examined all aspects of human, animal and plant morbidity, and environmental aspects in the Bonnybridge area. The Inquiry Report published in February 1985 dismissed links between morbidity at Bonnybridge and incinerator emissions.[34] A Welsh Office report published at the same time found that the incidence rate for all births with congenital malformations for Torfaen (the Local Authority area) was lower than would be expected for Wales as a whole. Increased levels of two particular eye malformations, anencephalus and polydactyly, were reported but the study found no statistically significant increase in the eye defects considered by the Lenihan inquiry.[35]

Since this time there has been continuing public concern over perceived cancer risks from the plant. In 1991 the Welsh Office commissioned an investigation into concentrations of PCBs and associated compounds in the Pontypool area, to determine if concentrations are in any way unusual and if so identify their source and assess their relative contribution to human intake. The first stage of the project was completed in August 1991. The report suggests some levels of PCBs in soils above background in the area and some evidence of elevated concentrations in small numbers of eggs from a local premises. However, it concludes that the data are insufficient to provide a complete characterisation of the area and all sources within it and are inadequate for the purposes of assessing population exposure. Further work is in progress.[36]

3.5.2. *Shipham.* The village of Shipham in Somerset was the subject of a comprehensive study into the health effects of heavy metal contamination caused by lead and zinc mining between the 16th and 19th centuries. No history of health problems had been associated with the village and no complaints had been lodged although it was known that the soil in the village was affected by the metals.

Following a nationwide survey of metal concentrations in stream sediments which identified high concentrations of cadmium and other metals in the Shipham area, the Department of the Environment, the Department of Health, the Ministry of Agriculture and the local authority initiated a study including diet studies and health checks. The survey concluded that whilst contamination was high, there were no associated health problems. However, residents were advised about general hygiene practices and against eating vegetables in substantial quantities.[27]

4. PUBLIC PERCEPTIONS OF RISK

In technical terms 'risk' is the probability of a hazardous event occurring. In the context of this chapter such an event might be the release of methane from an abandoned waste disposal site, or failure in an incinerator scrubbing system which resulted in emission of products of incomplete combustion. An objective analysis of risk attempts to quantify this likelihood. However, in any decision, issues of public concern and interest, such as the siting of a waste facility, will affect the risk evaluation process, often overshadowing the technical analysis of the risk. The effectiveness of any risk management system where public concerns and interests are represented will be highly dependent upon the ability of the risk assessor to communicate with the public in a meaningful way. To do this risk managers must comprehend the social context of the risks being evaluated and understand that they and the public will not share a common conceptual understanding of risk. Risk experts have a tendency to regard the public's judgements as irrational and ill-informed. However, public concerns must be seen as entirely rational in the light of their experience, the information available to them, and their knowledge. There is often little resemblance between what the public want to know about risks and what experts want to tell them.

4.1. *Psychometric and Sociological Evidence of Risk Perceptions*

A number of psychological and sociological studies have attempted to establish the determinants of public attitudes to, and perceptions of, risk, in order to assist in the understanding of the lack of commonality between expert and public.

Tversky and Kahneman[37] identified a number of heuristic principles which people use in making risk judgements. One such heurisitic is representativeness, which could be used to try and assess the probability of a hazardous event occuring. Basically this involves making a judgement as to how representative the problem is of problems in general. Hazardous waste sites would be judged to be highly representative of environmental problems in general, despite, in the UK, the lack of evidence of major pollution incidents involving such sites.

The availability heuristic is also used and involves assessment based on the availability of examples of the event having occurred, the ease with which something can be recalled. This is an area where the influence of the media is important. Most 'news' is bad news, dramatic coverage of pollution incidents or threats of such incidents, will provide for increased awareness which will lead to heightened perception of risk. Risks associated with memorable events, such as Bhopal and Love Canal, are considered to be more risky. Whilst an original estimate of risk may be subsequently adjusted, the original perception will be highly persistent and slow to change. This has considerable implications for the way in which information is provided in the first instance. Such cognitive biases that influence risk perceptions will of course be relevant also to the expert, who will have his own set of biases which must be recognised. How then do such heuristic principles affect perception?

Psychometric studies have been used to produce meaningful representations of risk attitudes and perceptions. A number of generalisations can be drawn from this literature (e.g. references 38–40):

a. Perceived risk is greater for hazards whose adverse effects are considered to be uncontrollable; unfamiliar (compare waste in a household dustbin with waste from a chemical plant); catastrophic; fatal; delayed and therefore represent a threat to future generations; generated by man, i.e. not natural; dread; and not offset by compensating benefits.
b. These characteristics of risk are highly correlated with each other. For example, risks that are regarded as voluntary (e.g. driving a car) are also regarded as controllable and understood. Risks regarded as involuntary are often also per-

ceived as catastrophic and a threat to future generations. Factor analysis of these interrelationships has been succesful in reducing this broad set of characteristics to two or three higher-order characteristics or factors (Figure I). Most important is the factor 'dread risk.' The higher a hazard's score on this factor the higher its perceived risk, the more vociferous the public's demands for control, regulation and reduction of the risk.

c. Experts tend to apply equal weight to consequences and probabilities, whereas the public tend to put more weight on consequences.

d. Proximity to an existing or proposed risky activity increases public resistance.

Sociological studies have provided further evidence of the factors which influence risk perceptions. 'Social context' and 'cultural bias' theory[41] stress the importance of looking not at the actual risks to explain any public reaction, but at the relationship between the risks and their social setting. Response to hazards can be caused by social influences for example the opinions of friends, work colleagues, family, etc. Socio-economic explanations of response to risk identify the differing values of the working classes and middle classes, of those with differing educational standards, of the young and the old, of men and women, etc.[42]

4.2. *Influences on Aversion*

The limitations of psychometric studies in helping us to understand actual behaviour and how to deal with public concern have also been recognised. Psychometric studies have often been based on hypothetical situations, involving surveys of small, discrete groups of people, with problems defined as social-psychological ones rather than analysing the incompatability of goals and values and behaviours of all the parties involved in decisions.

Sandmann[43] has attempted to explain people's fears and concerns about industrial hazards as a product of outrage rather than of hazard, and identified the public's lack of trust in responsible authorities and industries as one of the important sources of outrage. The risk communication literature has explored this issue of trust and credibility further and identified the need to tailor risk communication and management strategies to specific problem issues and to allow in the decision-making process for debate and argument not only about technical assessments of risk and prevention control procedures, but also about costs and benefits of industrial activities and ethical and moral questions about imposed hazards, threats to the environment and future generations, etc.[44]

Recent studies in the USA on various technologies[45] suggest that individuals have a need to delegate responsibility for assessing risks to both competent and trustworthy persons. If trust is lost it cannot be easily be regained. Linked to this issue of trust and competence is the openness and fairness of the decision-making forums themselves. Kemp has observed that in relation to radioactive waste disposal the final test of public acceptability relates to the question of the legitimacy of the decision-making process.[46] The public prefer management strategies which recognise that risks can occur and which provide for rapid detection of problems and implementation of corrective action. Experts have had a tendency to stress the prevention of risks based on technological solutions which are simply not trusted by the public.

However, Portney in a review of the NIMBY syndrome in relation to the siting of hazardous waste treatment facilities in the US[13] casts some doubt upon a

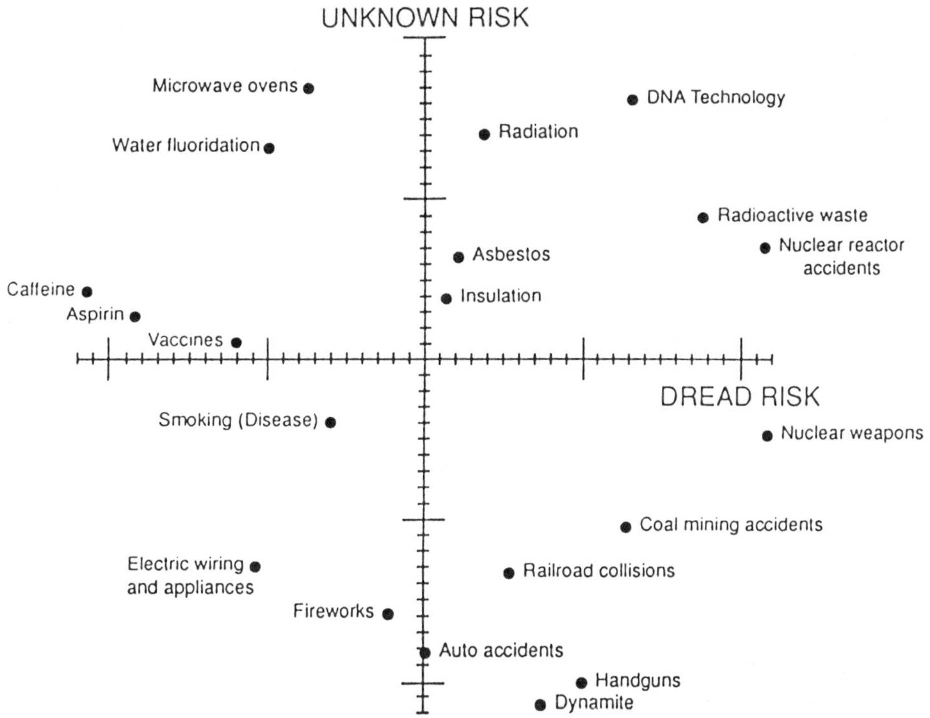

Figure 1 Hazard locations on 'Unknown Risk' and 'Dread Risk' factors adapted from Slovic *et al.*,[40] 1984.

number of these theories. In particular he suggests that there is no known way to build trust in political and social institutions, stressing that siting decisions actually raise a number of fundamental value conflicts such as scepticism of science and technology's ability to solve problems, the unfair distribution of risks in society, and the right of the individual to protect his interests. He sees improved risk communication and public involvement in decisions as unlikely to be successful in allaying the NIMBY syndrome, and instead suggests as an alternative the need for risk substitution.

4.3. *Dealing with Aversion: Compensation*

A study by Swartzman *et al.*[47] raised the question of the use of compensation in overcoming opposition. It is suggested that one of the reasons people object to having a waste disposal facility located near them is that they see it as providing risks without any benefits. There is some evidence that individuals are willing to condone a risk imposing activity if they are compensated either by direct financial inducements or providing non-monetary benefit, such as community facilities, or—more significantly—employment. The risk literature has seen a number of conflicting articles on the value of compensation in helping to alleviate public concerns and particularly as an aid to acceptance of new hazardous facilities. Proponents of the 'reverse Dutch auction'[48] suggest that communities should be allowed to bid for facilities, thus setting their own levels of compensation. However, others have stressed that communities prefer risk reduction strategies to compensation payments.[49]

4.4. *Dealing with Aversion: Comparing Risks*

Placing the risk of chemical exposure in relation to other daily risks has been the approach favoured by the UK's Health and Safety Executive (HSE) in its discussion of risk acceptability.[20] In common with most discussions of major hazard risk the HSE uses a comparative approach in risk communication. The HSE's discussion document included a table of examples of other risks of death against which the proposed land-use planning criteria may be compared (see Table III). Unfortunately the table shows risks from a number of 'voluntary' and occupational activities; these are exactly those types of activities for which public perceptions have been shown to be very different from concerns over major hazards. Covello[50] has summarised a number of the limitations and difficulties associated with risk comparisons which have been extensively discussed in the risk acceptability literature:

a. Failure to emphasise the uncertainties involved in the calculation of comparative risk estimates.
b. Failure to consider many quantitative factors including risk data for sensitive or high risk groups; annual probability of injury or disability, spatial extent, persistence, recurrence, delay, expected environmental damage, etc. In the HSE's criteria, which are based on risk of receiving a dangerous dose, there is the obvious problem that the comparative data refer only to risk of death and therefore do not assist in putting into context the non-fatal risks inherent in the dangerous dose.

Table III Examples of individual risks

Causes	Risk of death per million per year
All causes (mainly illnesses from natural causes)	11 900
Cancer	2 800
These figures vary greatly with age	
All violent causes (accident, homicide, suicide, etc.),	396
Road accidents	100
Accident in private homes (average for occupants only	93
Fire or flame (all types)	15
Drowning	6
Gas incident (fire, explosion or carbon monoxide poisoning)	1.8
Excessive cold	8
Lightning	0.1
Accidents at work—risks to employees	
Deep-sea fishing (UK vessels)	880
Coal extraction and manufacture of solid fuels	106
Construction	92
All manufacturing industry	23
Offices, shops, warehouses, etc. inspected by local authorities	4.5
Leisure—risks to participants during active years	
Rock climbing (assumes 200 hours climbing per year)	8 000
Canoeing (assumes 200 hours per year)	2 000
Hang-gliding (average participant)	1 500

c. Failure to consider the broad set of qualitative dimensions that underlie people's concerns about the acceptability of risks and technologies.
d. Failure to consider alternatives to the technology or product proposed, or legal constraints on these actions; or the social consequences of risk decisions such as loss of privacy, the generation of social conflict or loss of civil liberties
e. A tendency to draw on data from diverse sources that vary considerably in quality.

If risk comparisons are to be used to address risk perceptions their limitations must be openly addressed and the choice of comparative data must be targetted to the specific audience taking into account their needs, concerns, and level of knowledge.

5. WHAT IS STRESS AND WHAT ARE ITS EFFECTS?

Once people believe that they are living in situations of risk they can experience psychological and social effects, the most important of which are stress and social conflicts. In psychological terms, *stress* is a complex state deriving from a person's appraisal of their failure to adapt to the demands of their environment and the extent to which that environment meets their needs. Cox[51] has suggested that this process of appraisal takes account of at least four factors:

a. the demands on the person, matched against
b. their ability to meet those demands (personal resources);
c. the constraints that they are under when coping, and
d. the support received from others in coping.

The psychological effects of stress include quite large numbers of subjective, behavioural and cognitive effects. There appears to be an immediate response to the perception of a stressful situation in the form of a negative emotional experience. This emotional experience of stress is often accompanied by changes in the person's psychological, behavioural or physiological functions. Some responses are more controlled than others and are intended to remove or deal with the cause of stress. However, where increased internal arousal levels are maintained for long periods, these in turn lead to exhaustion and collapse.

The discovery of land contamination or the proposed siting of a new hazardous waste facility has the potential to provide a wide range of effects that can, and often do, act as stressors. It is important to distinguish between the influences of social and psychological factors *per se* and the influences of health and financial factors and the methods used in handling the situation. As discussed earlier, the effect of stress on health both directly and in pathophysiological consequences can result in reported health effects for which no causal relationship with the hazardous waste site can be established.

It is difficult to quantify the levels of stress or the levels of psychological disturbance due to stress which are elicited specifically by hazardous waste or contaminated land. However, it is possible to obtain an indication of the factors involved and the responses from the experiences of individuals situations. The important stressors relating to hazardous waste and contaminated land were identified in the study for the European Foundation:[27]

- Living in a situation of perceived risk—this usually centred around fears about health problems, in particular to children, resulting from exposure to chemicals or fears of a large scale disaster.
- Disturbance to environment—including air pollution (dust, odours), water pollution (discolouration, taste) or ecological effects (vegetation problems, pests)
- Disturbance to social organisation—caused particularly if evacuation or partial evacuation of an area occurs. Other disturbances to social functioning can arise where restrictions are placed on certain activities, for example reduced access to recreational areas.
- Financial problems—relating for example to loss in property value and resulting problems in moving house. This has been seen much more frequently at contaminated land sites rather than at waste disposal facilities.
- Feelings of anger and frustration—often these are caused by the way in which situations are handled, particularly where no one is apparently responsible, or will accept responsibility, or provide a solution.

One of the most important features of stress resulting from hazardous waste is that the public find themselves having to cope with situations that have been forced upon them and over which they have no control. It has been shown that in situations of uncontrollable stress the stress responses can become very damaging both psychologically and physically. This loss of control was illustrated at Loscoe in Derbyshire following the landfill gas explosion which demolished a home and injured the owners.[52] Here residents reported that they found themselves trapped and unable to escape from a situation which was not of their own making. The 'coping strategies' or 'flight or fight' were simply not available to the residents.[53] They could not take direct action by preparing themselves against harm, nor could they deny to themselves that the threat existed. The absence of any way of coping leads to feelings of apathy and depression. The methods available for the expression of community concerns and the way in which these are managed also have a significant influence on the social and psychological impact experienced by the community.

6. HOW DO THE PUBLIC RESPOND?

The effect of the stress factors, which is exacerbated if the individual or group feel that they have no control over the situation, is usually seen in a response pattern which starts with uncertainty and questions, followed by members of the community joining together and the identification of a figure to blame. The formation of a local action group is the next step with appeals to public figures for support and newspaper and television coverage. The result can be confrontation.[54] The following case studies examine these responses of the public in relation to three specific sites.

6.1. *Waste Treatment and NIMBY in the UK*

In the UK, NIMBY reactions to proposals for new incinerators, and also some chemical treatment plants, represent a complex response to a number of different yet interrelated concerns. Examination of local community responses to five recent incinerator applications indicates four main areas of concern or stress:[18]

a. The health effects of air emissions
b. Potential accidents involving 'toxic' waste
c. Adverse impact on the quality of life
d. The management of industrial waste

Subsumed within these four areas are a multitude of other concerns ranging from the type of risk perceptions common to many industrial activities as discussed in Section 4 through to specific concerns about the handling of hazardous wastes and the safety of operation of treatment plant. Subjects which commonly raise particular emotions are 'dioxins', 'toxic waste', 'PCBs', 'waste imports.'

The general NIMBY reactions appear to have their source in a few key areas (although in specific locations a number of local issues and concerns will also influence reactions—e.g. experience of a specific plant or accident; proximity of plant; socio-economic characteristics of residents):

a. The extremely poor image of the waste industry generally, and doubt as to its professionalism.
b. Mistrust in the state of knowledge about the risks from emissions.
c. The failure of environmental, and risk, assessments to communicate adequately the nature and extent of the impacts.
d. A lack of any strategic planning for waste disposal facilities.

6.1.1. *Case Study: NIMBY and Seal Sands.* Ocean Environmental Management (OEM) (now Cory Environmental) submitted an application to Cleveland County Council for a 30,000 tonnes per annum chemical incinerator in December 1988. The application was accompanied by an Environmental Statement, the result of an Environmental Assessment carried out for OEM by Environmental Resources Limited (ERL). The application related to a greenfield site on Seal Sands, Billingham, within the UK's largest agglomeration of heavy chemical and petrochemical industry, and an area designated in the county Structure Plan for special industry or capital intensive industry.

The proposal immediately met with strong and vociferous local opposition. A reported 300 residents took part in a protest march through the centre of Middlesbrough. The case made the national press and television. Public concern focused on the potential health effects. Insufficient conclusive evidence on 'safe' limits of human exposure to air emissions and conflicting views on the significance of individual health studies undertaken, and on their relevance to the proposed incinerator, were referred to by the Director of Economic Development and Planning in his report to the Planning Committee. However, he felt that he could not uphold these concerns in terms of a reason for refusal of permission.

The Council had received three separate reports from independent consultants which all in general terms supported the findings of the assessment by ERL. The Environmental Assessment had included a quantified public health risk assessment, one of the first undertaken in Britain to accompany an application for a waste treatment proposal.

Alongside the concern about health risks there was evidence of public concerns about importation of waste, the possible impact of traffic accidents involving hazardous waste, the effectiveness of the control system for the plant, and the competency of the regulatory authorities to monitor air emissions. Two interesting indicators of public concern were apparent. Firstly, another company, Fine Organics, was successful in obtaining permission for an in-house incinerator whilst OEM's proposal was being considered. There did seem to be a difference in the public concern over OEM's proposal that provided for merchant capacity which could attract wastes from outside of the area and Fine Organics' project which provided only for the handling of its own chemical waste arisings. Secondly, during discussions in the Council, and in the findings of an opinion poll commissioned by the County after the Committee's decision, there were indications that, in the minds of the local community, an incinerator run by the Council itself (i.e. a public authority) was preferable to an incinerator operated by the private sector.

Against the recommendation of the Director of Planning and his officers the planning application was refused by the full Council in July 1989. The Council concluded that it was 'not satisfied that such an incinerator would not endanger the health of the residents of Cleveland and would not have a detrimental effect on the quality of life and the environment in the County.'

The Seal Sands case is unique in UK land-use planning terms in that, following the refusal and subsequent appeal by the Company, the Secretary of State agreed to the County's request that a form of Planning Inquiry Commission be set up. The Inquiry was given the powers to consider three applications for incinerators in the area at the same time in order to examine the question of strategic need for such facilities and the technologies involved. This inquiry commenced in the autumn of 1990 followed by a public local inquiry into the individual applications. In November 1992 the Secretary of State upheld OEM's appeal and granted permission.

6.1.2 *Case Study: Incinerator Proposal at Doncaster.* An application by Leigh Environmental Ltd. for the provision of a waste treatment facility, incorporating a 26,000 tonnes per annum capacity incinerator, on Kirk Sandall Industrial Estate Doncaster, South Yorkshire, has been refused on appeal by the Secretary of State for the Environment.[55] The planning application was refused by Doncaster Metropolitan Borough Council on officer recommendation in June 1989. Eight reasons for refusal were given, relating to incompatability with the local plans; sensitivity of the aquifer at the location of the proposal; the need for such a facility; traffic hazards; and the 'additional risks to the health, safety and well being of the existing communities in the area'.

The application had generated considerable public concern, including representations from local trades unions, local doctors, and residents. A petition submitted at the public inquiry contained over 100,000 names, a major indicator of the strength of this concern. At the inquiry representatives of the action group stressed that they were not taking a NIMBY stance in that they did not oppose waste incineration in principle. Their objections focused on:

a. The socio-economic effects of the siting of such a facility in the area, including adverse impact on local industry and on the attractiveness of the area to new businesses.
b. Potential accidents on the plant and on roads in the neighbourhood which included a large residential area.
c. Health risks to the population downwind and the inadequacy of the assessment of these in the Environmental Statement accompanying the application. The ES submitted had been criticised by the Council's own independent con-

sultants in this respect. The inadequacy of the assessment was included in the reasons for refusal and extensively analysed during the public inquiry.

The findings of the inquiry inspector are interesting particularly in relation to the socio-economic effects and perceived risk. On these issues he concludes:

'I am convinced that many people in Kirk Sandall, because of the proximity of their homes to the site would feel constantly ill-at-ease if this development were to proceed, their fears being aggravated by constant reminders of the plant from passage of vehicles and the inevitable publicity for incidents at the site. Despite all assurances of safety which may be given or information programmes intended to assuage doubts and fears, the public would not know, for example, if long-term injury to health were taking place or whether control standards, some of which are less restrictive than elsewhere were adequate.'

He goes on to comment on the possible perceptions of those doing business with the food manufacturers in the area and the fact that they may choose not to do business with companies located near to a toxic waste plant.

The inquiry examined in detail all the reasons for refusal. The Secretary of State's decision letter, in considering the inspector's conclusions, primarily emphasises the unsatisfactory location of the proposal in terms of the sensitive aquifer. However, he also notes that 'rightly or wrongly, many members of the public living close to the appeal site have real fears that the proposal could have long-term health effects' and stresses that he has taken these issues into account insofar as they are relevant to the land use planning issue. On the traffic issue he does not uphold this as a significant reason for refusal considering that the objections are based more on fears of effects following an accident rather than actual evidence of a significant increase in traffic.

6.2. *Case Study: NIMBY and an Operational Treatment Plant*

In 1977 planning permission was granted to Polymeric Treatments Ltd. (a subsidiary of Leigh Environmental Services Ltd.) for an extension to full-scale operation of a plant to solidify hazardous waste by use of the 'Sealosafe' process at Aldridge, Birmingham, West Midlands. This patented method of wastes fixation and solidification involves polymerisation of the material with fly ash and cement. The polymerised material is disposed in clay pits and capped with clay. It solidifies, the resulting solid mass having a number of potential advantages over direct landfill of waste, including reduced leaching of materials into the environment and a more stable and inert fill which should enable reclamation of the site for after-use in a shorter space of time.

From the time full-scale operations commenced there were complaints from local residents, commencing with complaints about smells and developing into more serious concerns about health problems being experienced by residents. One of the main concerns of those living in the vicinity appeared to be possible airborne pollution resulting from vaporization of the semi-solid material, as well as possible water pollution. In 1981 County Council Environmental Health Officers investigated four potential licence violations including reports that arsenic and cyanide were present in waste alleged to have leaked from the site. The report concluded that the waste and contaminated land were being cleaned up. The following year saw a vehement campaign and protests by local residents when a prosecution alleging that the company had failed to prevent noxious emission in 1980 failed and a not-guilty verdict was returned. A fire at the site in 1983 and the authorised filling-in of a Site of Special Scientific Interest (SSSI) by the company in 1985 served to keep the protests alive.

Positive coverage of the operations was evident in 1988 after the company invited local schools to visit the site. However, during the same month the *Birmingham Evening Mail* carried a report that local health officials were to investigate Leigh's activities following a 1500 names petition. A study examined evidence relating to childhood cancers in the area, but no adverse health risks associated with the disposal site were established. Another epidemiological study examined reports of respiratory symptoms in children, but found no evidence of increased incidence in the area.[16]

At the annual conference of the British Medical Association in July 1989 a two-year study into the consequences of the importation of hazardous waste on public health was announced. Leigh at the time was importing some waste each year primarily for solidification by the 'Sealosafe' process. A report of the conference in the *Daily Telegraph* of July 4th 1989 indicated that a doctor, who practised in the area near the Leigh site, reported that local GPs were seeing an increasing number of people with sore throats, streaming eyes and persistent coughs. She suggested that the complaints were linked to the activities at the site. A second doctor told the conference that cases of asthma were above what was expected in the area. It was reported that the local Council was working with the Walsall Health Authority to examine death certificates for any possible links with hazardous waste. The BMA's report on hazardous wastes notes in relation to this anecdotal evidence that such symptoms are difficult to attribute with any degree of certainty to a single cause.[16]

In May 1989 the local action group, Community Action Against Toxic Waste, lodged a complaint with the European Commission. The complaint suggested the possibility of breaches of the EC Directive on Toxic and Dangerous Waste which required that waste must be disposed of without posing a danger to health or harming the environment. In June 1989 Walsall BC refused a licence application by Leigh to extend its operations at Aldridge. The decision came after councillors had seen a report by the Harwell Laboratory which suggested that claims for the 'Sealosafe' process could not be fully endorsed.

The House of Commons Select Committee on the Environment's inquiry into Toxic Waste,[3] concluded that the operations of Leigh were 'one of the most controversial public issues' to emerge during its nine-month study. The Committee concluded that such NIMBY responses were potentially damaging to the industry. In recognising the growing public awareness of environmental and waste management issues it also concluded that 'Greater sophistication of analysis and increased knowledge of toxicological effects will all raise public awareness of possible health effects.' The Committee's recommendations for dealing with public concern were that, firstly the industry must ensure that their operations are environmentally sound, thorough, and based on adequate research, and secondly that it must be open with the public in explaining activities and providing information.

7. HOW CAN THE MANAGEMENT OF RISK BE IMPROVED? SOME CONCLUSIONS

7.1. *Nature of Concerns and Risk Issues*

Table IV provides a summary of the main issues of concern identified from study of sites in the UK involving hazardous wastes and contaminated land which have

Table IV Problem site issues

Nature and extent of hazards	Social and economic
Health and toxicological effects	Liabilities—who pays?
Safety of drinking water safety of property Environmental effects partcularly from odour and dust	Blight effects—personal, financial, land-use Loss of property values difficulty selling property
Persistence of effects	Competency of authorities competency of industry
Likelihood of serious damage particularly arising from accidents, including traffic	Stress
Duration of effects particularly long-term effects, e.g. genetic effects	Ill-health leading to loss of earnings Effects on children
Extent of unknown hazards Uncertainty in investigation and assessment	Effect on local amenity Social and physical Disruption

caused public concern. The most important point to note is that the concerns cover a range of issues from the nature of the hazards to social and economic issues. Furthermore, the concerns focus on both effects on the individual as well as societal effects in terms of local community impact.

In relation to any specific 'problem' site (whether an operational site perceived as a problem, or a contaminated site discovered after development, or the site of a proposed new plant) a number of factors influence the public response:

a. Apparent, as well as actual, deficiencies in knowledge and understanding of the potential risks amongst the perceived experts. This lack of understanding can be compounded by apparent disagreements between experts, e.g. over the use of specific hazard and risk assessment models.
b. The timing of information provision—premature disclosure before the issue is fully understood being just as bad as late provision of information.
c. Failure of the risk assessors and/or competent agencies to address issues of local public concern due to a lack of understanding of these concerns.
d. Failure to involve the public in development of policy and strategy with regard to waste management, i.e. the public are only involved when the strategy is implemented.
e. Limited technical understanding amongst local agencies.
f. Tendency of industry and competent agencies to stress regulatory control as a means of reducing risks.
g. Apparent lack of a responsible party, i.e. someone to blame and who can take action and pay for any mitigation measures.

7.2. *Improving Risk Management*

From our understanding of public perceptions, of stress, and of actual responses to hazardous waste issues, it is possible to identify some key components of an improved management strategy. Fundamentally it is clear that there is no single answer to the problem of public stress and its resulting impact, particularly

NIMBY. Improving the risk management system is going to require attention to:

- Risk communication.
- Control of waste management operations and involvement of the public in the control mechanism.
- Understanding of the health effects of waste treatment and disposal.
- Public discussion of alternative management strategies.

7.2.1. *Risk Communication.* Risk communication improvements must be planned and effective systems which can relieve situations of stress must be set up. These improved systems must focus on the primary objective of providing for informed debate, i.e. providing for the two-way discussion of risk issues between 'experts' and the 'public'. Experience indicates that it can be very difficult to overcome or change an initial perception of risk. Therefore, the strategy overall has to be to reduce the original biases. In order to assist and promote informed debate there must be general education of the public as to the nature of waste, how wastes are handled and managed, and how experts make risk decisions. This could involve both improved environmental education within the formal education system, and informal education through the media.

Communicators of risk information must be trained to understand the risk issues and to communicate effectively. This could include local doctors who have to deal directly with local concerns and stress; local authority officers who are often the first source of information in a community; and industrial operators. Effective communication includes not only attention to the terminology used and the manner in which complex technical subjects are discussed, but also attention to the timing of information provision and the onset of communication. A study for the Department of the Environment has looked at risk communication requirements in relation to potentially contaminated land, in the light of the possible blight and public concern which could be raised by identification of sites.

Effective communication must allow for the public to be involved in consideration of specific issues as early as is possible, e.g. during site selection for a new facility and during scoping of the impacts to be considered in an environmental assessment. Then this involvement must be maintained, e.g. throughout the life of an installation or the duration of an issue such as remediation of a contamination problem. Many authorities and companies have a policy of involving the public in their activities beyond the statutory requirements, by, for example, committees, presentations, positive and proactive relationships with the press, and open days.

7.2.2. *Control of Waste Management Operations.* If waste management operations are to be seen as a solution to hazardous wastes rather than as a problem then they must be controlled adequately. Stronger enforcement and monitoring of operations by statutory agencies is essential. Importantly this aspect of control must be seen to be focusing on preventative measures, as promoted under integrated pollution control.

However, responsibility to control lies also with the operators themselves. Under the Environmental Protection Act, 1990, the concept of the 'duty of care' of anyone who handles waste provides for a new focus of responsibility. The industry cannot be expected to be accepted if it operates unprofessionally. It must learn to regulate itself.

The third group of people who should be important in any control system are the public themselves. Direct involvement of representatives of local communities in the monitoring of facilities, and the publishing of monitoring data and reports for public consumption and criticism are both ways in which local communities can contribute to the control process. Both provide an opportunity for early detection of problems and discussion of appropriate solutions.

7.2.3. *Understanding of Health Effects.* Fear of health effects is one of the dominant influences on public concern over hazardous wastes and a potential source of stress reactions. In a situation where the possibility of health effects from specific sites is often most openly explored in the first instance by the media, who are often seen to search for an 'expert' to pass comment on the likelihood of damage in relation to a situation in which they are not involved, then the potential for distortion, misrepresentation, and generation of stress is obvious.

The review of existing evidence indicates that the risk is probably small, but that it is difficult to quantify. There is an urgent need to understand better the nature and degree of potential and actual effects; to improve assessment methodologies; and for independent monitoring of claims of adverse impact and publication of findings of such studies. The BMA Report[16] recommends the need for more epidemiological data from well-designed and executed research on human populations, including waste workers and vulnerable populations. The BMA stresses the need for further training of medical practitioners in investigation of diseases which may be associated with environmental hazards.

A key area in the management of hazardous wastes has to be in the assessment of potential adverse health impacts before specific options are selected and then before specific sites are developed. This requires considerable attention to public health risk assessment techniques, in terms of development of, and publicity and guidance on, appropriate methodologies, and the validation of assessment models by the relevant independent control agencies. The Department of the Environment, Contaminated Land Branch, has a research programme which is looking at a framework for risk assessment in the management strategy for contaminated land.

7.2.4. *Public Discussion of Alternative Management Strategies.* The perceived imposition of an involuntary risk which creates imbalance in risk distribution and imposes unfairly on individuals and specific communities and their property, is the second issue which lies at the heart of public stress over hazardous wastes. The latter are not seen as a product of a consumer society, but rather as the fault of industry. Portney argues that decisions as to waste control and management are inherently political decisions, and that the social, cultural, and psychological construction of opposition to hazardous waste (particularly facility siting decisions) reflect general opposition to any political decision which impinges on the individual.[13]

This argument perhaps suggests there can be no solution to public stress. However, attention to the involvement of the public in political decisions on waste management, and particularly involvement of local communities in decisions as to alternate strategies which affect their individual rights, should be explored. Strategic planning for the siting of waste facilities needs to be strengthened and this planning system opened to public involvement and consultation.

Open examination of the best practicable environmental options for dealing with specific hazardous wastes, and central guidance to those managing wastes as to the safest and most appropriate form of disposal and treatment of wastes and contaminated land, is important.

8. CONCLUSIONS

Public stress over hazardous wastes is real and can be acute in specific situations. This stress if often caused by a combination of paucity of understanding, of information, and of communication. Stress can impact on decision-making at the local level particularly with regard to the siting of facilities. Stress can also impact on health and on the confidence of the public in the management of hazardous wastes. There is a need for proactive attention to measures by which stress can be ameliorated. There is no single solution to the problem. Attention to the whole risk management system (risk identification, assessment, evaluation, decision-making, control and audit) is required. The overall theme is the need to understand the risk and to involve communities in decisions about the management of these risks. The responsibility to improve understanding and communication lies with everybody involved in the management system.

Acknowledgements

The author would like to thank Judith Denner and Roger Smithson of the Contaminated Land Branch, Department of the Environment, for very useful comments on this paper.

References

1. Central Statistical Office, *Social Trends 20* (HMSO, London, 1990).
2. Department of the Environment Waste Management Paper No 4, *The Licensing of Waste Facilities* (HMSO, London, 1988).
3. House of Commons Environment Committee Second Report, *Toxic Waste* (HMSO, London, 1989).
4. Department of the Environment, *The Government's Reply to the Second Report from the Environment Committee Toxic Waste* (HMSO, London, 1989).
5. Her Majesty's Inspectorate of Pollution, *Third Annual Report* (HMSO, London, 1991).
6. The Hazardous Waste Inspectorate, *Third Annual Report* (HMSO, London, 1988).
7. Clayton, Bostock, Hill and Rigby Ltd., "Evidence to the House Commons Select Committee on the Environment" **In**: *House of Commons Environment Committee First Report Contaminated Land* (HMSO, London, 1980) p. 37.
8. Department of the Environment, *This Common Inheritance: Britain's Environmental Strategy* Cmnd. 1200. (HMSO, London, 1990).
9. Royal Commission on Environmental Pollution Twelfth Report, *Best Practicable Environment Option*, Cmnd. 310. (HMSO, London, 1988).
10. B. Croft and D. Campbell, "Characterisation of 100 UK Landfill Sites" **In**: *UK Atomic Energy Authority Environmental Safety Centre: Proceedings of 1990 Harwell Waste Management Symposium* (AEA, Harwell, 1990) p. 1.
11. Interdepartmental Committee on the Redevelopment of Contaminated Land (59/83) (2nd Edition), *Guidance on the Assessment and Redevelopment of Contaminated Land* (Department of the Environment, London, 1983).
12. Commission of the European Communities, Proposal for a Council Directive on the Landfill of Waste. *COM (91) 102 Final* (Brussels, 22 May 1991).
13. K. E. Portney, *Siting Hazardous Waste Treatment Facilities. The NIMBY Syndrome* (Auburn House, New York, 1991).
14. J. Petts, "The Challenge of Contaminated Sites: Developing an Integrated Approach," **In:** Pro-

ceedings of the Environment Northern Seas Conference, Stavanger, Norway, August 1990 **Vol. 3** (ENS, Stavanger Forum, Norway, 1991) pp. 17–29.
15. S. Wells and S. Caldwell, "Overview of the revised Hazard Ranking System (HRS)" **In**: *Proceedings of Superfund '90, Washington D.C. 1990, Hazardous Materials and Research Institute, Silver Spring MO, USA* (1990) pp. 71–76.
16. British Medical Association, *Hazardous Waste and Human Health* (Oxford University Press, Oxford, 1991).
17. S. M. Brett *et al.*, "Assessment of the public health risks associated with the proposed excavation of a hazardous waste site," **In**: *The Risk Assessment of Environmental and Human Health Hazards* (D. J. Pasutenbach, ed.) (John Wiley & Sons, New York, 1989) pp. 427–458.
18. J. Petts, "Incineration Risk Perceptions and Public Concern—Improving Risk Communication, Experience in the U.K." *Waste Management and Research* **10**(2) 169–182 (1992).
19. G. H. Eduljee, "Health risk assessment of incinerator emissions" **In**: *Proceedings of Seminar 'Incineration: An Environmentally Acceptable Means of Waste Disposal'* 13 December 1989 (Institution of Mechanical Engineers, London, 1989)
20. Health and Safety Executive, *Risk Criteria for Land Use Planning in the Vicinity of Major Industrial Hazards* (HMSO, London, 1989).
21. J. W. Grisham (ed.), *Health Aspects of the Disposal of Waste Chemicals* (Pergamon, New York, 1986).
22. T. Cox, *Stress* (Macmillan, London, 1978).
23. L. S. Zeyans, "Stress and the development of somatic disorders" **In**: *Handbook of Stress: Theoretical and Clinical Aspects* (L. Goldberger and S. Breznitz, eds.) (Free Press, 1982).
24. B. Jemmott and S. E. Locke, "Psychosocial factors, immunologic mediation and human susceptibility to infectious diseases: how much do we know" *Psychological Bull.* **95**, 78 (1984).
25. L. S. Sklar and H. Anisman, "Stress and Cancer" *Psychological Bull.* **89**, 369 (1981) pp. 369–406.
26. P. Martin, "Psychology and the Immune System" *New Scientist* (9 April, 1987) pp. 46–50.
27. J. M. Denner *et al.*, *Hazardous Wastes and the Public: United Kingdom* (Working Paper No. WP/91/52/EN/European Foundation for the Improvement of Living and Working Conditions, Dublin, 1991).
28. Department of Health, *Investigation of the Possible Increased Incidence of Cancer in West Cumbria* Report of the Independent Advisory Group (Chairman, Sir Douglas Black) (HMSO, London, 1984).
29. Warren Spring Laboratory, *Review of Municipal Solid Waste Incineration in the UK* Report LR 776 (PA) (Warren Spring Laboratory, Stevenage, 1991).
30. S. Openshaw *et al.*, "Geological analysis machine for the automated analysis of point data sets" *Int. J. Geographical Information Systems* **1**, 335–358 (1987).
31. P. J. Diggle *et al.*, "Modelling the Prevalence of Cancer of the Larynx in Part of Lancashire: A New Methodology for Spatial Epidemiology" **In**: *Spatial Epidemiology* (R. W. Thomas, ed.) (Pion, London, 1990) pp. 35–47
32. A. C. Gatrell and A. A. Lovett, "Burning Questions—Incineration of Wastes and Implications for Human Health" **In**: *Waste Location: Spatial Aspects of Waste Management, Hazards and Disposal* (M. Clarke *et al.*, eds.) (Routledge and Kegan Paul, London, 1992).
33. E. G. Knox *et al.*, "The Quality of Notification of Congenital Malformations" *J. Epidemiology and Community Health* **38**, 296–305 (1984).
34. J. Lenihan, *Scottish Home and Health Department Report: Bonnybridge/Denny Morbidity Review* (SHHD, Edinburgh, 1985).
35. Welsh Office, "The incidence of congenital malformations in Wales, with particular reference to the district of Torfaen, Gwent," *A Welsh Office Review of Notifications Made to the Office of Population Censuses and Survey* (Welsh Office, Cardiff, 1985).
36. Environmental Risk Assessment Unit, "Panteg Monitoring Project," *First Interim Report to the Welsh Office, August 1991* (University of East Anglia, Norwich, 1991).
37. A. Tversky and D. Kahneman, "Judgement under uncertainty: heuristics and biases," *Science* **185**, 1124–1131 (1974).
38. P. Slovic *et al.*, "Facts and fears: understanding perceived risk" **In**: *Societal Risk Assessment: How Safe is Safe Enough* (R. C. Schwing and W. Albers, eds.) (Plenum Press, New York, 1981) pp. 181–216.
39. S. Lichtenstein *et al.*, "Judged frequency of lethal events" *J. Experimental Psychology (Human Learning and Memory)* **4**, 551–578 (1975).
40. P. Slovic *et al.*, "Behavioural decision theory: perspectives on risk and safety" *Acta Psychologica* **56**, 183–203 (1984).
41. M. Thompson and A. Wildavsky, "Towards a cultural theory of risk" **In**: *The Risk Analysis Controversy* (H. Kunreuther and E. Ley, eds.) (Springer-Verlag, Berlin, 1982) pp. 145–161.

42. P. Prescott-Clarke, *Public Attitudes Toward Industrial Work-Related and Other Risks* (Social and Community Planning Research, London, 1982).
43. P. M. Sandmann, "Hazard versus outrage: a conceptual framework for describing public perception of risk" **In**: *Themes and Tasks of Risk Communication* (H. Jungermann *et al.*, eds.) (KFA Jülich, Germany, 1988) pp. 211–219.
44. G. Cvetkovich and P. M. Wiedemann, "Results of the working group: trust and credibility in risk communication" **In**: *Themes and Tasks of Risk Communication* (H. Jungermann *et al.*, eds.) (KFA Jülich, Germany, 1988) pp. 83–86.
45. W. W. Lowrence, "Stewardship of chemical production risks" *Paper presented at the First IUPAC Workshop on Safety in Chemical Production, 9–13 Sept. 1990, Basel, Switzerland* (1990).
46. R. V. Kemp, "Risk tolerance and safety management" *Reliability Engineering and Systems Safety* **31**, 345–353 (1990).
47. D. Swartzman *et al.*, "Reducing aversion to living near hazardous waste facilities through compensation and risk reduction" *J. Environmental Management* **20**, 43–50 (1985).
48. H. Inhaber, "A market-based solution to the problem of nuclear and toxic waste disposal" *J. Air Waste Manage. Assoc.* **41**(6), 808–816 (1991).
49. M. L. P. Elliot, "Improving community acceptance of hazardous waste facilities through alternative systems for mitigating and managing risk" *Hazardous Waste* **1**, 397–410 (1984).
50. V. T. Covello, "Informing people about risks from chemicals, radiation, and other toxic substances" **In**: *Prospects and Problems in Risk Communication* (W. Leiss, ed.) (University of Waterloo Press, Canada, 1989) pp. 1–50.
51. T. Cox, "The nature and measurement of stress" *Ergonomics* **28**, 1155–1163 (1985).
52. G. Ryan *et al.*, *Report of the Non-Statutory Public Inquiry into the Gas Explosion at Loscoe, Derbyshire, 24 March 1986* (2 Vols) (Derbyshire County Council, 1988).
53. J. D. Tankard, "The Plight of Loscoe, Derbyshire" *Wastes Management* **77**(3), 172–175 (1987).
54. J. DeBoer, "Community response to soil contamination—risk and uncertainty" **In**: *Contaminated Soil* (Assink and Van der Brink, eds.) (Kluwer Academic Publishers, Dordrecht, 1986) pp. 211–219.
55. Department of the Environment, Appeal by Leigh Environmental Ltd. Appeal decision 11 November 1991. Reference APP/F4410/A/89/126733 (Department of the Environment Yorkshire and Humberside Regional Office, Leeds, 1991).

9. STRESS AND THE ENVIRONMENT: THE IMPACT OF THE MOTOR VEHICLE

J. McHARRY*

Motor vehicles create substantial environmental impacts "from cradle to grave". One of the most serious issues is the pollutants found in vehicle emissions; these are responsible for accelerating other forms of environmentally-related stress, affecting up to one fifth of the UK population. Traffic fumes also rank high on the scale of public disturbance. Road traffic remains the largest single source of noise nuisance in the UK. Road pricing, traffic calming and improved public transport systems are only partial solutions if the rising number of vehicles on our roads is not curbed.

1. INTRODUCTION

Human health, the environment and the economy are inextricably linked. The state of the environment determines not only the health and safety of current generations but also future generations. Human activities are exerting non-sustainable stresses upon the global ecosystem—such pressures will undoubtedly have knock-on consequences for humankind.

This chapter will discuss various aspects of stress arising from transportation—and in particular the motor vehicle. In 1970, Alvin Toffler's *Future Shock* described the "shattering stress and disorientation that we induce in individuals by subjecting them to too much change in too short a time". The impact of new technology upon our lives is one cause of such stress, another is the growing predominance of the car.

Stress is an inescapable part of life. If the pressures felt by the individual are managed effectively, stress can provide the necessary motivation to overcome all sorts of obstacles blocking the path to goals and aims. However some types of stress are overwhelming and, if allowed to get out of control, problems will occur. In the individual this can manifest itself in a variety of ways, but reduced performance at work and in undertaking everyday tasks is a common result. Another frequent symptom is lowered resistance to disease, so sickness occurs. In extreme cases, death might occur if the nature of the stress is significant and prolonged. These effects make it vital that symptoms of stress are heeded as an early warning sign.

But it is not just humankind that suffers stress. Over the last couple of decades scientists have, with increasing voracity, sounded warnings about global systems out of equilibrium. Human activities have stretched many of the earth's systems to their very limits; the results are clearly evident—pollution, loss of biodiversity, and now the threat of climate change through global warming. Pollutants act

*Environmental Research and Information, P.O. Box 893, London E5 9RU, UK

alone or together—synergistically—to increase natural stresses upon the environment. The dominance of the motor vehicle in creating unsustainable pressures is not disputed.

2. THE MOTOR VEHICLE AS AN AGENT OF STRESS

"Life in the fast lane" is a phrase commonly used to describe the western experience of life, particularly in urban areas. Conjuring up visions of noise, glare, pollution, fast movement, competition—all these images actually add up to a picture of transport today.

Transport, especially the motor vehicle, is a major source of environmental disruption. It is also a major consumer of raw materials and energy. In the developed world, transport accounts for 30% of all energy consumption. Our quality of life is profoundly influenced by transport—a poor transport policy will severely undermine the quality of life for many people, especially for those with no access to a motor vehicle. Pedestrian danger, visual intrusion, noise, vibration, dust and dirt, fumes and smoke, and physical irritation are the major disturbances to the public from road vehicles.

The growth in road traffic is now coming up against real environmental limits—with increasing pollution, congestion and conflicts in land use. Despite policies and techniques aimed at cleaning up emissions from vehicles, air pollution has reached serious levels in a number of cities in the developed world, and is a major problem in most developing countries. The biggest culprit is the motor vehicle—generating more air pollution than any other single human activity. In many cities, traffic exhausts have become the number one problem, responsible for a wide variety of health problems and building soiling.

Vehicle emissions are responsible for photochemical smogs—the haze visible over many cities in the summer. This sort of air pollution places increased stress on the most vulnerable members of the population in both developing and developed countries. Up to one fifth of the population in the UK fall into this category—the elderly, babies and young children, pregnant women, and people suffering from respiratory or circulatory problems.[1] Ill health is a result of physical stressing agents, and may be seen as a failure to adapt to such conditions.

Beyond vehicle emissions, transport exerts major stresses elsewhere. Cancelled trains, crowded public transport systems, long-distance commuting, noisy car horns all take their toll upon the individual. Urban areas, unable to cope with increasing volumes of traffic suffer congestion and noise pollution. For pedestrians there is the added problem that, traditionally, urban areas have been planned for vehicles, with inadequate provision for pedestrian safety. Other stress-provoking factors involve bad planning in transport facilities. Safety of individuals while travelling or on foot is a growing issue, along with accessibility to transportation systems. Whilst the environmental and health effects of motor vehicle pollution have been well documented in many scientific papers and reports, there is little information available regarding the nuisance caused to the public from vehicle-derived pollutants. Studies from the Social and Community Planning Research (SCPR) indicate that disturbance, which could be interpreted as stress, generally increases with traffic flow.[2] A nation-wide survey conducted by SCPR back in 1972 showed that traffic fumes ranked equal second with noise in importance as a public disturbance.

Transport is a contributory factor to the global problem of "acid rain". This is often described as the "silent killer", insidious in its effects, placing significant stress upon natural systems. Irreversible damage to forests has already occurred in Germany and many parts of eastern Europe. Within urban areas, pollution from vehicle emissions causes continuous and costly damage to buildings and structures. Unsightly soiling by diesel particulates, as well as direct damage from gaseous nitrogen pollutants, is affecting the stonework on many urban buildings. Cleaning and renovation work is extremely expensive.

The greatest environmental impact occurs when products from exhaust emissions chemically interact. For instance, ozone is formed when pollutants such as nitrogen oxides and hydrocarbons react together in the presence of sunlight. Ozone at ground level is a secondary pollutant, causing significant damage to trees and vegetation, including loss of agricultural productivity through damage to crops. Throughout the European Community the annual cost of pollution damage to crops is calculated at over US$1 billion.

There are now 408 million cars in the world and this figure is expected to reach half a billion by the end of the century. Beyond the sheer impact of such a number of vehicles, the supporting infrastructure, for instance, large-scale road building and widening programmes create substantial impacts upon the environment. Finally, the eventual disposal of this growing number of vehicles is an issue that has been given too little attention yet is starting to cause increasing environmental impacts. Although the iron and steel content of vehicles makes salvage worthwhile, plastic is increasingly being substituted. Currently the use of mixed plastic components, textiles, rubber, and specialised glass makes recycling uneconomic.

With rising waste disposal costs and tightening legislation, car manufacturers are being forced to acknowledge the substantial impact of disposal. In Europe, landfill operators are increasingly reluctant to accept unsalvageable car parts which could total 20% over the next five years.

3. VEHICLE EMISSIONS AS AGENTS OF STRESS

Vehicle emissions are of particular concern because of the rising concentrations being monitored at street levels, and in densely populated areas. Since motor vehicles generate more air pollution than any other single human activity, their potential role in accelerating environmentally-related stress cannot be ignored. The air in our cities frequently does not meet the guidelines laid down by the World Health Organisation (WHO) for pollutants such as carbon monoxide and nitrogen dioxides. Both these are waste products emitted through the exhausts of vehicles.[1]

In OECD countries, road transport produces 75% of carbon monoxide emissions, 48% of nitrogen oxides and 40% of hydrocarbons. In the US in 1985, transport was responsible for 70% of the carbon monoxide, 45% of the nitrogen oxides and 34% of the volatile organic compounds in the air. In Europe, a similar burden of pollution exists. For 1991 it was estimated that motor vehicles in Europe alone would emit approximately 5.5 million tonnes of hydrocarbons, 28 million tonnes of carbon monoxide and 6.5 million tonnes of nitrogen oxides.[1] The average car produces more than a quarter of a tonne of pollution every year.

What is this doing to people's health? Research by Friends of the Earth, based on WHO official information, reveals that 20% of the population is at risk from polluted air in Britain today. A similar figure has been calculated for the US by the American Lung Association. Groups most sensitive to air pollution are children under two years of age, pregnant women, the elderly and people suffering from asthma, bronchitis, other respiratory ailments, angina and other coronary heart diseases.[3]

These effects are global. In Athens, renowned for its summer car-induced smogs, between six to ten people, mainly the elderly or very young, are killed each day in summer by the smog or *nefos*. Doctors offer the parents of a new baby a straight choice "fit your child with an oxygen mask or leave town".[4] In Bombay the air is so polluted that a day's breathing is equivalent to smoking half a pack of cigarettes.

Every day a reasonably active person will inhale between 10,000 and 20,000 litres of air over 24 hours—if vigorous exercise is undertaken this figure will be higher. With this volume of air being taken into the human body, it is easy to see how damage from pollutants may occur. Within the lungs, air is exchanged through tiny air sacs known as alveoli, which are in close contact with small blood vessels. At any one time, the total surface area of blood vessels within the alveoli exposed to the atmosphere is as large as half a tennis court.[1]

Young children are particularly at risk from air-borne pollutants as their respiratory systems are not fully developed. Some pollutants, for instance lead, damage the nervous system. The greatest impact occurs at times of maximum brain development—in other words when children are young. Children absorb lead more quickly and easily than adults—the position of a car exhaust pipe is exactly at the level of a baby in a pushchair or young children walking.

Over 10,000 different substances are found in vehicle exhausts. Many contribute to environmental stresses such as acid rain and the "greenhouse effect".

3.1. *Sulphur Dioxide*

The most serious health effects occur when sulphur dioxide is present with increased levels of particulates, or smoke. Sulphur dioxide can be absorbed onto the surface of the particulates and if these are inhaled, they will be carried deep into the lungs. Moisture in the lungs creates the right conditions for the formation of sulphuric acid.[3] Short term exposure to sulphur dioxide, in combination with particulates, is linked with increased mortality at times of heavy pollution. In London, airborne levels of SO_2 regularly exceed the one-hour World Health Organisation guideline of 350 micrograms per cubic metre.

3.2. *Nitrogen Oxides*

OECD research confirms that the primary source of most NO_x emissions is road transportation. Nitrogen oxides include nitric acid and nitrogen dioxide. A range of health risks is associated with nitrogen dioxide, including increased susceptibility to viral infections and lung irritation. Asthmatics are particularly susceptible to nitrogen dioxides.

New research from St Bartholomew's Hospital in London reveals that car exhausts may increase the incidence of hay fever. Studies show that nitrogen dioxide kills the cilia, the microscopic hairs that line the nose and windpipe. These hairs play a crucial role in helping to expel mucus and foreign bodies from the human body—expelling it through the nose or mouth. If the cilia are killed by nitrogen dioxide, the study argues that allergens from pollen will stay longer on the cells of the nose and diffuse more easily into the body—increasing the likelihood of an allergic response.[5]

Nitrogen dioxide levels from traffic fumes have been increasing since the 1970s. Nitrogen oxides play an important role in the formation of photochemical smog, responsible for the haze that hangs over many cities, particularly in the summer. Nitrogen oxides are implicated in the formation of acid rain. Vehicle drivers should be aware that nitrogen oxide emissions are six times greater at speeds of 80mph than at 30mph.

3.3. *Carbon Monoxide*

This odourless and colourless gas is produced by the incomplete combustion of fuel. 85% of the UK's emissions come from road traffic. An average car emits five lungfuls of poisonous carbon monoxide gas per mile.

Carbon monoxide reduces and impairs the oxygen-carrying capacity of the blood and aggravates heart and lung conditions. It will dull people's perception, as well as causing headaches. Carbon monoxide in the urban atmosphere has been linked to loss of worker productivity and general discomfort. A study in 1963 revealed that people's senses become dulled by an hour's exposure to carbon monoxide levels of around 55 parts per million (typical rush hour level). A carbon monoxide level of 30 parts per billion has been found to impair the central nervous system in that part of the brain which controls perception and response.[6] With these effects it can be assumed that atmospheric levels of carbon monoxide must be taken as a significant contributory cause towards misjudgements prior to accidents.

The highest concentrations of carbon monoxide are emitted from vehicles with cold engines and vehicles stationary at road junctions. Overall, and in high concentrations, carbon monoxide is the most directly toxic gas found in vehicle emissions.

People identified as being most at risk include pregnant women and young children, the elderly, and sufferers of heart disease and respiratory complaints such as chronic bronchitis. Carbon monoxide may retard an unborn child's growth and mental development.

3.4. *Volatile Organic Compounds (VOCs)*

These consist of a large number of compounds including hydrocarbons, which arise from a variety of industrial processes and from the evaporation of liquid fuels, solvents and organic chemicals. VOCs are directly implicated in the formation of ground-level ozone and photochemical smogs—it is vital that stringent controls are introduced to reduce the public health hazard and potential damage to vegetation. Of particular concern is benzene, a constituent of petrol, and known to have carcinogenic effects.

Emissions of VOCs in the air have been rising steadily since 1970 and figures from the Department of the Environment show that the increase is a direct result of emissions from road transport. These have increased by over 70% since 1970.[7]

3.5. *Ozone*

Ozone is formed when hydrocarbons from incomplete combustion in vehicle emissions and other sources chemically react with nitrogen oxides and VOCs in sunlight. The result is corrosive ozone gas. It can irritate the eyes, nose and throat, as well as causing catarrh and headaches. Ozone aggravates chronic respiratory diseases such as asthma and bronchitis. At high concentrations, ozone will damage the tissues of the lung and can impair the human body's defences against bacteria and viruses.[3] Ozone is responsible for the famous photochemical smogs of Los Angeles and Mexico City.

Research in the US shows that concentrations of ozone—lower than formerly believed—cause short-term breathing difficulties and long-term lung damage. A Canadian study confirms that there are more admissions to hospital of children suffering from asthma when ozone levels are high. During the summer months more than half of all Canadians are exposed to concentrations of ground-level ozone, or smog.[8]

3.6. *Diesel and Airborne Particulates*

One of the main hazards from diesel-fuelled vehicles is the emission of large quantities of particulate matter. This mixture of pollutants in solid form is a chemical stressor—with the ability to carry poisonous heavy metals deep into the most sensitive parts of the human lung.[9] Here heavy metals and certain complex organic compounds, such as polyaromatic hydrocarbons, are potentially carcinogenic.

The higher incidence of lung cancer in urban areas is well known. It is not a minor problem—some researchers believe that a mix of sulphate particles (transformed from SO_2) can mix with air and liquify into an aerosol form capable of bringing toxic metals and gases into the lung. This mix could be responsible for as many as 50,000 deaths in the US each year. Estimates from the World Health Organisation and the United Nations Environment Programme suggest that more than one billion people—one in five people in the world—are exposed to excessive levels of suspended particulates.[10]

3.7. *Lead*

Lead is a dangerous neurotoxin (brain poison). There is no known safe level of exposure; it has the ability to accumulate in the environment as well as in the human body. At high levels, lead is known to cause damage to various organs in the body, including the liver and kidneys, and will affect the function of the brain. The hazards of lead at lower levels has become of great concern in recent years. It can cause seizures, behavioural changes, mental retardation, irritability and clumsiness.

Lead in petrol is responsible for 80% (but this proportion is falling) of the lead in the air we breathe, and for large quantities of lead particles in urban dust. Concentrations are higher in urban areas and near busy roads but the impact of the motor vehicle means that lead distribution is widespread. Available data now shows that lead in petrol at the scale of use in the 1970s produced significant environmental lead contamination and was responsible for increasing the average blood/lead levels in the general population.

Lead in petrol has been the source of much controversy for many years. Since the 1920s lead compounds, especially tetra-ethyl lead, have been added to petrol to prevent "knocking"—to ensure that petrol burns rather than explodes. Much of the lead ends up in vehicle exhausts.

Studies linking children's body/lead levels with reduced intelligence emerged in the 1970s. At the time such a link was highly controversial, but world-wide research has confirmed that such a link does exist.

A major Scottish study published in 1987 by Dr. Mary Fulton and colleagues at Edinburgh University gave the results of tests upon 500 Edinburgh children aged between six and nine. Reports in the medical journal *The Lancet* (May 1987) reveal that the children showed a clear dose-related link between low levels of lead in the bloodstream and the ability to read and work with numbers. The study took into account various social factors which might affect the children's performance, yet the study found that the difference between the 10% of children with highest lead levels and the 10% with lowest only averaged six points on the British IQ intelligence scale. This infers that most children in Britain have enough lead inside them to hinder intellectual development.

But one of the most worrying aspects of this survey is the fact that 98% of the children had levels that fell well within the Department of Health and Social Security's definition of "normal". Such studies reveal that the effects of lead as an environmental contaminant are severe. A similar picture is obtained from international studies. Research in Mexico City found that seven out of ten newborns had lead levels in their blood in excess of World Health Organisation guidelines.[9]

Air pollution is a major problem in developing countries. Rapid industrialisation in Thailand has led to immense traffic congestion problems in Bangkok, bringing worsening health for its inhabitants. The city's 2.3 million vehicles (a figure that is increased each day by 800) daily pour out a cocktail of pollutants from their exhausts. Vehicle pollution in Bangkok is now so severe that 41% of traffic police surveyed were found to have respiratory problems. Children in Bangkok are born with levels of lead in their blood exceeding 18 milligrams per decilitre—brain damage by lead can occur at lower levels of 10–15 mg/dl.[12]

In an effort to ameliorate some of these problems, technology transfer from the developed countries must be made available to allow pollution control equipment to be developed. At present less-polluting vehicles are either few and far between or they are priced beyond the reach of the ordinary citizen.

3.8. *Smog Alerts*

Smog intensifies air pollution problems by holding pollutants in the air. This invariably leads to an increased "dose" reaching individuals—this is significant if individuals fall within the vulnerable population groups.

Smog or pollution alerts are already common in many countries in Europe. Pollution alerts are issued through the media in the US, Japan, Germany, the Netherlands, France, Sweden and Denmark. People at risk are advised to avoid strenuous exercise or to stay indoors. When pollution is particularly bad, local authorities in some countries may close down polluting factories or halt the use of cars. Overall, however, the smog problem needs to be tackled by far tougher enforcement of national air quality standards—in line with World Health Organisation guidelines, and backed by an expanding monitoring network and public data registers.

3.9. *Air Quality Standards*

Mandatory air quality standards have been introduced in a number of countries around the world to improve air quality in polluted areas. They are also used in areas where the air remains relatively clean in an attempt to maintain air quality. Protecting human health and the environment demands an integrated approach—one which will both tackle pollution at source as well as establish national air quality standards.

In 1987 the World Health Organisation published air quality guidelines for Europe. Their primary aim is to "provide a basis for protecting public health from adverse effects of air pollution and for eliminating, or reducing to a minimum, those contaminants of air that are known or likely to be hazardous to human health and well-being".[13] If these guidelines are exceeded, it does not mean that ill effects occur; but the levels of pollution could present a risk, particularly to susceptible members of the public.

Serious attention needs to be given to the fact that air pollution can present a hazard to all sectors of the population. Air pollution is a stressing agent, acting physically and chemically upon the body. Even in healthy individuals, the effects can be measured. A study undertaken by doctors from the Institute for Medical Psychology at the University of Munich monitored 22 healthy women for the effects of air pollution on their physiological state and mental functioning. The women, living in polluted and non-polluted regions of Bavaria, were assessed over a two-month period. The results indicated that impairments in reaction time to visual stimuli, and the inability to concentrate, appeared to be associated with increased concentrations of pollutants. Other tests revealed the effects of low-level ambient air pollution on impaired physiological well-being.[14]

3.10. *Controlling Emissions*

Technical fixes such as catalytic converters and legislation are necessary to curb the impact of air pollution—yet these will do little more than maintain the status quo if the underlying cause, the huge increase in vehicles, is not tackled. It is here that increased expenditure on public transportation systems is vital; incentives to get people to abandon their car for journeys into city centres, or to promote car pooling, are well under way in many parts of Europe and the USA.

Legislation designed to reduce emissions includes the Sofia Protocol Concerning the Control of Emissions of Nitrogen Oxides or their Transboundary Fluxes, signed in November 1988, becoming legally binding in February 1991. Signatories to the Protocol agreed a package of measures designed to freeze nitrogen oxide emissions from 1994. The Protocol is based upon the "critical load" concept which provides a scientific rationale for emission reductions.

An EC Directive for controlling emissions from small cars has already been adopted. In addition, legislation targeted at medium and large cars requires that emissions of nitrogen oxides (NO_x) from passenger cars is, by the end of 1992, reduced by 75%. All new petrol-engined cars will have to be fitted with catalytic converters to attain such a reduction. Amendments to the EC Directive on emissions from large diesel vehicles will reduce NO_x, as well as smoke. Diesel vehicles are currently responsible for half the NO_x from road transport in the UK. Unfortunately those features of diesel engines which make them fuel-efficient rule out the use of catalysts to remove NO_x.

Legislation to set up air quality standards for ozone is being tackled by the European Commission, which published draft guidelines in June 1991. Existing international guidelines, designed to protect human health and vegetation from ozone damage, are routinely exceeded throughout Europe. The draft Directive proposes to establish common procedures for ozone monitoring and information exchange.

It is important that all legislation designed to curb air pollution is based upon four principles:

- The Polluter Pays.
- The Precautionary Principle—prevention is better than cure. Taking action now will reduce the impact upon the environment, on people's health and will, in the longterm, be more economical.
- Adoption of the Best Available Technology for pollution control. The UK's Environmental Protection Act 1990 and its system of Integrated Pollution Control is based upon BATNEEC—Best Available Technology Not Entailing Excessive Cost.
- Freedom of Information. Available data on pollution levels and limits, as well as monitoring results, should be publicly available and open to inspection.

4. NOISE AS A STRESS FACTOR

"It is possible that the hearing of the population as a whole has begun to be affected by the rise in ambient noise in cities through the growth of the mechanisation of transport", warned *British Medical Journal* in September 1965.

Noise is one of the least recognized and regulated of all pollution sources—yet it is a major public health issue. Noise acts as a physical stressor and, at intense levels, unwanted or unpleasant noise can contribute to ill health. Constant noise causes fatigue as the mind seeks to minimize the intrusion. It is thought that the body still reacts to sound even when it is below the level of conscious awareness. Short-term physical effects to noise include increased muscle tension and fatigue or headaches, whilst longer-term stress may lead to changes in the amount of hormones secreted into the blood stream. This can lead to physiological damage: typical long-term effects are insomnia, irritability and fatigue.

"The Environmental Protection Agency estimates that more than 70 million Americans now live in neighbourhoods with noise levels high enough to interfere with communication and to cause annoyance and dissatisfaction", *Psychology Today* reported in October 1981.[14]

Traffic noise has become a major source of pollution in urban areas. Road traffic remains the largest single source of environmental noise nuisance in the UK (although the proportion of people bothered by it has been overtaken by noise from neighbours). Complaints to Environmental Health Officers about road traffic noise has increased by nearly 50% over the last ten years.[15]

In OECD countries about 130 million people are exposed to "unacceptable" noise levels—above 65 decibels—caused by road, rail and air traffic. In addition, over 50% of the OECD population are exposed to noise levels in excess of 55 decibels—a level which needs to be considered as unsatisfactory. Despite these statistics, measures to combat noise are often given low priority.[16]

Sound is measured in decibels with the threshold of hearing registering zero decibels. A quiet street registers about 50 decibels, a noisy street 70, and at 80 (the danger level) the noise is equivalent to the noise inside a small car. Beyond this, a heavy truck registers 90 decibels, whilst the noise from a jet engine 30 metres away registers 130–140 decibels.

According to a World Health Organisation working group, there is no precise critical sound level below which nobody is annoyed and above which everyone will be annoyed. Generally levels of less than 50 decibels outdoors in daytime are thought to cause little or no nuisance to the general community. At night the outdoor level needs to drop to about 45 decibels.

4.1. *Controlling Noise*

UK legislation for controlling noise usually follows guidelines set by the United Nations Economic Commission for Europe (UNECE). Maximum noise limits for new vehicles and all vehicles in use are laid down by law. Provisions are also made for ensuring that vehicles are fitted with exhaust silencers; that road users should not make excessive noise or run the engine unnecessarily when stationary. Controls over noise from car horns on both moving and stationary vehicles also exist. Offenders can be dealt with by warnings, fixed penalty notices or prosecution. In October 1986 most police forces implemented Vehicle Defect Registration Schemes (VDRS). These allow the motorist to remedy vehicle defects or to scrap the vehicle as alternatives to being reported for prosecution.[15]

Overall, obtaining the evidence for successful prosecution of noise offenders is extremely difficult. Figures from the UK Home Office relating to noise from motor vehicles, actually show that convictions for noise offences fell by 28% between 1986 and 1989.

As the amount of traffic on the roads increases, a further problem has emerged. Whilst some compensation and financial grant aid is available for noise insulation work such as double glazing, this is only available for residents living alongside new or substantially altered roads; other residents are left unprotected. This loophole is felt most keenly in areas where roads are now carrying substantially increased volumes of traffic.

Traffic management schemes may be one solution to alleviate some of the noise and resultant stress upon nearby residents, but these are not always possible. Transport planning involving new roads must be designed either to route roads away from noise-sensitive locations, or to introduce construction remedial measures into the planning process.[17] Such measures include those proposed by the OECD in 1988. In addition to compensation, they recommend the use of quieter surfacing materials on major roads—for example asphalt rather than cobblestones. Green architectural measures such as willow walls and tree planting have also been tried in an attempt to reduce noise. Otherwise measures are aimed specifically at traffic—the introduction of pedestrian precincts, traffic cells, stricter controls on noise-emission levels from moving vehicles, and controls to rationalise the use of vehicles. Regulating heavy lorry traffic remains a controversial issue.

5. ENVIRONMENTAL IMPACT

Environmental damage is an inescapable consequence of an increase in the number of cars on the road. Difficulties experienced in Europe and Japan demonstrate how environmental improvement will always be a problem whilst the number of vehicles keeps growing. Catalytic converters and unleaded petrol are technical fixes which reduce some of the cocktail of pollutants emitted from vehicles, but there are real limits as to how far improvement can be made.

There are now over 400 million cars in the world—this figure is expected to rise to half a billion by the end of the century. In 1989 the UK Government's White Paper "Roads to Prosperity" forecast an increase in traffic of 142% by the year 2025 if the economy grows at a rate of 3% until that year. This would lead to an extra 27.5 million vehicles on our roads. It is calculated that these extra vehicles would form a queue 104,000 miles long. Such a jam could be accomodated, if stationary, on a new motorway stretching from London to Edinburgh—but only if it were 257 lanes wide.[18]

Besides the sheer impact of these vehicles, there is also the whole supporting infrastructure. Large-scale road building and widening schemes create a significant impact upon the environment. All roads cause some environmental damage. This is felt most acutely when proposals for new roads are likely to cause damage to important wildlife habitats or heritage sites. Research by Friends of the Earth in 1986 identified 110 officially designated Sites of Special Scientific Interest (SSSI) damaged by transport programmes, and 23 other important wildlife habitats damaged, destroyed or under threat from road schemes. Such threats have not abated. In 1988, for instance, it was announced that the proposed East London River Crossing would cut through Oxleas Wood, an 8,000 year old ancient woodland and SSSI in south east London.

Road building proposals such as bypasses are intended to re-route unnecessary traffic from certain areas to relieve congestion. However the gains made by such schemes can be short-lived due to the increasing volume of traffic on our roads. And of course these proposals can themselves threaten areas of environmental significance.

5.1. *Damage to Ecosystems*

Vehicle emissions from transport are recognized as a major contributor to the phenomenon known as "acid rain". Acid deposition is responsible for placing severe stress on many ecosystems, although the precise inter-relationships have yet to be fully understood. Acid rain has increased acidity in lakes in Scandinavia, north-east America and south-east Canada, as well as causing widespread tree damage in Scandinavia and eastern European forests. But acid rain is not just a problem for the industrialized countries; significant pollution is happening in the developing countries too.

Pictures of dead and dying trees have been a familiar sight in the media over the last seven or eight years, but signs of actual damage were occurring long before that time. Theories as to how acid rain produces such pronounced effects are still the subject of considerable debate—but one theory is regarded as the most plausible. The "stress hypothesis", suggested by Professor Schutt and colleagues at the Forestry Department of Munich University, proposes that the cocktail of chemical pollutants in the air disturbs the essential life-support systems occurring within the tree—including photosynthesis.

Pollutants in the air act synergistically to weaken the trees, making them less resistant to disease. The leaves or needles become affected, making them unable to help the tree build up reserves of nutrients. The constant stress from pollutants affects the vital functions of the tree; it becomes less resistant to various climatic extremes such as frost or drought. The overall effect is widespread tree damage and death.

5.2. *Pollution and Resource Use*

Beyond vehicle emissions, other aspects of transport create substantial environmental impacts—all forms of pollution are stress agents upon natural ecosystems. Vehicles produce volumes of waste products that have to be disposed of somewhere. For example, although it is possible to recycle waste lubricating oil, every year millions of litres of oil are dumped or poured down the drain. Used oil contains substances that are toxic, potentially carcinogenic and highly polluting. Just four litres of oil, the amount from one car change can contaminate an area of water twice the size of a football pitch. If the oil enters a sewage works, it may kill bacteria needed for the anaerobic digestion of sewage sludge. Car products including antifreeze, brake fluid, and batteries are frequently discarded by illegally dumping them.

The ever-growing number of used car tyres in the world presents a major environmental headache. The US discards a staggering 250 million tyres a year, compared to 25 million from the UK. There is a growing reluctance by landfill operators to accept large numbers of tyres for disposal—they have a tendency to make landfill sites unstable, as well as being bulky items occupying a lot of space. Used tyres represent a potential source of energy and valuable chemical products, but mishandling them leads to enormous problems. A massive fire at a Canadian tyre dump in 1989 resulted in 15 million tyres going up in toxic black smoke. The fire is expected to smoulder on for years.

The manufacture of vehicles is energy-intensive and consumes large amounts of non-renewable resources. In 1987 the US Vehicle industry consumed (as a proportion of the total US consumption) 74% of the natural rubber, 54% of lead, 46% of synthetic rubber, 43% of iron, 39% of platinum, 16% of aluminum, and 15% of steel. Making one car uses a total amount of energy ranging from 11 Gigajoules in Japan to 32 Gigajoules in the USA. 98% of transport energy comes from non-renewable resources, such as petroleum oil. Improved engine efficiency is a clear need.

A new problem is emerging—what to do with the growing volume of worn-out, unsalvageable vehicle waste. Whilst steel and other metals are recoverable as scrap, the escalating amount of plastic, a non-renewable resource, is leading to disposal problems. Over 1.35 million tonnes of plastic were used by western car manufacturers in 1989. By 1990, 8–14% of a car was plastic. Several leading car manufacturers predict this figure will rise to as much as 25% by the end of the century as lighter materials, including aluminium, are substituted in place of heavier iron and steel.

In an attempt to ameliorate this growing disposal problem, a research centre is being set up in Germany to investigate the best means of dismantling vehicles for recycling. Methods of identifying the plastics used, by labelling or bar coding schemes, are on trial.

6. ENDING DOMINATION BY THE MOTOR VEHICLE

Space for roads is a scarce resource, but seldom are its real effects taken into account in traffic forecasts. Road pricing is a measure that goes some way towards recognition of the social and economic costs of transport—this entails charging motorists extra to drive in congested areas. As well as discouraging more vehicles from travelling in an already over-congested space, revenue gained from such a scheme can be directed towards improving public transportation systems and improving the local environment.

The dominance of traffic and its needs has led to a decline in the quality of life in urban areas. Noise and pollution are present—and often in large quantities. Back in 1976 the Noise Advisory Council estimated that over 6 million people in the UK lived in homes where traffic noise was above recommended limits. Planning measures are usually defined with road users in mind and not pedestrians.

6.1. *Safety*

"We value speed more highly than we value human life. Then why not say so, instead of every few years having one of these hypocritical campaigns (at present it is 'Keep Death off the Roads'. . .), in the full knowledge that while our roads remain as they are, and present speeds are kept up, the slaughter must continue?"

George Orwell, "As I please" *Tribune*, 8 November 1946

Does it have to be like this?

Some of the negative impact and stress from motor vehicles is removed by improved planning to make local environments safer. Traffic calming is such a practice—already popular in Norway, Sweden, Netherlands, Germany, Switzerland, and now in France. Traffic calming involves the introduction of engineering measures such as road humps, chicanes and street furniture as a means of slowing down traffic. These are usually combined with legal measures to give pedestrians priority over motor vehicles.

Traffic calming, as an idea, was first developed in the Netherlands in the early 1970s. The first Dutch "woonerf" abandoned the traditional division between carriageway and pavement. One surface was created upon which all road participants mixed and had equal rights. By law the maximum vehicle speed was restricted to walking pace.[19]

"Woonerven" were created as a result of pressure from local residents in the inner parts of the city who felt that traffic was dominating their lives to an unacceptable extent. The "woonerf" concept was originally applicable to residential areas, where the needs of local residents have clear priority over the interests of traffic. This idea is now spreading to whole cities. In these areas pedestrians can walk anywhere on the road and children may play anywhere. All cars, mopeds and cycles must be driven at walking pace. Anyone driving within "woonerfen" must not impede pedestrians—but equally pedestrians should not obstruct vehicles.

There are clear environmental and safety benefits to be gained from traffic calming. There is a reduction in the level of air pollution and noise, which results in improvements to the local environment. Forty per cent of car trips are around 3.2 kilometres or less. In the city such journeys are often "stop-start" driving—which is polluting, as fuel is burnt inefficiently. The level of air pollution drops if drivers drive more calmly; acceleration, slowing down and braking are all driving practices which increase the amount of air pollution a vehicle emits.[19] In Scandinavia traffic calming measures are used as cheap alternatives to bypasses—the negative environmental effects of traffic in urban areas are largely countered by simply slowing down the traffic.

As well as environmental benefits, the safety of pedestrians is improved. In

Britain we make three times as many journeys on foot as we do on buses or trains. Almost 40% of all journeys are made on foot or bicycle, yet there is a long-term underspending on footways. The condition of pavements is often poor, creating further pedestrian hazards. In Britain half of all pedestrian casualities are aged sixty or more. Additionally, pedestrians in cities regularly experience kerbside pollution levels above guidelines set by the World Health Organisation.

In the vulnerable group of road users, 90% of accidents take place in urban areas. The UK child pedestrian fatality rate is 31% above the EC average. As Figure 1 demonstrates, one of the most striking facts is the extent to which children are injured as pedestrians or cyclists on minor roads in built-up areas. These are precisely the sort of roads which would benefit from traffic calming into 20 mph zones.

In many European cities, the general 50 mph speed limit is acknowledged to be too high. At this speed, a pedestrian involved in an accident with a car has a 37% likelihood of receiving fatal injuries whilst with a speed limit of 30 mph this likelihood is reduced to 5%.[19] Traffic calming measures address important issues such as accessibility. By making speed reduction a priority, city streets become considerably more pleasant places in which to walk.

In Denmark and Germany, where large areas of their cities have traffic calming measures in operation, the drop in road accident casualties has been dramatic—in some cases by up to 50%. A MORI poll commissioned by Friends of the Earth in May 1987 asked respondents to comment on the statement "Lower speeds should be introduced into residential areas to reduce the number of accidents on our roads". Altogether 56% of the population agreed with the statement, whereas only 35% opposed it. Women and elderly people were particularly

CHILD ROAD CASUALTIES

Number five in the DTp's glossy Accident Factsheets series. The accompanying DTp press note does not highlight the fact that the UK child pedestrian fatality rate is 31% above the EC average; it focuses on the overall fall in numbers of children killed on the roads. The most striking effect shown in this chart is the extent to which children are injured as pedestrians or cyclists on minor roads in built-up areas—the roads where they live, which could and should be traffic-calmed into 20 mph zones.

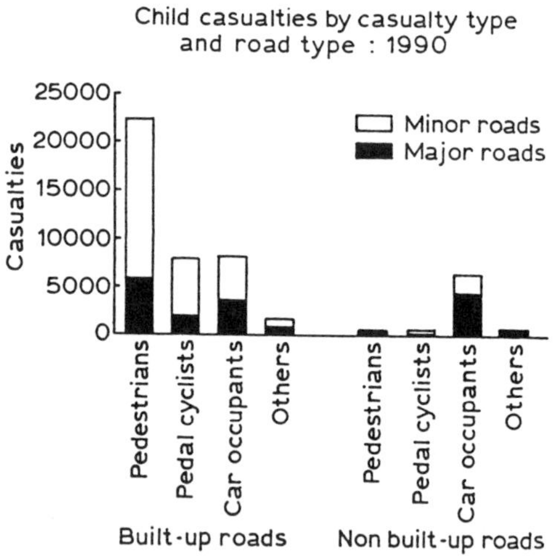

Child Casualties In Road Accidents 1990,
Accident Fact Sheet 5, Dept of Transport,
Rm B648, Romney House, 43 Marsham St,
London SW1 3PY

Figure 1 Child Road Casualties.

in favour of reduced speeds. Similar results were found by a poll commissioned for *Which*? magazine ("Right up your Street?" *Which?* October 1988.)

7. CONCLUSIONS

From the issues raised, it is clear that the motor vehicle is a major stressor. Vehicle emissions directly affect people's health—vulnerable groups now include 20% of the UK's population. Besides the physical stress arising from air pollution, from dust suspended in the air we breathe and from noise, vehicle-induced pollution also has the ability to act as a chemical stressor within our bodies. Noise pollution is regarded by the public as a major nuisance which can contribute to ill health—yet as the noise from city traffic grows, this issue is not high on the agenda of pollution control. Projected growth in traffic and accompanying road building programmes will only add to this existing burden of unacceptable stress. Transport policies must broaden to improve local environments; provision for pedestrians must not be the forgotten component. Ending the domination of the motor car by improved public transportation systems and better planning is vital for future sustainable development.

References

1. R. Read and C. Read, "Breathing can be hazardous to your health" *New Scientist* 34–37 (23 Feb., 1991).
2. I. D. Williams, "Road traffic pollution and public nuisance" *Clean Air J.* National Society for Clean Air and Environmental Protection **21**, No. 3, 123–131 (1991).
3. C. Holman, *Air Pollution and Health* (London, Friends of the Earth, 1989).
4. G. Davies, "The road to ruin" *Radio Times* (28 Dec., 1989).
5. J. Webb, "Car exhausts may cause hay fever" *New Scientist* (22 June, 1991).
6. J. Schulte, "Effects of mild carbon monoxide intoxication" *Archives of Environmental Health* (November, 1963).
7. Department of the Environment, *Digest of Environmental Protection and Water Statistics* (London HMSO, 1991).
8. *Canadian Green Plan* (1991).
9. H. French, "Clearing the Air" *State of the Earth 1990* (Worldwatch Institute, Washington, D.C., January, 1990).
10. UNEP and WHO Assessment of Urban Air Quality: in *State of the Earth 1990* (Worldwatch Institute, Washington, D.C., January, 1990).
11. H. French, *op. cit.*
12. A. Usher, "Slow Death in the City of Angels" *Panoscope*, (London, July, 1991).
13. Department of the Environment, *Digest of Environmental Protection and Water Statistics* (London, HMSO, 1989).
14. S. Cohen, "Social effects on behaviour" *Psychology Today* (New York, October, 1981).
15. Department of the Environment (1991), *op. cit.*
16. OECD, *Environmental Effects of Automotive Transport* (Paris, 1986).
17. National Society for Clean Air and Environmental Protection, *Pollution Handbook* (Brighton, 1989).
18. J. Adams, "Car Ownership Forecasting. Pull the ladder up or climb back down?" *Traffic Engineering and Control* (March, 1990).
19. Friends of the Earth, *Illustrated Guide to Traffic Calming* (London, 1990).

INDEX